Fundamentals of Metal Joining

Dheerendra Kumar Dwivedi

Fundamentals of Metal Joining

Processes, Mechanism and Performance

 Springer

Dheerendra Kumar Dwivedi
Department of Mechanical and Industrial
Engineering
Indian Institute of Technology Roorkee
Roorkee, India

ISBN 978-981-16-4821-2   ISBN 978-981-16-4819-9   (eBook)
https://doi.org/10.1007/978-981-16-4819-9

This Springer imprint is published by the registered company Springer Nature Singapore Pte Ltd.
The registered company address is: 152 Beach Road, #21-01/04 Gateway East, Singapore 189721,
Singapore

This book is based on fundamental understanding of subject matter, research and teaching experience of subjects in the area of joining at postgraduate level and research publications. I would like to dedicate this work to my father Late Shri Mahabir Prasad Dwivedi and mother Late Shrimati Shanti Dwivedi.

Preface

The need for welding of metals and dissimilar metals is growing for fabrication of energy-efficient, lightweight systems having high reliability. The book entitled *Fundamentals of Metal Joining: Processes, Mechanism and Performance* is expected to provide fundamental understanding on technological aspects related to arc welding, heat flow, relevant metallurgical transformations and quality assurance methodology joints, keeping in mind requirements of all those who are keen to pursue UG/PG studies, research and development broadly in the area of welding. Undergraduate and postgraduate students of mechanical engineering, production engineering and metallurgical engineering discipline will get fundamental of physics of welding arc, arc welding processes, heat flow in welding, welding metallurgy, design of welded joints, and inspection and testing methods of welded joints. This work will also be useful for researchers in two ways: (a) understanding the ways by which arc welding processes and their parameters affect the soundness and mechanical performance of the joint joints, (b) significance of heat flow in welding and its effect on mechanical and metallurgical properties of welded joints and (c) how to establish the structure–property relationship for welded joints. Quality control and shop floor engineers and managers dealing with challenges of controlling the weld defects and enhancing performance and reliability will find it very handy information on what can be done to ensure quality of welded joints. Reason for authoring a book in the area of joining technologies includes (a) to consolidate the different areas (arc welding processes, heat flow, design of weld joints, welding metallurgy, weldability of metals, inspection and testing, advancement in field of joining) of importance for welding especially for PG and research students, and (b) to present advances in the area of fusion and solid state and hybrid joining approaches. This book consists of matter from four different areas: (a) fundamental understanding of the subject matter, (b) findings of R&D activities of students under author's supervision, (c) recent research publications and (d) literature available in public domain in the form of books, handbooks and Web. The author with thanks acknowledges the support provided by NPTEL and MHRD for developing content and then permission to publish the same.

Roorkee, India   Dheerendra Kumar Dwivedi

About This book

The need for welding of metals and dissimilar metals is growing for fabrication of energy-efficient, lightweight systems having high reliability. The book entitled *Fundamentals of Metal Joining: Processes, Mechanism and Performance* is expected to provide fundamental understanding on technological aspects related to arc welding, heat flow, relevant metallurgical transformations and quality assurance methodology joints, keeping in mind requirements of all those who are keen to pursue UG/PG studies, research and development broadly in the area of metal joining. Undergraduate and postgraduate students of mechanical engineering, production engineering and metallurgical engineering discipline will get fundamental of physics of welding arc, arc welding processes, brazing and soldering, heat flow in welding, welding metallurgy, design of welded joints, and inspection and testing methods of welded joints and weldability of metals. This work will also be useful for researchers: (a) understanding the ways by which arc welding processes and their parameters affect the soundness and mechanical performance of the joints, (b) significance of heat flow in welding and its effect on mechanical and metallurgical properties of welded joints and (c) how to establish the structure property relationship for welded joints. Quality control and shop floor engineers and managers dealing with challenges of controlling the weld defects and enhancing performance and reliability will find it very handy information on what can be done to ensure quality of welded joints. Main highlighting points of the book on *Fundamentals of Metal Joining: Processes, Mechanism and Performance* making it unique include:

(a) Use of detailed coloured schematics to demonstrate the effect of heat temperature gradient on structure, properties and formation of different zone
(b) Integrating different areas of metal joining (joining processes, heat flow, design of welded joints, welding metallurgy, inspection and testing, weldability of metals) which are needed to develop a comprehensive in UG, PG and research students
(c) Presenting few advances as well in the area of fusion and solid state metal technologies.

This book has been developed considering (a) fundamental understanding of the subject matter, (b) findings of R&D activities of students under author's supervision,

(c) recent research publications and (d) literature available in public domain in the form of books, handbooks and Web.

Contents

About the Author

Dheerendra Kumar Dwivedi, Ph.D. Professor in the Department of Mechanical and Industrial Engineering, Indian Institute of Technology Roorkee. He has completed his Bachelor of Engineering (mechanical engineering) from Government Engineering College Rewa in 1993, Master of Engineering (welding engineering) from University of Roorkee (now IIT Roorkee) in 1997 and Doctorate (metallurgical engineering) from MNIT Jaipur in 2003. He has been involved in teaching, research and development, industrial consultancy for the last 25 years broadly in the area of manufacturing technologies in general and casting, welding and surfacing modification for improved mechanical properties in particular.

Recipient of Binani Gold Medal Award-2001 (IIM), and recognized in top 1% Global Scientists in Materials Domain by Elsevier-Stanford (2020). Five best research paper awards in National/International conferences.

He has developed more than 200 h video lectures in area of manufacturing technologies under MOOCS/NPTEL program of MHRD Government of India. He supervised 16 Ph.D. thesis and more than 50 M.Tech. dissertations.

He has published in more than 148 research papers in peer reviewed SCI/SCIE indexed International Journal with h factor 34 and i-10 index 84 with total citation more than 3650 and RG Score 36.62.

He has published two books namely *Production and Properties of Cast Al-Si Alloys* with New Age International, New Delhi in 2013 and *Surface Engineering for Enhanced Tribological Life of Component* with Springer nature in 2018.

He has executed more than 20 Research and Development project and 50 industrial consultancy project. Filed three Indian patents on technologies developed in area of A-GTAW and FSW.

He has undertaken eight bilateral international collaborative research projects with reputed university namely Chemnitz University, Germany, Technical University Munich, Germany, Institute of Metal Research Shenyang, China, University of Belgorod, Russia, University of Coimbra, Portugal, University of Uberlandia, Brazil, University of Zacatecas, Mexico and Physical Technical Institute, Minsk, Belarus. Author has undertaken research projects in the area of friction stir welding, welding bonding, activated flux GTAW, oxy-fuel flame and high velocity oxy-fuel spraying for improved abrasive and erosive wear resistance, laser cladding of none-cobalt base alloys for improved cavitation resistance, laser assisted nitriding and ion implantation cast martensitic stainless steel for improved erosion resistance, friction stir processing of cast Al-Si alloys, Ni-Al-Bronzes.

Part I
Basics of Metal Joining

Chapter 1
Metal Joining: Need, Approaches and Mechanisms

1.1 Basics and Need of Joining

The joining of metals is needed to realize complex shapes more commonly for large size components, ships, automotive, pressure vessels, gas cylinders, etc. Moreover, small size components like electronic components also need joining primarily to facilitate the flow of current. Joining of metals can be needed for establishing connections between (a) similar metals and (b) dissimilar metals. Joining between similar metals is found to be easier than dissimilar metals because response (behaviour) of both the faying surfaces of the components of the similar metals being joined to heat, pressure or both applied during joining becomes same. Similar metals joining under the influence of heat/pressure/both result in almost identical expansion/contraction, melting/solidification, chemical interactions and deformation. Therefore, similar metal joining offers somewhat fewer issues in the form of weld defects and discontinuities, residual stress and distortion problems than dissimilar metal joining. Issues in dissimilar metal joining primarily occur due to difference in chemical, metallurgical and physical properties.

1.1.1 Fundamental Approaches of Metal Joining and Joint Capability

Metals can be joined using heat, pressure and chemical reactions depending upon the method to be employed. The joint is developed (as per process/approach) through (a) metallurgical continuity (through fusion and diffusion), (b) mechanical interlocking and (c) chemical/physical forces. The fundamental approaches based on above mechanism are shown schematically in Fig. 1.1a–g.

Metallurgical continuity is realized through fusion of the flying surfaces or diffusion across the mating/faying interfaces of the components to be joined, and this approach results in the high joint strength with efficiency sometimes even greater

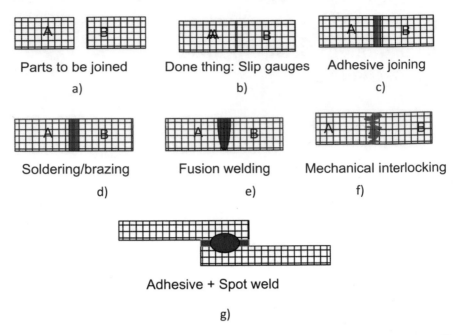

Parts to be joined   Done thing: Slip gauges   Adhesive joining

a)   b)   c)

Soldering/brazing   Fusion welding   Mechanical interlocking

d)   e)   f)

Adhesive + Spot weld

g)

Fig. 1.1 Fundamental approaches for joining of metals showing schematically **a** parent metallic components, **b** weak joint formed in case of slip gauges, **c** adhesive joining, **d** soldering and brazing, **e** fusion welding, **f** mechanical interlocking based on plastic deformation and **f** hybridization of joining approaches involving adhesive joining and resistance spot welding

than 100% suggesting joint is stronger than its base metal. A combination of chemical/physical interactions and weak mechanical interlocking approaches becomes active in case of adhesive joining and sheet metal joining; this approach usually results in lower joint strength (say 20–50% of based metal). A hybrid joining approach like weld bonding and adhesive mechanical joining combines the advantages of both metallurgical joining and adhesive joining and adhesive joining and mechanical joining like crimping. Joints produced through chemical bonding (adhesive joining) only offer low joint strength. These are usually used for compressive loading, maintaining the relative position of mating components, and are generally not preferred as load carrying joint.

In view of above, it is therefore preferred to develop joints through metallurgical continuity by either fusion- or diffusion-based approaches. The development of the joints through fusion-based approaches of the faying surfaces is easier and faster than those based on diffusion. The diffusion-based approach is slow, tedious and time consuming as it requires very high surface finish of faying surfaces of components to be joined and controlled atmosphere in the form of inert gas/vacuum during joining. The diffusion-based approaches are still preferred when fusion-based approaches for joining of dissimilar metals are not feasible due to metallurgical incompatibility and large difference in physical properties.

1.1.2   Relevance of Heat and Pressure in Metal Joining

The heat and pressure either singly or in combination are commonly used for achieving the metallurgical continuity depending upon the joining process. All the fusion-based joining processes apply heat primarily for melting of the faying surfaces of the components to be joined, while solid state joining processes use heat primarily for the thermal softening of the faying surfaces of base metals to facilitate the local-ized plastic deformation and facilitate metallic intimacy and metallurgical continuity (Fig. 1.2). The application of heat increases diffusion and results in the formation of joint in solid state joining process like diffusion bonding. Diffusion across the joint interface in both solid state joining, and solid–liquid joining processes signifi-cantly determine the joint strength. Diffusion plays a significantly important role in ensuring metallurgical continuity across the joining interfaces in solid state joining methods, development of microstructure, favourable/unfavourable microstructural transformations like reversion/dissolution of precipitates, change of grain sizes in weld and HAZ. The presence of impurities in the form of oxides and other traces interferes with diffusion which in turn lowers the joint strength due to reduction in metallic continuity.

Effect of Heat

Further, heating of surfaces of components being joined facilitates cleaning by removing surface impurities through thermal decomposition, evaporation of organic compound and fracture of hard and brittle oxides. However, rise in temperature due to heat supplied/generated (during joining) under ambient condition increases the oxidation and nitride formation tendency as per affinity of workpiece metal.

Effect of Pressure

Pressure in case of solid state metal joining process governs soundness of the joints. Pressure facilitates metallic intimacy between the faying surfaces of the components

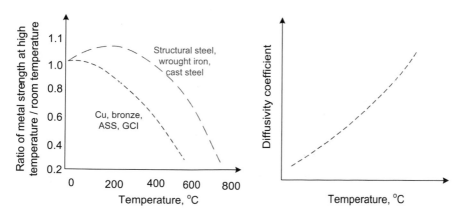

Fig. 1.2   Effect of temperature on **a** strength of metal and **b** diffusivity coefficient

being joined by (a) fracturing brittle oxide layers present at the faying surfaces, (b) squeezing out the surface impurities and (c) plastic/elastic deformation of asperities present at the faying surfaces as per joining process. The plastic deformation can be at micro-level or macro-scale as per joining process. Micro-level deformation typical of 1–3 μm is observed at the joint interface, while it can be in range from millimetres to centimetres in case of macro-level deformation processes. Surface layer deformation of mating surfaces results in mechanical interlocking, atomic level metallic intimacy to facilitate diffusion and strain hardening of the deformed metallic layers at the joint interface. Micro-level deformation primarily affects the interfacial metallic intimacy, mechanical interlocking and so diffusion, while macro-level diffusion mainly governs the mechanical interlocking, strain hardening, microstructural modification and so the joint strength.

1.1.3 Sources of Heat for Metal Joining

Heat for joining can be applied using various sources like combustion of oxygen–fuel gas mixture, exothermic chemical reaction, welding arc, interfacial frictional heat between mating components, electrical resistance heating, application of radiation energy in the form of laser and electron beam and hybridization of heat sources by bringing in two or more heat sources together in action for thermal softening or melting the faying surfaces as per the need of joining approach. Since each heat source (associated with a joining process) applies heat within a limited range of power density (W/mm^2), the amount of the heat supplied for thermal softening/fusion by different joining processes becomes different. Increase in power density of the heat source associated with a process reduces the heat required for the joining purpose (Fig. 1.3).

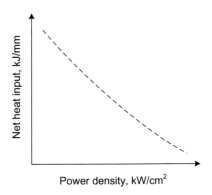

Fig. 1.3 Relationship between power density and net heat input in fusion welding processes

1.1.4 Heat and Characteristics of Metallic Joint

Heat supplied to the base metals for joining purpose significantly affects the soundness, quality, metallurgical and mechanical properties, residual stress and distortion tendency of the joint. In general, it is always preferred to develop a sound joint with required properties using minimum possible heat input so as to reduce adverse effect of heat on the base metal properties in the form of heat affected zone formation, residual stress development, unfavourable metallurgical transformation and chemical interaction in weld and the heat affected zone (Fig. 1.4). Too low or high heat input deteriorates the characteristics but due to different factors. Low heat input may cause weld discontinuity and defects in fusion weld joint, while high heat input adversely affects the microstructure, mechanical properties and residual stresses especially in the heat affected zone. Higher the heat input for joining, greater will be weld volume and wider heat affected zone and so more related issues. The heat affected zone is a region near the joint interface or weld fusion boundary which experiences the change in the microstructure and that mechanical properties due to the application of the heat during the joining process. The microstructure variation in the heat affected zone determines the mechanical properties and the corrosion resistance. Therefore, it is important that microstructure of the weld and heat affected zone is properly controlled/developed. Thus, there should be an optimum level of heat input in case of fusion welding.

Further, the heat applied for the joining purpose increases the temperature (at high rate of heating) of the base metal followed by (comparatively slow) cooling. Rise in temperature of base metal causes the fusion of the faying surfaces and thermal expansion and contraction of the nearby base metal (heat affected zone) due to the heating/cooling cycle called weld thermal cycle. The localized thermal expansion and

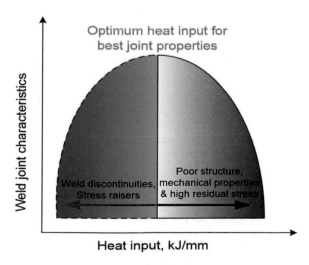

Fig. 1.4 Effect of heat input on weld joint characteristics

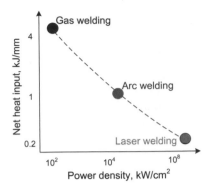

Fig. 1.5 Effect of power density and net heat input for different fusion welding processes

contraction of the base metal near the joint interface produces residual stresses and so distortion tendency. The development of the tensile residual stress in joints adversely affects the performance in many ways: (a) reduces the tensile and fatigue strength under tension, (b) increases stress corrosion cracking tendency and (c) undesirable dimensional and shape change in components being joined in the form of distortion. Therefore, efforts are always made to develop the joints using minimum possible (a) heat input, (b) volume of weld and (c) thin layer in vicinity of faying surface of the base metal which is heated so as to minimize the width of the heat affected zone. Power density of the fusion-based joining process being used to develop a joint directly determines above-mentioned three aspects. Low power density joining process like gas welding results in high heat input (>4/5 kJ/mm), large volume of the weld metal and very wide heat affected zone; on contrary, high power density joining process like laser welding develops joint using very low heat input (<0.3 kJ/mm), small volume of the weld metal and very narrow heat affected zone (Fig. 1.5).

1.2 Mechanisms of Joining

Joining of similar and dissimilar metallic components, metal and non-metal combination to achieve the desired shape can be obtained using a range of methods such as mechanical joining, fusion weld joining, solid state weld joining, adhesive joining and a combination of above methods. In case of mechanical joining, components to be joined are held together using suitable force with the help of nuts and bolt, and rivets. There is no interaction between components joined using mechanical joining. However, other joining approaches such as fusion weld joining, solid state weld joining and adhesive joining involve some kind of metallurgical and chemical interactions either with filler metal or with mating components. These interactions are largely governed by temperature variation during joining as a function of time, i.e. weld thermal cycle. In the following section, four mechanisms of developing joints are fusion, plastic deformation, chemical and metallurgical reactions. Diffusion is inherently part of all these mechanisms, and it plays a major role in development of joint using a solid state joining process called diffusion bonding/joining.

1.2.1 Fusion

Joining of components by this mechanism is primarily based on fusion/melting of faying surfaces of the components to be joined using suitable heat source. Heat may also be used for melting filler metal or electrode depending upon the type of joining process. However, a part of heat applied for fusion is dissipated to the underlying base metal (according to thermal conducting behaviour). This heat alters the microstructure, mechanical and corrosion properties of the zone next to the fusion boundary (HAZ) and sometime even weld zone itself unfavourably. Invariably, the region next to the fusion boundary (towards the base metal) that experiences the effect of heat applied for realizing fusion is called heat affected zone (Fig. 1.6). Heat affected zone is subjected to a range of phenomenon such as partial melting, recrystallization and grain growth, dissolution of precipitates and formation of both precipitate-free zone and new precipitates. These microstructural changes in heat affected zone deteriorate mechanical properties in the form of either softening or hardening according to the strengthening mechanism of the metal system being joined and degrade the corrosion resistance. Additionally, molten metal produced during joining is subjected to (a) interaction with gases (oxygen, nitrogen, hydrogen) present in vicinity of molten pool and (b) solidification. Interaction of the atmospheric gases with molten pool (if not protected properly) leads to inclusion and porosity in the joint, while unfavourable solidification can lead to solidification cracking, shrinkage porosity and coarse grain structure (Fig. 1.7). Since fusion mechanism of developing joints uses maximum heat input (for melting faying surfaces) as compared to the other two mechanisms, namely plastic deformation and chemical reaction heat, it also leads to the greatest effect on base metal properties and widest heat affected zone.

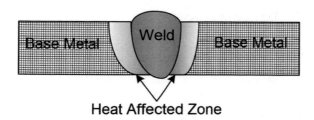

Fig. 1.6 Schematic of fusion weld joint showing weld zone and heat affected zone

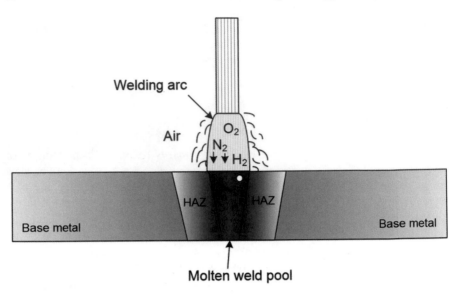

Fig. 1.7 Schematic of arc welding and interaction effect of atmospheric gases leading to pores and inclusion due to poor shielding

1.2.2 Plastic Deformation

Joining of metallic components by this mechanism is based on realizing of either micro-level plastic deformation or macro-scale deformation using suitable compressive force coupled with/without any external/internal heating. Micro-level (usually < 10 μm) plastic deformation facilitates (a) metallic intimacy between faying surfaces for diffusion across the interface and (b) mechanical interlocking and (c) even breaking of surface oxide layers.

Micro-plastic deformation

The micro-scale plastic deformation at interface can be used for joining of both thin sheets and thick plates using joining processes like ultrasonic welding and explosive welding, respectively (Fig. 1.8). The extent of flow of metals from the faying surfaces into each other (to cause mechanical and metallurgical bonding) during joining is determined by yield strength and ductility of the components being joined and force applied to facilitate interfacial plastic deformation. Low strength and high ductility metals flow more than high strength and low ductility metal system at the joint interface. Limited plastic flow and so less interfacial mechanical locking reduce strength and ductility of the joint. Plastic deformation of metals near the joint interface increases the hardness due to work hardening if the effect of heat generated/applied is limited on thermal softening. Therefore, joint interface may exhibit higher hardness than the respective base metals.

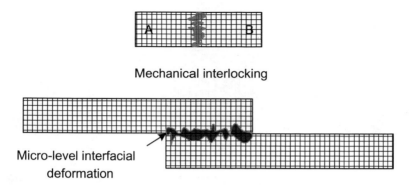

Fig. 1.8 Schematic of joint interface developed in processes involving micro-level interfacial plastic deformation like ultrasonic welding and explosive welding

Macro-plastic deformation

Joining processes like riveting, clinching and friction stir welding (FSW) using macro-scale plastic deformation involve flow of the metal from one zone to another to obtain a joint continuity. The macroscopic plastic deformation in case of riveting and clinching is primarily used to give the desired shape to joining members so as to achieve mechanical interlocking (Fig. 1.9), while in case of other macroscopic plastic deformation-based joining processes like FSW, friction welding, etc., a metallurgical joint is created.

In case of joining processes like friction stir welding wherein a rotating tool having shoulder and pin facilitates plastic deformation to develop a metallic joint using solid state joining approach (Fig. 1.10a–c). This macro-scale deformation is coupled with fracturing of both hard and brittle micro-constituents, and the matrix usually leads to significantly grain refinement in joint area due to dynamic recrystallization. Plastic deformation and heat (generated due to friction and plastic flow) in combination determine the microstructure and mechanical and corrosion properties of weld, joint interface and heat affected zone. Metal systems joined using macro-scale plastic deformation-based processes therefore show large variation in hardness/strength distribution across the joint interface including heat affected zone. Precipitation-strengthened metals (like 6061 aluminium alloys) usually show softening, while transformation hardening and work hardening metals (like steel) show hardening embrittlement according to their strengthening mechanisms.

1.2.3   Chemical Reactions

Joining of the metallic components using chemical reaction is based on principle occurrence of certain chemical and reactions at the faying surfaces. The typical joining methods where joining is realized through chemical reactions and inter-metallic formation at the joint interface include adhesive joining and low melting

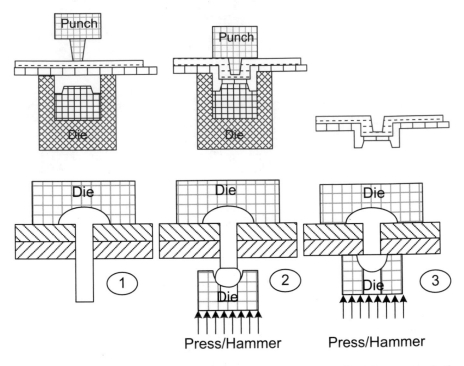

Fig. 1.9 Schematics showing sequential step of joining processes involving macro-plastic deformation **a** clinching and **b** riveting

point brazing and soldering. Chemical reaction-based joining is found suitable for those metals where both fusion and deformation of faying are not feasible to develop joints due to soundness, poor strength, embrittlement and metallurgical issues. However, joints developed using chemical reaction-based mechanism usually suffer from limited joint strength and applicability in specific environment and unsuitable for high temperature conditions. Adhesive joining involves application of suitable chemical mixture on the cleaned faying surfaces which after curing offers required peak joint strength (Fig. 1.11a). Chemical reactions like cross-linking polymerizations in adhesives and at adhesive–metal interface during curing provide high joint strength. However, application of adhesive joining is limited by low stability of adhesive at high temperature and in wet environment. Depending upon the soundness and relative strength (load carrying capacity) of adhesive itself and adhesive–substrate interface joint, the failure of adhesive joint may occur from either adhesive–substrate interface or cohesive failure of adhesive itself whichever is weaker (1.11b). Chemical reaction-based joining methods do not affect properties of components being joined like there is no HAZ formation as observed in case of joints developed using fusion and plastic deformation-based mechanisms.

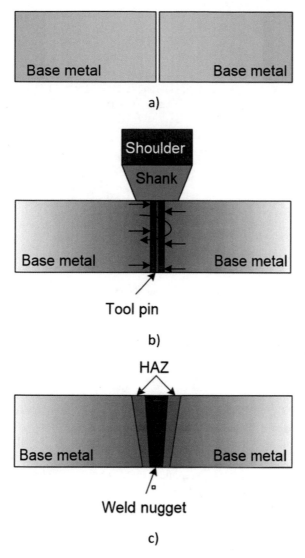

Fig. 1.10 Schematic of FSW **a** base metals to be joined, **b** FSW tool in action during welding and **c** weld joints exhibiting weld nugget and HAZ

1.2.4 Metallurgical Reactions

Joining of metals purely based on metallurgical reactions (without any fusion or deformation of faying surfaces) involves formation of intermetallic compounds at faying surfaces due to metallurgical reactions between faying surfaces of the base metal and molten low melting point metals. The faying surface of the base metal

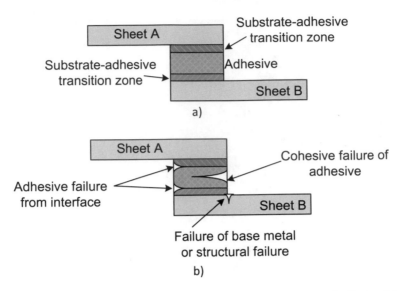

Fig. 1.11 Schematic related to adhesive joining showing **a** zones/parts of adhesive joint and **b** common failure modes determining adhesive joint strength

remain is solid state, while low melting metal (solder/brazing metal) is brought to the molten state to facilitate the metallurgical reaction besides mechanical interlocking (Fig. 1.12). Solid–liquid–solid joining processes like brazing and soldering fall in this category. The components to be joined are heated to a temperature high enough (as per the metal system from 250 to 1000 °C) so as to ensure melting of the low melting point brazing and soldering metal, while base metals during these processes remain in solid state. Metallic interactions between the molten brazing/soldering filler at the mating interface of components being joined result in intermetallic. The solidification of low melting metals later facilitates mechanical interlocking too. Metallurgical reactions and mechanical interlocking both determine the joint strength. However, strength of low melting filler metal and that of intermetallic compounds formed at joint interface predominantly dictate the joint efficiency.

1.3   Choice of the Method of Joining

The fabrication of engineering systems frequently needs joining of simple components and parts to get the required size and shape. Three types of joining methods, namely mechanical joining (nuts and bolts, clamps, rivets), adhesive joining (epoxy resins and Fevicol) and welding (fusion welding, brazing and soldering) are commonly used for manufacturing variety of engineering products/systems. Each type of joint offers unique load carrying capacity, reliability, compatibility in joining of similar or dissimilar metals, suitability for application in different environments

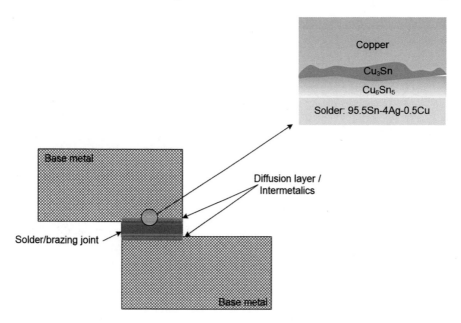

Fig. 1.12 Schematic of soldering of copper with details of layers of compounds formed at substrate–solder interface due to metallurgical reactions

and cost. Consideration of following points will be appropriate for selection of a suitable method of joining for an application.

1.3.1   Type of Joint

1.3.1.1   Temporary Joint

These joints are used for applications where frequent assembly and disassembly of parts being joined is needed like nuts and bolts and clamps (Fig. 1.13).

1.3.1.2   Semi-permanent Joints

These joints are used for applications where disassembly may be needed later in future. These joints allow disassembly with little efforts like adhesive joints and solder joints (Fig. 1.14).

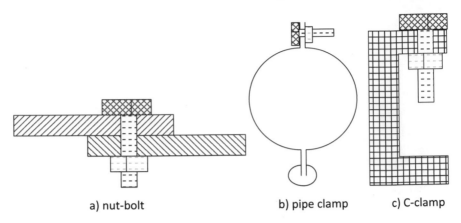

a) nut-bolt   b) pipe clamp   c) C-clamp

Fig. 1.13 Schematic of different types of temporary joints: **a** nut–bolt, **b** pipe clamp and **c** C-clamp

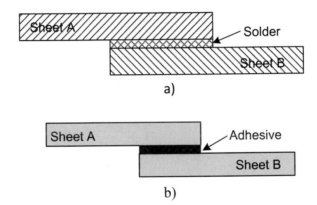

Fig. 1.14 Schematic of common semi-permanent joints: **a** solder joint and **b** adhesive joint

1.3.1.3 Permanent Joint

These joints are used for applications where a permanent connection is needed between components. These joints are permanent in nature and take lot of effort in disassembly like fusion weld joint, resistance weld joints, riveted joints and many other solid state weld joints (Fig. 1.15).

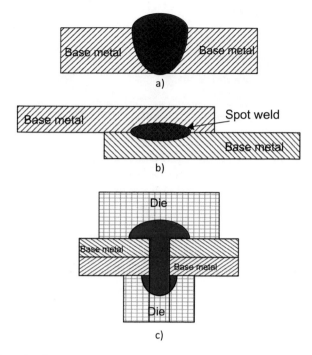

Fig. 1.15 Schematic of common permanent joints: **a** fusion weld joint, **b** spot weld and **c** riveted joint

1.3.2 Type of Metallic Combination

1.3.2.1 Similar Metals

Joining of similar metal systems needs technological consideration with regard to the choice of (a) joining process and (b) filler/electrode metal. The choice of joining process is governed by (a) thickness of the section to be welded and (b) thermal sensitivity of metal of components to be joined. Metals sensitive to thermal/heat experience problem in weld and heat affected zone after the joining. The undesirable mechanical and thermal damages may lead to increased cracking tendency, loss of corrosion resistance and embrittlement. For example, fusion welding of cast iron and high carbon steels may cause cracking and embrittlement of weld and HAZ, while brazing can avoid these issues if strength requirement is satisfied. Likewise, fusion welding of Al-Cu alloy and stainless steel leads to severe corrosion attack in heat affected zone in corrosive environment. Therefore, thermal sensitive metals need to be joined using low heat input processes like (a) high power density joining processes, e.g. plasma arc, laser and electron beam, and (b) solid state joining processes. Therefore, sensitivity of the metal systems to thermal and mechanical damages by potential joining process even in case of joining of similar metal systems must be given

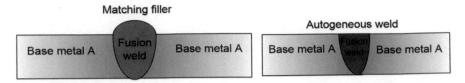

Fig. 1.16 Schematic of fusion weld joints of similar base metals developed using **a** matching filler/electrode and **b** autogenous welding without filler

through consideration while choosing a joining process. Further, autogenous weld or weld joints developed using filler/electrode of composition similar to base metal also do not necessarily offer sound and good weld joint due to (a) high solidification cracking tendency of weld metals having a wide solidification temperature range (say > 100 °C) and (b) requirement of special weld properties like enhanced corrosion resistance, high toughness, high temperature/oxidation resistance and high wear resistance. Under these situations, it is required to use such electrode/filler (metal composition) different from base metal which can offer weld metal properties required in the light of the target application (Fig. 1.16).

1.3.2.2 Dissimilar Metals

Joining of dissimilar metals certainly affects the choice of joining process in a big way as it directly determines the volume fraction of both the metals being joined will be contributing towards the development of joints. The volume fraction of both the metals (in the developed joint) must be kept a minimum possible level using appropriate selection and control of joining process parameters in order to reduce issues related to difference in physical, thermal and mechanical properties of dissimilar metal components being joined. These differences in properties of dissimilar metal systems (to be joined) frequently cause problems of metallurgical incompatibility and heterogeneity/deterioration in respect of chemical, mechanical and corrosion properties of the joints. Therefore, the selection of a suitable filler/electrode for fusion welding of dissimilar base metals becomes crucial as it should be compatible either with both the base metals or one base metal and buttering layer being applied to isolate one of the base metals primarily to reduce dilution in the weld metal so as to reduce the incompatibility-related issues such as IMC formation. Issues in dissimilar metal joining are more pronounced with fusion welding processes than solid state joining processes and solid/liquid state joining processes. Moreover, development of dissimilar metal joints using fusion welding processes needs additional efforts in terms of application of buttering layers on the faying surfaces of components to be joined, use of special filler/electrode metals compatible to both dissimilar metals, use of multi-pass joining technique to take advantage of auto-bead tempering and grain refinement, use of low heat input using high power density joining processes like laser and electron beam welding (Fig. 1.17). Solid state joining processes like friction stir welding, diffusion bonding and ultrasonic welding produce dissimilar

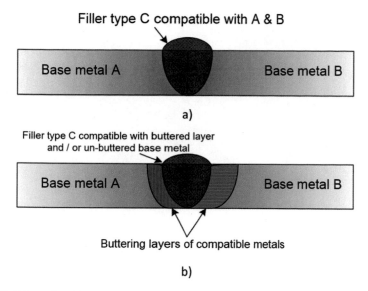

Fig. 1.17 Schematic of fusion weld joints of dissimilar base metals developed using **a** suitable different type of filler metal compatible to both the base metals and **b** buttering layer of different electrode/filler metal(s) compatible to one/both the base metal and weld metal

metal joints of much better performance in service primarily due to the absence of fusion/solidification-related issues and low heat input.

1.3.3 Workpiece Metal

1.3.3.1 Physical Properties

Thermal expansion coefficient, melting temperature and solidification temperature range are few very important physical properties of metals which determine the ease of welding. High thermal expansion coefficient metals must be heated over the smallest possible regions if required during joining in order to reduce the extent of thermal expansion and contraction experienced by them due to weld thermal cycle so as to minimize residual stresses and distortion tendency of joints. Melting temperature of base metals (to be joined) has limited influence on the ease of fusion for welding purpose and selection of joining process. High melting temperature metals like steel (in thickness section) are found to be difficult to join using low power density gas welding processes as the maximum temperature generated by oxy-fuel gas mixture flame is limited to about 3100 °C, while melting temperature of metals being joined by other fusion welding processes (like arc welding, plasma welding, laser welding) may not be an important consideration as these offer maximum temperature (from 6000 to 25,000 °C) much higher than melting temperature of base metals

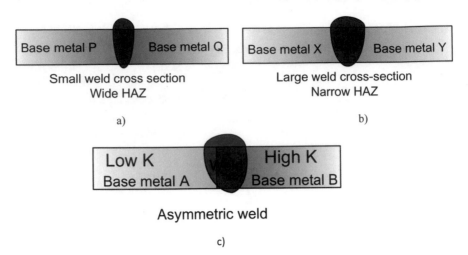

Fig. 1.18 Schematic of fusion weld joints of dissimilar base metals: **a** symmetric weld of high K base metals, **b** symmetric weld of low K base metals and **c** asymmetric weld of different K and other base metal properties

itself. However, longer solidification temperature range (despite low melting temperature like in case of few Al alloys) increases the solidification cracking tendency of weld metal; therefore, joining of such metals may be realized either through solid state joining processes, adhesive joining process, solid/liquid joining processes or use proper filler metal in case of fusion welding processes so that solidification temperature range can be reduced (Fig. 1.18).

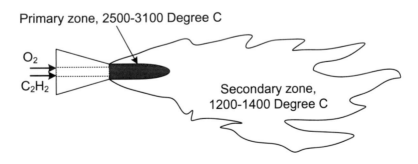

1.3.3.2   Chemical Properties

Chemical composition, solubility of gases in liquid and solid state, affinity of metal to ambient gases at high temperature and electro-chemical nature are few factors which must be considered for selection of suitable joining process. Chemical composition of metal significantly determines the cracking tendency, and mechanical, physical and

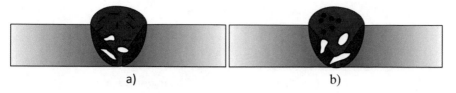

Fig. 1.19 Schematic of fusion weld joints showing weld discontinuities like porosity, inclusion in the form of oxides, nitrides formed due to molten weld metal and atmospheric gas interactions

metallurgical properties of the metal. The presence of impurities like S and P in steel increases the solidification, liquation and under-bead cracking. Further, metal with large difference in liquid and solid state solubility to gases promotes porosity in weld metal. Such metals need proper control of welding conditions to avoid entrapment of gases in the weld metal. Metals having high affinity to ambient gases (oxygen, nitrogen, hydrogen) promote oxidation, nitride and inclusion formation tendency, so joining of such metals may be performed using either (a) solid state joining processes or (b) fusion welding processes (GTAW, GMAW, EBW) providing excellent protection to the metal being heated from atmospheric gases (Fig. 1.19). Electro-chemical behaviour of base metal after joining may get changed appreciably w.r.t. weld and heat affected zone (in a corrosion-sensitive environment) which may promote localized pitting and galvanic corrosion. Such deterioration in electro-chemical behaviour is directly linked with metallurgical transformation in the weld and heat affected zone and chemical heterogeneity developed in weld after joining due to weld thermal cycle experienced. Increase in heat input in general increases such deteriorations in electro-chemical behaviour; therefore, efforts are made to select low heat input/high power density joining processes.

1.3.3.3 Metallurgical Properties

During the joining, metal systems experience many undesirable metallurgical transformations in the form of formation of hard and brittle phases, intermetallic compounds, carbide, dissolution of precipitates, presence of precipitate-free zone (PFZ) in the joint and heat affected zone due to weld thermal cycle and plastic deformation. These changes degrade the mechanical and corrosion performance of the joint during service or make weld fabrication difficult. For example, the formation of chromium carbide in heat affected zone (called sensitization) during welding of austenitic stainless steel due to sensitization reduces the corrosion resistance and promotes the intergranular cracking and stress corrosion cracking in halide environments (Fig. 1.20). Solid state joining and solid/liquid joining process can reduce the issues commonly related to high heat input and weld thermal cycles associated with fusion welding. However, solid state joining processes can be applied in metal systems having enough ductility and low yield strength as high-strength

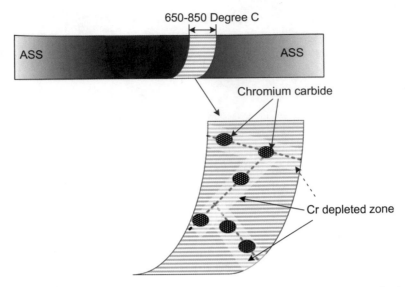

Fig. 1.20 Schematic of fusion weld joints of austenitic stainless steel showing mechanism of sensitization due to unfavourable metallurgical transformation

metals make plastic flow more difficult during joining which in turn results in defective joints. However, need/importance of optimization of process parameters related to a joining process is mandatory and need not be overemphasized.

1.3.4 Service Environment

Many BCC and HCP metals and their weld joints exhibit the very poor impact resistance and brittle behaviour at low temperature. The low temperature brittle behaviour is commonly studied in terms of ductile to brittle transition temperature (DBTT). The DBTT is a median temperature and is obtained from the temperature range over which sharp drop in toughness/impact resistance of a given metal or weld joint takes place (Fig. 1.21). Carbon steel and their weld joints show very poor impact resistance (5–20 J) at low temperature. Low DBTT metals/welds are preferred over high DBTT. The DBTT of carbon steels in general increases with increase of carbon content from 0.05 to 0.8%, while addition of manganese and nickel in steel lowers the DBTT. Moreover, DBTT of weld metal can be adjusted using suitable electrode/filler metal (composition) selection. However, HAZ-related issues can only be reduced using development of proper welding procedure comprising grove preparation, preheating, welding process, filler metal selection and post-weld heat treatment. For example, the corrosion resistance of the fusion weld joints can be realized using suitable filler/electrode but HAZ due to undesirable metallurgical transformation

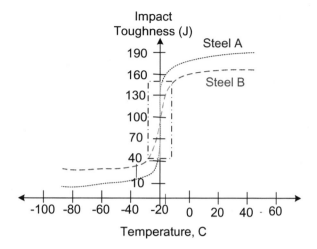

Fig. 1.21   Plot showing relationship between impact toughness (J) and temperature of two different steels

frequently exhibits poor corrosion resistance. Low heat input welding processes like pulse GTAW and GMAW, laser and electron beam welding help in reducing HAZ-related issues.

1.3.5   Reliability

Joints are developed for fabrication of components and systems for a range of applications from simple structural joints used in construction industry to very critical joints for nuclear reactors, aircraft components involving huge risk property and life. Accordingly, reliability in performance expected from a joint varies from moderate (reliability) for non-critical joints to extremely high (reliability) for critical joint. Reliability of the joint performance is dictated by (a) soundness of the joints and (b) extent of degradation in base metal properties caused by thermal and mechanical factors during joining. In general, most of the joints for general applications are developed by shielded metal arc welding process, while joints for semi-critical application like pressure vessel and materials handling machines are developed using gas metal arc welding and submerged arc welding and those for critical application like nuclear reactor joints, aircraft, spacecraft are developed using gas tungsten arc welding, plasma arc welding, pulse variants of gas tungsten arc and gas metal arc welding, laser and electron beam welding, and solid state joining processes. Moreover, suitability of these joining processes is significantly governed by quality desired, productivity and economics.

It would be the best (from the reliability and performance point of view) if a joint is developed between two members without doing anything (literally without doing anything) as it will not be causing any thermal/mechanical damage to base metal properties; however, it is not possible. Therefore, we look for alternatives which will be less damaging (thermally/mechanically) to the base metal properties like use of high power density processes, laser and electron beam welding, solid state joining processes. Adhesive bonding, soldering, brazing, etc., are few such examples which cause least thermal/mechanical damage to base metal properties; however, load carrying capacity and ability to work in hostile service conditions (like high temperature and corrosion) of such joints become very limited. These joints may be suitable for light service conditions. Therefore, application of joints significantly affects their reliability.

1.3.6   Nature of Loading

The nature of load during the service does not affect choice of joining process appreciably, but still it needs little consideration. The extent of influence of nature of load factor consideration on selection of suitable joining processes is primarily determined by the severity of stress concentration induced (after developing a joint) by a joining process. Few types of joints (lap, corner joint, etc.) inherently cause high stress concentration; therefore, such joints promote premature failure through fast crack nucleation and growth under fluctuating load conditions. Joining processes like adhesive bonding, weld bonding, brazing and soldering are primarily used for developing lap joints; therefore, these should not be considered for developing joints which are expected to take fluctuating loads during the service. Joining processes which are able to develop butt joint (offering minimum stress concentration) need to be considered for fluctuating and impact loads. Joints for static loading can be developed using any suitable joining processes as static loading is considered to be least troublesome for reliability and performance of the joints (Fig. 1.22).

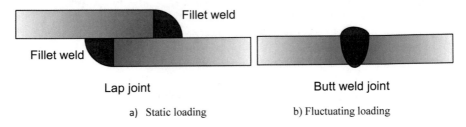

a)   Static loading   b) Fluctuating loading

Fig. 1.22 Schematic of preferred weld joints as per loading conditions

1.3.7 Type of Stress

Similar to the nature of loading during the service, type of stress also does not influence the selection of joining process appreciably; however, still it needs little consideration. The influence of type of stress on choice of joining processes depends on crack nucleation and propagation tendency. Joints subjected to tension, shear and bending load cause easy fracture through cracking; therefore, joining processes that are able to develop sound, free from stress raisers, and minimum metallurgical/mechanical damage should be considered. On contrary, joints primarily used for positioning of connecting members under compression can survive even with moderate discontinuities, stress raisers, metallurgical and mechanical damages; hence, the choice of joining process for fabricating joints for compressive loads becomes somewhat less critical.

1.3.8 Economy of Joining

Economy of joining is measured primarily in terms of cost per unit length of joints developed (Rs./m). For a given set-up of machines, manpower in an established industry economy can be measured in terms of quantity of weld metal deposited per unit time period (kg/day). Economy associated with a joining process certainly affects selection for fabrication of joints in the mass manufacturing industries significantly. However, economy may not be a very important consideration for selection of joining process if joints are to be developed for critical applications or manufacturing of capital goods where quality is of prime importance. High deposition rate process like submerged arc welding and gas metal arc welding processes are considered to be more productive and economical for high volume productions, while shielded metal arc welding processes are popular and economical for depositing weld metal in smaller quantities for general applications. However, high-quality weld joints are not necessarily produced economically; therefore, if quality is of utmost importance, then we should consider for gas tungsten arc welding, plasma arc welding and radiation-based laser and electron beam welding processes for selection. These processes produce quality weld joints but high cost due to use of costly skilled welder requirement, costly equipment and machines, consumable requirement besides low weld metal deposition rate (Fig. 1.23).

1.4 Manufacturing Processes and Welding

The manufacturing technology primarily involves sizing, shaping and imparting desired combination of the properties to the material so that the component or engineering system being produced to perform indented function for design life. A wide

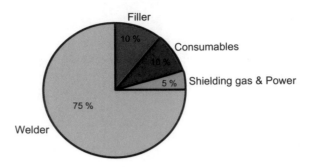

Fig. 1.23 Tentative proportion of various factors determining the cost of welding

range of manufacturing processes has been developed in order to produce the engineering components ranging from simple to complex geometries using materials of different physical, chemical, mechanical and dimensional properties. There are four main manufacturing processes, i.e. casting, forming, machining and welding (Fig. 1.24). Selection of suitable manufacturing process for a produce/component is dictated by complexity of geometry of the component, number of units to be produced, properties of the materials (physical, chemical, mechanical and dimensional properties) to be processed and economics. Based on the approach used for obtaining desired size and shape by different manufacturing processes, these can be termed as positive, negative and/or zero processes.

- Casting: zero process
- Forming: zero process
- Machining: negative process
- Joining (welding): positive process.

Casting and forming are categorized as zero processes as they involve only shifting of metal in controlled (using heat and pressure singly or in combination) way from one region to another to get the required size and shape of product. Machining is considered as a negative process because unwanted material from the stock is removed in the form of small chips during machining for the shaping and sizing of a product purpose. During manufacturing, it is frequently required to join the simple shape components to get desired product. Since simple shape components are brought together by joining in order to obtain desired shape of end useable product, joining is categorized as a positive process.

1.5   Welding and Its Uniqueness

Welding is one of the most commonly used joining techniques for manufacturing engineering components for power, fertilizer, petrochemical, automotive, food processing and many other sectors. Welding generally uses localized heating during

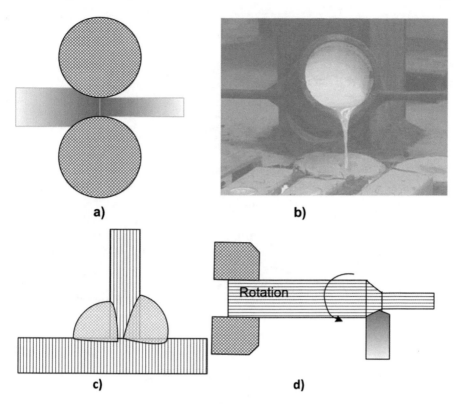

Fig. 1.24 Schematics showing different manufacturing processes: **a** forming, **b** casting, **c** joining and **d** machining

common fusion welding processes (shielded metal arc, submerged arc, gas metal arc welding, etc.) for melting the faying surfaces and filler metal. However, localized and differential heating and cooling experienced by the metal during welding makes it significantly different from other manufacturing techniques:

- Residual stresses are induced in welded components (development of tensile residual stresses adversely affects the tensile and fatigue properties of workpiece).
- Simple shape components to be joined are partially melted.
- Temperature of the base metal during welding in and around the weld varies as function of time (weld thermal cycle).
- Chemical, metallurgical and mechanical properties of the weld are generally anisotropic.
- Little amount of metal is wasted in the form of spatter, run in and run off.
- Process capabilities of the welding in terms of dimensional accuracy, precision and finish are poor.

- Weld joints for critical applications generally need post-weld treatment such as heat treatment or mechanical working to get desired properties or reduce residual stress.
- Problem related to ductile to brittle transition behaviour of steel is more severe with weld joints under low temperature conditions.

1.5.1 Selection of Welding as a Route for Manufacturing

A wide range of welding processes is available to choose from. These were developed over a long period of time. Each process differs in respect of their ability (power density) to apply heat for fusion, protection (shielding) of the weld pool and soundness of weld joint. Therefore, performance of the weld joint developed by a process differs from those developed using other processes. However, selection of a particular process for producing a weld joint is dictated by the size and shape of the component to be manufactured, the metal system to be welded, expected quality of weld joint, availability of consumables and machines, precision required and finally economy. Moreover, irrespective of welding process selected for developing a weld joint, it must be able to perform the intended function for designed life.

1.6 Advantages and Limitations of Welding

Welding is mainly used for the production of comparatively simple shape components. It is the process of joining the metallic components with or without application of heat, pressure and filler metal. Application of welding in fabrication offers many advantages; however, it suffers from few limitations also. Some of the advantages and limitations are given below.

   Advantages of welding are enlisted below:

1. Permanent joint is produced, which becomes an integral part of workpiece.
2. Joints can be stronger than the base metal if good quality filler metal is used.
3. Economical method of joining.
4. It is not restricted to the factory environment.

Disadvantages of welding are listed also below:

1. Labour cost is high as only skilled welder can produce sound and quality weld joint.
2. It produces a permanent joint which in turn creates the problem in dissembling if sub-component required.
3. Hazardous fumes and vapours are generated during welding. This demands proper ventilation of welding area.

4. Weld joint itself is considered as a discontinuity owing to variation in its structure, composition and mechanical properties; therefore, weld joints being of low reliability are usually not commonly recommended for critical applications if avoidable.

1.7 Applications of Welding

Welding processes with their broad field of applications are given below:

- Resistance welding: Automobile and electronic industry
- Thermite welding: Rail joints in railways
- Tungsten inert gas welding: Aerospace and nuclear reactors
- Submerged arc welding: Heavy engineering and ship building
- Gas metal arc welding: Joining of metals (stainless steel, aluminium and magnesium) sensitive to atmospheric gases where reasonably good quality weld joints needed for manufacturing.

The welding is widely used for fabrication of pressure vessels, bridges, building structures, aircraft and spacecraft, railway coaches and general applications besides shipbuilding, automobile, electrical, electronic and defence industries, laying of pipe lines and railway tracks and nuclear installations. Specific components need welding for fabrication.

1. Transport tankers for transporting oil, water, milk, etc.
2. Welding of tubes and pipes, chains, LPG cylinders and other items
3. Fabrication of steel furniture, gates, doors and door frames, and body
4. Manufacturing white goods such as refrigerators, washing machines, microwave ovens and many other items of general applications.

The requirement of the welding for specific area of the industry is given in the following section.

1. Oil and gas: Welding is used for joining of pipes, during laying of crude oil and gas pipelines, construction of tankers for their storage and transportation. Offshore structures, dockyards, loading and unloading cranes are also produced by welding.
2. Nuclear industry: Spheres for nuclear reactor, pipe line bends and joining of pipes carrying heavy water require welding for safe and reliable operations.
3. Defence industry: Tank body fabrication and joining of turret mounting to main body of tanks are typical examples of applications of welding in defence industry.
4. Electronic industry: Electronic industry uses welding to limited extent; e.g. joining leads to special transistors, but other joining processes such as brazing and soldering are widely used.
5. Soldering is used for joining electronic components to printed circuit boards (PCBs).

6. Robotic soldering is very common for joining of parts to printed circuit boards of computers, television, communication equipment and other control equipment, etc.
7. Electrical industry: Components of both hydro and steam power generation systems, such as penstocks, water control gates, condensers, electrical transmission towers and distribution system equipment, are fabricated by welding. Turbine blades and cooling fins are also joined by welding.
8. Surface transport.
9. Railway: Railway uses welding extensively for fabrication of coaches and wagons, repair of wheel, laying of new railway tracks by mobile flash butt welding machines and repair of cracked/damaged tracks by thermite welding.
10. Automobiles: Production of automobile components like chassis, body and its structure, fuel tanks and joining of door hinges require welding.
11. Aerospace industry: Aircraft and Spacecraft—Similar to ships, aircraft were produced by riveting in early days but with the introduction of jet engines welding is widely used for aircraft structure and for joining of skin sheet to body.
12. Space vehicles: These systems encounter frictional heat as well as low temperature requirement of outer skin and other parts of special materials. These materials are welded with full success for achieving safety and reliability.
13. Ship industry: Ships were produced earlier by riveting. Welding found its place in ship building around 1920, and presently, all welded ships are widely used. Similarly, submarines are also produced by welding.
14. Construction industry: Arc welding is used for construction of steel building structures leading to considerable savings in steel and money. In addition to building, huge structures such as steel towers also require welding for fabrication.

Questions for Self-assessment

*   How does fusion weld joint offer different performances from those developed by solid state joining and solid/liquid joining?
*   Does the metal system of two components to be joined affect the choice of joining process?
*   Elaborate the conditions where non-fusion joining processes would be more appropriate.
*   How does difference of mechanisms of developed joints differ from each other?
*   How is welding different from other manufacturing processes?
*   What are the factors affecting the selection of manufacturing processes?
*   Do the properties of the metals to be joined dictate the selection of the suitable joining process?
*   Explain the factors to be considered for selection of a joining method.
*   How can common manufacturing processes be compared in the light of their fundamental principles?
*   Compare the welding with other common manufacturing processes.

Further Reading

American Welding Society (1983) Welding handbook, 7th edn, vol 1 & 2. USA

Dwivedi DK (2013) Production and properties of cast Al-Si alloys. New Age International, New Delhi

Dwivedi DK (2018) Surface engineering. Springer, New Delhi

Groover MP (2010) Fundamentals of modern manufacturing: materials, processes, and systems. Wiley, USA

Little R (2001) Welding and welding technology. 1st edn

Chapter 2
Classification of Joining Processes

Numerous joining processes have been developed over a period of time. These processes help in development of joints using different mechanisms like fusion, deformation, diffusion, chemical and metallurgical reactions, and accordingly, the performance of these joints also varies significantly. Classification of these joining processes in certain group/category on the basis of their fundamental characteristic/feature/approach helps in number of ways: (a) increased understanding about the joining processes (in same group) based on similar fundamental feature/approach/mechanism/heat source, (b) ease of giving a name for new/hybrid joining process based on approach/similarity, (c) grouping various joining processes based on fundamental similarity in their nature, (d) making communication easier in joining by having just one broad name for each process and (e) better organizing information about joining in better way. Joining processes can be grouped on the basis of many technological factors, approaches and the unique ways by which joints are developed. Some of these factors are given below.

2.1 Classification Based on Approach of Joining Processes

Classification based on the mechanism/approach of developing joints: Primarily, the way metallic pieces are united together can be done as follows: (a) fusion of faying surfaces of the components to be joined is realized for developing a joint; therefore, all such joining processes are termed as *fusion weld process*, (b) molten metal is made available (from outside/other source) between components to be joined similar to that of casting so these joining processes are called *cast weld processes*, (c) heating of metal faying surfaces primarily to plasticize and then to forge them together using suitable pressure; thus, all processes based on this approach are called *resistance weld processes*, and (d) plastic deformation at the faying surfaces of component mating (interfaces) using suitable compressive, bending, shear force to produce a joint in solid state only is called *solid state joining process*.

© The Author(s), under exclusive license to Springer Nature Singapore Pte Ltd. 2022   33
D. K. Dwivedi, *Fundamentals of Metal Joining*,
https://doi.org/10.1007/978-981-16-4819-9_2

Further, there are many other joining processes aimed at developing (a) joints and (b) surfaces with desired properties for enhanced performance/life during the service. These processes are termed as allied joining processes. Some of the allied joining processes used for developing joints include adhesive bonding, brazing, soldering, weld bonding, crimping, while those used for improving surface properties are metal depositing processes, weld surfacing, thermal spraying, coating.

There are no criteria, parameters, and approaches, which are perfect for classification all the joining processes without doubt as there are always few technical difficulties encountered in these classifications. These classifications have been described with such technical difficulties associated with such categorizations:

2.1.1 Fusion Weld Processes

Those welding processes in which faying surfaces of components to be joined are brought to the molten state by applying heat and followed by solidification during cooling are classified as fusion weld processes. Cooling rate experienced by weld metal in these processes is much higher than casting. The heat required for melting can be produced using electric arc, plasma, laser and electron beam and combustion of fuel gases (Fig. 2.1). Probably, this is undisputed way of classifying few welding processes. Still electro-slag and electro-gas welding processes find place in other classification/grouping of the joining processes due to few peculiar technological characteristics of these processes which will be described later in the chapter. Common fusion welding processes are given below:

- Shielded metal arc welding
- Submerged arc welding
- Gas metal arc welding
- Carbon arc welding
- Gas tungsten arc welding
- Plasma arc welding
- Electro-slag welding
- Electro-gas welding
- Laser beam welding
- Electron beam welding
- Oxy-fuel gas welding.

2.1.2 Cast Weld Processes

The cast weld processes are those in which either molten weld metal is supplied from external source or metal is melted and solidified at very low rate during solidification like casting process. The following are two common welding processes that are grouped under cast welding processes:

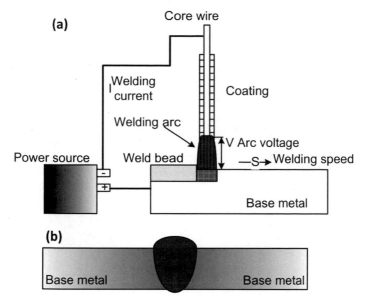

Fig. 2.1 Schematic of **a** shielded metal arc welding and **b** typical fusion weld joint

- Thermite welding
- Electro-slag welding
- Electro-gas welding.

In thermite welding, the molten metal is produced externally using heat generated by exothermic chemical reactions and then fed between the components to be joined (Fig. 2.2). While in case of electro-slag and electro-gas welding processes, the weld metal is produced through fusion of faying surfaces of components (to be joined) and melting of the electrode by electrical resistance heating and followed by very slow cooling of the weld pool to complete the solidification (almost similar to that of the casting process).

This classification, however, is true for thermite welding where like a casting process molten melt is supplied from external source and subsequently weld metal is solidified at comparatively low cooling rate. While in case of electro-slag/electro-gas welding processes, weld metal is obtained by melting of both electrode and base metal and molten metal is not supplied from outside. Therefore, this classification is not perfect.

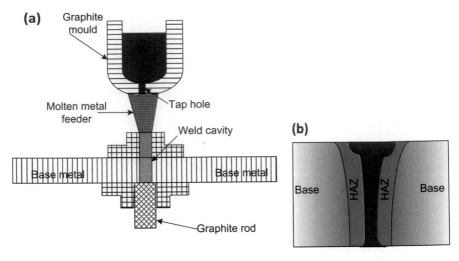

Fig. 2.2 Schematic of **a** thermite welding set-up and **b** weld cross section of thermite weld joint

2.1.3   Resistance Weld Processes

The welding processes in which heat required for thermal softening or partial melting of base metal is generated by electrical resistance heating followed by application of pressure for consolidation to develop a joint are grouped as resistance welding processes. Spot welding and seam welding are the most common resistance welding processes (Fig. 2.3). Heat is generated by flow of high current (few tens of thousands of amperes) for very short period, and interfacial contact resistance between the components to be joined becomes of great importance. This criterion fits good for grouping/classification of the most of resistance welding processes perfectly except flash butt welding. In case of flash butt welding, the process begins with sparks between components (to be joined) during welding and so heat of sparks is used in this joining process and there is not heat generation due to electrical resistance heating.

- Resistance welding processes
- Spot welding
- Projection welding
- Seam welding
- High-frequency resistance welding
- High-frequency induction welding
- Resistance butt welding
- Flash butt welding
- Stud welding.

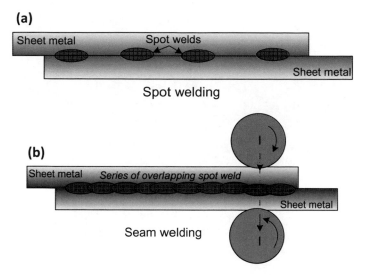

Fig. 2.3 Schematics showing cross section of **a** spot weld joints and **b** seam weld joints

2.1.4 Solid State Welding Process

The solid-state welding processes are those in which weld joint is developed mainly by application of pressure and heat through various mechanism such as mechanical interlocking, large scale interfacial plastic deformation and diffusion etc. fall under this category. Heat is primarily used for thermal softening and increases diffusion to accelerate the joining process and develop a sound joint. Depending up on the amount of heat generated/used during the solid-state joining, these processes are further categorized as high or low heat input solid state joining processes.

- Low heat input solid state joining processes

  Ultrasonic welding
  Cold pressure welding
  Explosion welding

- High heat input solid state joining processes

  Friction welding
  Forge welding
  Diffusion welding.

The flow chart (below) is showing classification of welding and allied processes for better understanding of nature of a specific process.

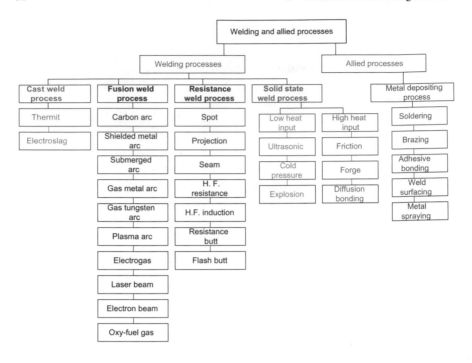

2.2 Classification Based on Technological Factors of Welding Processes

Welding is a process of joining metallic components with or without application of heat, with or without pressure and with or without filler metal. A range of welding processes has been developed using singly or a combination of said factors, namely heat, pressure and filler. Welding processes can be classified on the basis of the following technological factors:

- Filler: Welding with or without filler material
- Heat source: Source of energy for welding
- Arc: Arc and non-arc welding
- Fusion: Fusion and pressure welding.

2.2.1 Welding With/Without Filler Material

A weld joint can be developed just by melting of edges (faying surfaces) of plates or sheets to be welded especially when thickness is lesser than 5 mm. A weld joint developed by melting the faying surfaces and subsequently solidification only (without

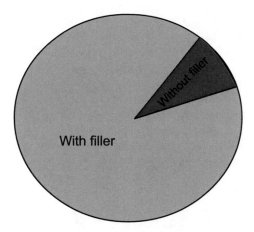

Fig. 2.4   Pie chart shows tentative industrial usage of welding processes with and without filler

using any filler metal) is called "autogenous weld". Thus, the composition of the autogenous weld metal largely corresponds to the base metal only except if there is any of loss of alloying element (a) due to evaporation (b) interactions with other elements or gases present in weld pool to form inclusion/slag. However, autogenous weld can be crack sensitive if solidification temperature range of the base metal (to be welded) is significantly high (50–100 °C). In general, the welding processes with application of filler dominate over without filler processes (Fig. 2.4) due to a very wide range of reasons. The following arc typical welding processes in which *filler metal is generally not used* to produce a weld joint for joining **thin section** components.

- Laser beam welding
- Electron beam welding
- Gas tungsten arc welding
- Plasma arc welding
- Resistance welding
- Friction stir welding.

However, for welding of thick section plates using any of the following processes filler metal can be used as per need. Autogenous fusion welds during joining of thick plates may result in concave weld or under fill like discontinuity in the weld joint. The composition of the filler metal used for welding of two metallic members can be similar to that of base metal or different (as per need to develop a sound joint with desired combination of properties); accordingly, weld joints are categorized as homogeneous or heterogeneous weld, respectively.

Solidification of the metals usually occurs in two stages: (a) nucleation and (b) growth to consume the entire molten metal to complete the solidification sequence. However, in case of autogenous and homogeneous welds, solidification occurs directly by growth mechanism without nucleation stage. This type of solidification is called epitaxial solidification. The solidification in heterogeneous welds takes place in conventional manner in two stages, i.e. nucleation and growth. The autogenous and homogeneous welds are considered to be lesser prone to the development of weld discontinuities than heterogeneous weld because (a) of more uniformity in composition of the weld metal and (b) solidification largely takes place at a constant temperature (with very narrow/no solidification temperature range). Metal systems having a wide solidification temperature range show issues related to segregation, solidification cracking, partial melting zone formation and liquation cracking tendency. The following are few fusion welding processes where *filler may or may not* be used for developing weld joints:

- Laser beam welding
- Electron beam welding
- Gas tungsten arc welding
- Plasma arc welding
- Gas welding.

Few welding processes are inherently designed to produce a weld joint by applying heat for melting the faying surfaces of both base metals and filler metal. Thus, the filler is bound to be applied (between the components to be joined) whenever the following processes are used. Therefore, processes of category are mostly used for welding of thick plates (usually section thickness > 5 mm) with comparatively high deposition rate.

- Metal inert gas welding
- Submerged arc welding
- Flux-cored arc welding
- Electro-gas/electro-slag welding.

Comment: During early days of welding technology development, the gas welding process was the only (then) available fusion welding process using which joining could be achieved with or without filler material. The gas welding performed without filler was termed as autogenous welding. However, later with the development welding technologies in the form of tungsten inert gas welding, plasma arc welding, electron and laser beam welding and many other such welding processes, classification of welding processes based on filler (with/without) created confusion as many welding processes were falling in both the categories.

2.2.2 Source of Energy for Welding

Almost all types of the weld joints are produced by applying energy in one or other form to develop atomic/metallic bond between metallic components being joined. The weld joints are realized using approaches like: (a) by melting the faying surfaces using heat, (b) by localized plastic deformation using pressure either at room temperature or high temperature (0.5–0.9 times of Tm where Tm is melting temperature

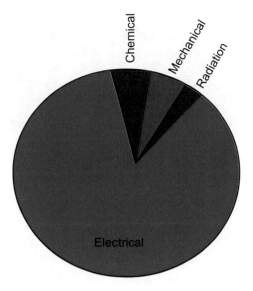

Fig. 2.5   Pie chart shows tentative use of type of energy in welding processes

of base metal in Kelvin), (c) diffusion across the mating interfaces using heating at high temperature. Based on the type of energy being used for creating metallic bonds between the components to be welded, welding processes can be grouped as under. The use of electrical energy dominates over the other energies for welding purpose (Fig. 2.5).

- Chemical energy: Gas welding, explosive welding, thermite welding
- Mechanical energy: Friction welding and ultrasonic welding
- Electrical energy: Arc welding, resistance welding, electro-slag and electro-gas welding
- Radiation energy: Laser beam welding and electron beam welding.

Comments: Weld joints are realized using energy in various forms, namely chemical, electrical, light, sound, mechanical energy. However, except chemical energy all other forms of energies are generated from electrical energy for welding. Hence, categorization of the welding processes based on the source of energy parameter also does not justify classification perfectly.

2.2.3   Arc and Non-arc Welding

Metallurgical continuity between metallic components can be developed either (a) by using heat for melting of the faying surfaces of member to be joined and then allowing it to solidify or (b) by applying pressure on the components to be joined (under ambient conditions or at elevated temperature) for developing joints through

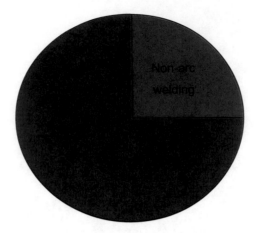

Fig. 2.6 Pie chart shows tentative use of arc and non-arc welding processes

mechanical interlocking, diffusion, etc. All those welding processes in which heat for melting the faying surfaces is provided by arc are classified as arc welding processes. The arc can be established either between the base plate and an electrode or between electrode and nozzle. Further, the use of arc welding processes dominates over non-arc welding processes (Fig. 2.6). According to this classification, rest of welding processes in which a weld joint is produced using either pressure or heat generated from sources other than arc, namely chemical reactions, friction, radiation, etc., are categorized as non-arc welding processes. Welding processes corresponding to each group are given below.

- *Arc welding processes*

    - Shielded metal arc welding: Arc between base metal and covered electrode
    - Gas tungsten arc welding: Arc between base metal and tungsten electrode
    - Plasma arc welding: Arc between base metal and tungsten electrode
    - Gas metal arc welding: Arc between base metal and consumable electrode
    - Flux-cored arc welding: Arc between base metal and consumable electrode
    - Submerged arc welding: Arc between base metal and consumable electrode

- *Non-arc welding processes*

    - Resistance welding processes: uses electric resistance heating
    - Gas welding: uses heat from exothermic chemical reactions
    - Thermite welding: uses heat from exothermic chemical reactions
    - Ultrasonic welding: uses both pressure and frictional heat
    - Diffusion welding: uses electric resistance/induction heating to facilitate diffusion
    - Explosive welding: involves pressure.

Comments: Arc and non-arc welding process classification leads to grouping of all the arc welding processes in one category, while all other welding processes are categorized as non-arc welding processes. However, welding processes such as electro-slag welding (ESW) and flash butt welding (FBW) were found difficult to be classified in either of the two categories as ESW process starts with arcing, and subsequently on melting of sufficient amount flux, the arc is extinguished and heat for melting of faying surfaces of the base metal is generated by electrical resistance heating by flow of current through molten flux/metal. While in case of flash butt welding, tiny arcs i.e. sparks are established during initial stage of the FBW process followed by forging/upsetting action. Therefore, classification of welding processes based on this factor is also found not to be perfect.

2.2.4 Pressure and Fusion Welding

Welding processes in which heat is primarily applied for melting the faying surfaces of the base metal to develop a weld joint are grouped as fusion welding processes, while other processes in which pressure is primarily applied (with little or no application of heat for thermal softening of metal up to the plastic state) for developing a weld joint are grouped as pressure welding processes. Since there is no fusion of faying surfaces, processes falling in group are also termed as solid state welding processes. Moreover, industrial application of fusion welding processes dominates over pressure welding processes (Fig. 2.7).

- Pressure/solid state welding process

    - Resistance welding processes (spot, seam, projection, flash butt, arc stud welding)
    - Ultrasonic welding
    - Diffusion welding

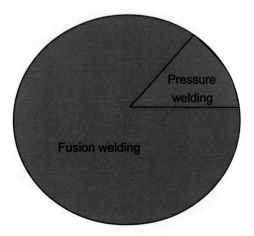

Fig. 2.7 Pie chart shows tentative use of fusion and pressure welding processes

- Explosive welding
- Fusion welding process
  - Gas welding
  - Shielded metal arc welding
  - Gas metal arc welding
  - Gas tungsten arc welding
  - Submerged arc welding
  - Electro-slag/electro-gas welding.

Comments: This factor (fusion welding or pressure welding) is one of most widely used classifications as it covers all processes in both the categories irrespective of heat source and welding with or without filler material. In fusion welding, all those processes are included in which molten metal solidifies freely, while in pressure welding, molten metal, if any, is retained in confined space (as in case of resistance spot welding or arc stud welding) and solidifies under pressure or semi-solid metal cools under pressure. This type of classification poses no problems, and therefore, it is considered as one of the best ways for the classification of the welding processes.

Questions for Self-assessment

- What are the factors on the basis of which welding processes can be classified?
- Describe difficulties encountered in classification of various welding processes?
- Which is the most preferred way of classifying welding processes?
- What is autogenous weld?
- How does filler metal affect the solidification of the weld?
- How can fusion welding processes be distinguished from pressure welding processes?
- What are fundamental mechanisms of developing weld joints by pressure welding processes?
- What are difficulties in classifying the following welding processes?

  - Electro-slag welding
  - Flash butt welding
  - Stud welding.

- What are cast weld processes?
- Explain the difficulty in putting electro-slag welding under cast weld processes?
- Define fusion welding processes and enlist fusion welding processes.
- What is rationale behind putting the flash butt welding and high-frequency induction welding process under resistance welding processes?
- Define solid state welding processes?
- What are fundamental mechanisms of developing weld joints using solid state processes?

Further Reading

American Society for Metals (1993) Metals handbook-welding, brazing and soldering, 10th edn, vol 6, USA

American Welding Society (1987) Welding handbook, 8th edn, vol 1 & 2. USA

Cary H (1988) Welding technology. 2nd edn. Prentice Hall

Little R (2001) Welding and welding technology. 1st edn. McGraw Hill

Nadkarni SV (2010) Modern arc welding technology. Ador Welding Limited, New Delhi

Parmar RS. Welding process and technology. Khanna Publisher, New Delhi

Chapter 3
Heat Generation and Protection of Weld

Heat is generally supplied to faying surfaces of the components to be joined by using either fusion welding or pressure/solid state welding to produce a sound weld joint. Heat can be supplied from external source like arc, flame, radiation beam, electrical resistive heating, molten metal or generated inherently during welding itself by friction, deformation, etc. Need of heat source during welding will be presented first, and thereafter, common sources of heat generation for welding, factors affecting heat generation and effect of heat generation on weld joint performance will be described. Metals become very reactive to surrounding atmospheric gases and impurities with increase of temperature. Heating of faying surfaces of base metals (during the welding) either to molten state or to a temperature high enough to cause thermal softening makes them active/sensitive enough for oxidation, decolourization and formation impurities like inclusion and slag. Thereafter, protection of molten metal pool during welding becomes crucial. Therefore, need for protecting the weld pool and approaches available for protecting the molten weld pool with the common welding processes will be described.

3.1 Need of Heat in Welding

Different mechanisms namely fusion, plastic deformation, diffusion and metallurgical reactions are used to realize the metallurgical continuity between two or more metallic components for development of a weld joint with application of heat. Depending upon the mechanism associated with a given welding process, the heat can be applied for (a) melting of faying surfaces of base metal in case of fusion welding processes, (b) thermal softening of the faying surfaces to plastic state in case of resistance welding and friction welding, (c) melting of the filler (solder/brazing metal) and heating base metal only to a high temperature high enough (but not for melting) during brazing, soldering and thermite welding and (d) facilitating diffusion across the mating interfaces during diffusion welding. Additionally, heating of the faying

D. K. Dwivedi, *Fundamentals of Metal Joining*,
https://doi.org/10.1007/978-981-16-4819-9_3

surfaces helps in getting the cleaner faying surfaces for welding through removal of absorbed gases present at the surfaces, decomposition of oxides and other compounds and evaporation of moisture and organic impurities like oil, paint, grease, etc. Thus, it can be noted that depending upon the welding process, heat is applied for different purposes.

3.2 Heat Generation

Principle of heat generation for welding heavily depends on welding process. Heat for the welding can be generated using following principles:

- Chemical reactions: gas and thermite welding
- Electrical resistive heating: resistance welding
- Frictional heating: friction and ultrasonic welding
- Electric arc heating: arc welding process
- Radiation heating: microwave, laser
- Interfacial friction or impact: ultrasonic, explosion welding.

3.2.1 Chemical Reactions

Gas welding

Oxy-fuel gas welding and thermite welding processes primarily involve exothermic chemical reactions by combustion of oxygen–fuel mixture and chemical reactions between mixture of oxides and metallic powders, respectively. Oxy-fuel gas welding process can use different type fuel gases as per requirement for amount of heat, temperature, economics and availability. The flame of oxy-fuel gas welding can be composed of two or three zones as per ratio of oxygen and fuel gas mixture. Accordingly, chemical reactions for combustion occurring in primary and secondary zones are given below and shown in Fig. 3.1. Heat generated and maximum temperature attained in the flame of different oxygen–fuel gas mixtures is given in Table 3.1. It can be observed that total heat generated (sum of heat in primary and secondary zones) by propane is higher than other fuel gases, while the peak temperature generated by flame using oxy-acetylene mixture is higher than other oxy-fuel gas mixtures. This is evident from the following chemical reactions and heat generated in respective zones.

Primary zone : $C_2H_2 + O_2 \rightarrow 2CO + H_2 + 448\,kJ/mol$ ($18.75\,MJ/m^3$ of acetylene)

Secondary zone: $4CO + 2H_2 + 3O_2 \rightarrow 4CO_2 + 2H_2O + 812\,kJ/mol$ ($35.77\,MJ/m^3$)

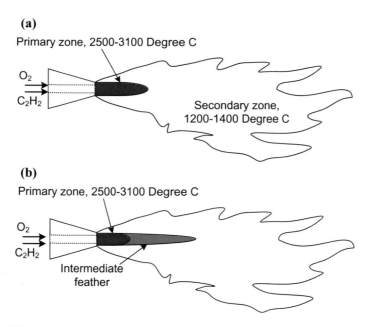

(a)

Primary zone, 2500-3100 Degree C

O_2

C_2H_2

Secondary zone,
1200-1400 Degree C

(b)

Primary zone, 2500-3100 Degree C

O_2

C_2H_2

Intermediate
feather

Fig. 3.1 Schematics showing different zones in **a** oxidizing/neutral flame and **b** carburizing flame

Table 3.1 Heat generation and peak temperature in different fuel gas flames

Fuel gas	Formula	Primary heat (MJ/m^3)	Secondary heat (MJ/m^3)	Total heat (MJ/m^3)	Peak temperature $(°C)$
Acetylene	C_2H_2	19.0	36	55	3100
Propane	C_3H_6	16.4	71.6	88	2500
Hydrogen	H_2	–	–	10	2390
Natural gas	$CH_4 + H_2$	0.4	36.6	37	2350

The heat generated in inner (primary) zone is found to be lower than outer (secondary) zone, while peak temperature observed in the two zones exhibits reverse trend, i.e. higher temperature in the primary zone than secondary zone. This is an unexpected trend/relationship between peak temperature and heat generated in different zones of oxy-acetylene flame which is primarily due to the difference in two conditions experienced by primary and secondary zones for transfer of heat conversely localization of heat. These two conditions affecting heat localization and so the peak temperature are (a) **surface area** of a zone affects the localization/transfer of heat directly, i.e. more surface area for secondary zone than primary zone which in turns results lesser heat localization (owing to higher heat transfer rate), and therefore, secondary zone despite of high heat generation shows up the lower temperature than primary zone and (b) **surrounding temperature** for primary and secondary zones affecting heat localization/transfer. Primary zone experiences more heat localization than secondary zone owing to lesser heat transfer rate as primary zone is surrounded by hot gases of secondary zone; while secondary zone is directly exposed to ambient conditions facilitating higher heat transfer rate than the primary zone, and therefore, secondary zone experiences lesser heat localization than the primary zone. To achieve high melting rate, the gas welding torch is therefore adjusted in such a way that the peak temperature/primary zone of the flame touches the faying surface of the base metal.

Peak temperature and heat generation both are important in welding as per following characteristics of the base metal (a) physical properties such as thickness, melting point, thermal expansion coefficient, (b) microstructure/heat sensitivity and (c) tendency for distortion and residual stress. Welding of high melting point metals requires an oxy-fuel gas mixture for gas welding process to produce flame of high peak temperature enough to achieve the melting of faying surfaces. In addition to required peak temperature, oxy-fuel mixture for gas welding of thick base metal of high thermal conductivity should generate heat high enough for melting of faying surfaces.

An oxy-fuel mixture having such volume (proportion) of oxygen which just enough for complete combustion of the fuel gas results in a neutral flame and corresponding ratio of oxy-fuel mixture is called stoichiometric ratio. The neutral flame (stoichiometric ratio) in case of oxy-acetylene gas mixture is obtained when there is equal proportion of acetylene and oxygen, i.e. ratio of volume fraction of oxygen and acetylene is 1. The proportion of acetylene greater and lesser than one results in carburizing and oxidizing flames, respectively. It is mostly preferred to use neutral flame for gas welding over carburizing and oxidizing flames because carburizing and oxidizing flames tend to alter the chemistry (chemical composition) of the weld metal, and so microstructure, mechanical and tribological properties are affected accordingly. A carburizing flame during welding steel increases the carbon content which in turn increases the fraction of hard micro-constituents in steel like pearlite, cementite and martensite, and so hardness, tensile strength, embrittlement and abrasive wear resistance are increased but at the cost of ductility and toughness. On the other hand, application of oxidizing flame for welding of steel tends to modify the chemical composition due to oxidation of elements having high affinity to the oxygen at elevated temperature and get removed as slag. Loss of alloying elements will be affecting microstructure and mechanical properties.

Moreover, peak temperature and heat input (kJ/mm) offered by a welding process largely have inverse relationship and are governed by power density of the welding process in consideration. In general, high-power density processes (like laser, electron beam welding) offer higher peak temperature and lower heat input (required for welding) than low energy density welding process (such as gas welding, shielded metal arc welding).

Thermite Welding

The heat for joining of steel rails by thermite welding is also generated by exothermic chemical reactions of metal oxides (iron oxide, copper oxide) and with reactive metals

$$3FeO + 2Al > Al_2O_3 + 3Fe + 880 \text{ kJ}$$

$$Fe_2O_3 + 2Al > Al_2O_3 + 2Fe + 850 \text{ kJ}$$

$$3CuO + 2Al > Al_2O_3 + 3Cu + 880 \text{ kJ}$$

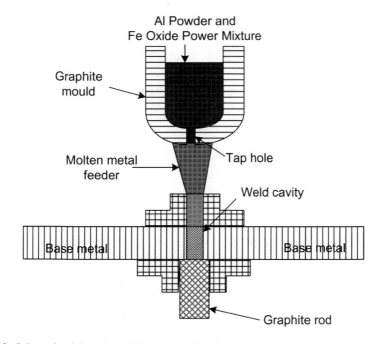

Fig. 3.2 Schematic of thermite welding system showing components

like aluminium and magnesium (Fig. 3.2). Temperature rise can vary from 1400 to 2500 °C. Thermite is a trade name, for a mixture of granular aluminium metal and powdered iron oxide. Usually mixture is ignited with ribbon of Mg, and it gives off large amounts of heat.

$$3FeO + 2Al > Al_2O_3 + 3Fe + 880 \text{ kJ}$$
$$Fe_2O_3 + 2Al > Al_2O_3 + 2Fe + 850 \text{ kJ}$$
$$3CuO + 2Al > Al_2O_3 + 3Cu + 880 \text{ kJ}$$

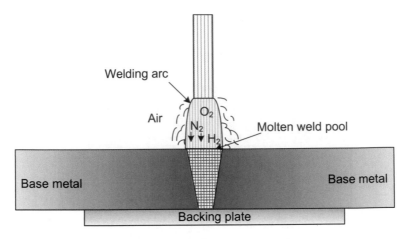

Fig. 3.3 Schematic of arc welding system showing common members

3.2.2 Electric Arc

Heat generation by electric arc depends on the welding parameters (current and arc voltage), shielding gases, fluxes, if any. Heat generated by welding arc (power of arc) is given by product of VI where V arc voltage (V) and I welding current (A) as shown in Fig. 3.3. Since during the welding, arc is moving, and therefore, net heat applied to the faying surfaces of components being joined is obtained from ratio VI/S in J/mm (where S is welding speed mm/s). However, upper and lower limits of arc voltage (10–100 V) and welding current (50–2000 A) can vary significantly with type of welding process. Few welding processes (GTAW, PAW) work at low current (50–250 A) and low voltage (10–20 V), while other welding processes like SMAW use high arc voltage (30–100 V) and high current (100–500 A). SAW works at high welding current (500–2000 A). Accordingly, heat generation and power density associated with each process is found to be different.

3.2.3 Resistance Heating

Heat is generated by electric resistance heating in case of all resistance welding processes like spot welding, seam welding, induction welding, projection welding, etc. Heat generated by electrical resistance heating during these welding processes is given by I^2Rt where I welding current (A), R contact resistance or electrical resistance at interface (ohms) and t is time (millisecond). Maximum contact resistance is obtained at interface of mating components to be joined. Therefore, for a given set of welding conditions (welding current and time), the maximum heat generation and so peak temperature are also realized at contacting interface which in turn facilitates thermal softening and even partial melting of mating surfaces for joining (Fig. 3.4).

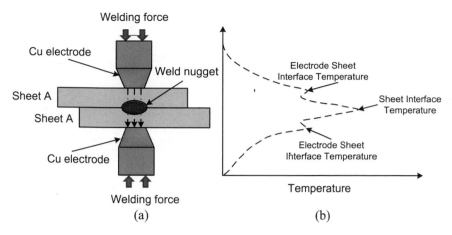

Fig. 3.4 Schematic shows **a** formation of nugget at contact interface in case of resistance spot welding process and **b** variation in temperature across the plates being joined

Pressure and surface finish of contacting surfaces affect the contact resistance so the heat generation for welding. In general, increase in pressure and surface finish increases the metallic contact at mating interfaces and reduces contact resistance, so heat generation due to electrical resistive heating is also reduced. Further, the heat generated at contact interface between the copper electrode and workpiece surface (although unavoidable) is considered to be harmful as it increases (a) tendency of electrode impression on the surface of the workpiece (due to electrode pressure) which adversely affects the appearance and tensile shear load-carrying capacity owing to stress concentration caused by impression and (b) increased electrode weld possibility due to thermal softening of electrode materials and frequent rubbing with workpiece surface during the welding.

3.2.4 Friction and Deformation Heat

Heat generation by friction in friction-based welding processes namely friction welding, friction stir welding, ultrasonic welding softens the metal at faying surfaces. The frictional heat generation rate (W/J/s) is obtained from: $\eta \times F \times v$ [Where η is the fraction of energy (energy lost in friction) converted into heat (varies from 0 to 1), F is friction force (N) obtained from product of normal force (N) friction coefficient (μ) and v is relative speed (m/s)]. Heat generated leads to thermal softening and so improved plastic flow behaviour which in turn assists in joining through consolidation under forging pressure (Fig. 3.5). Although during friction-based welding processes, in addition to friction, heat can also be generated due to severe plastic deformation of metal like in friction stir welding process and ultrasonic welding, if any.

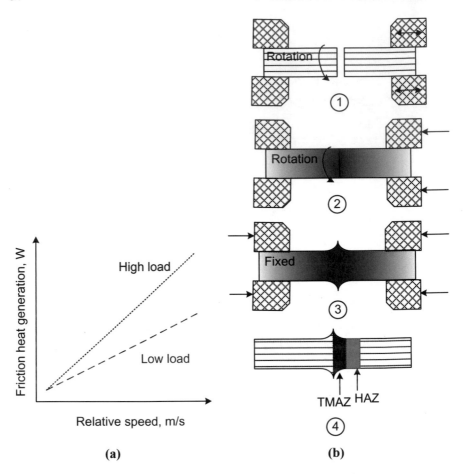

Fig. 3.5 Schematics show **a** effect of relative speed, normal load and **b** sequential steps of friction welding using frictional heat

3.3 Effect of Heat Generation and Weld Joint Characteristics

Application of heat during welding, apart from facilitating the melting of faying surfaces of base metal for developing a joint can cause (a) increased sensitivity of metal to react with atmospheric gases in arc zone, if any, (b) development of heat affected zone with undesirable characteristics and (c) unfavourable cooling rate leading to coarsening of grain structure, formation of undesirable micro-constituents resulting in embrittlement, cracking and loss of toughness of the weld joints.

Increase in temperature of metal (both in solid and molten state) with the application of heat during welding increases the chemical reactivity/affinity with gases present in and around the weld pool. This affinity can be observed in two ways (a)

increased solubility of gases in both liquid and solid states and (b) increased tendency for chemical compound formation like oxides, nitrides and carbides. These factors can lead to the development of discontinuities such as porosity and inclusions.

Usage of heat during welding causes many mechanical and metallurgical changes in base metals adjacent to the fusion zone (weld metal) called heat affected zone (HAZ). The HAZ may get softened/hardened depending upon its response to applied heat. Both hardening and softening of HAZ are considered undesirable as these deteriorate the mechanical performance and expected service life.

Cooling rate experienced by weld pool during the welding directly affects the solidification time. The solidification time of weld pool in turn determines (a) **production time** and so the productivity like in thermite welding, (b) **inclusion and gas entrapment:** tendency for natural rejection of impurities like gases, inclusions due to reduction in solubility with reduction in temperature during cooling (longer the solidification time lower will be inclusion and gas entrapment tendency), (c) **microstructure** of the weld metal and heat affect zone in respect of both phase and grain structure which in turn governs the mechanical properties of the weld joints and (d) **chemical homogeneity** (alloy segregation tendency) of the weld metal: short time of weld solidification results in less segregation tendency and so more uniformity in chemical composition of the weld metal which in turn affects the solidification cracking tendency (Fig. 3.6).

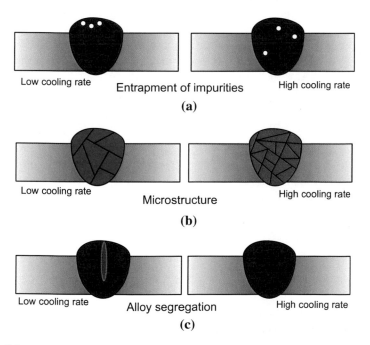

Fig. 3.6 Schematics showing effect of cooling rate on **a** entrapment of impurities **b** microstructure and **c** segregation

Additionally, heat generation can affect the chemical composition of the weld metal if base metal/filler metal are brought to molten state. Excessive heat generation increases the temperature, weld pool size and solidification time which in turn so possibility of loss of alloying element through evaporation, oxidation and slag formation. Few welding processes like gas welding, submerged arc welding and flux-cored arc welding are known to affect the weld metal composition appreciably due to carburization and oxidation in case of gas welding, oxygen enrichment in case of submerged arc welding and alloy addition as per need by flux cored arc welding. Alternation in chemical composition brought in by welding processes changes the microstructure and mechanical properties. These aspects must be kept in mind.

3.4 Protection of Weld Pool

In fusion welding, application of heat of the flame, arc, plasma and radiation results in melting of faying surfaces of the components to be welded. At high temperature, metals become very reactive to atmospheric gases such as nitrogen, hydrogen and oxygen present in and around the arc environment. These gases either get dissolves in weld pool or form their compound which in turn leads to development of weld discontinuities. These discontinuities adversely affect the soundness of the weld joint and mechanical performance. For example, increase in nitrogen concentration in the weld metal of steel increases the tensile strength but at the cost of ductility and toughness primarily due to the formation of hard and brittle needle shape iron nitride (Fig. 3.7). Similarly, excessive oxygen content in the weld metal increases the porosity and oxide inclusion formation tendency which in turn deteriorates the mechanical performance of the weld joint (Fig. 3.8). While hydrogen in weld metal is known to cause pin hole porosity in weld joint of aluminium alloys and hydrogen-induced cracking in hardenable steel weld joints. Following three approaches are commonly used to protect the weld pool from atmospheric gases.

- Developing a firm shroud/blanket/cover/envelop of protective gases around arc and weld pool so as to isolate molten weld pool metal from the atmospheric gases using either inactive gases like carbon dioxide during GMAW and SMAW or inert gases like argon and helium in case of GTAW, PAW and MIGW processes.
- Covering the molten weld pool with molten flux during SAW and ESW processes so as to avoid/reduce the possibility of interactions between atmospheric gases and molten metal
- Developing suitable vacuum environment for welding like in case of EBW.

3.4.1 Forming Envelop/shroud of Inactive or Inert Gas

Two types of gases are being used to protect the molten weld metal during fusion welding by forming shroud of (a) inactive gases such as carbon dioxide and (b)

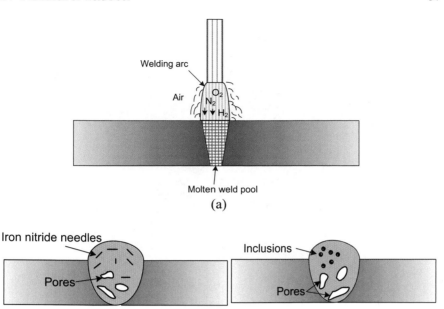

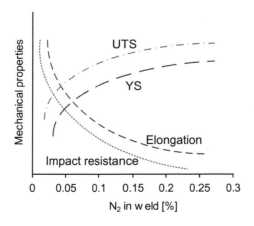

Fig. 3.7 Schematics showing **a** gases in arc environment, **b** formation of iron nitrides needles and iron oxides and **c** pores and inclusions due to molten weld metal-gas interactions

Fig. 3.8 Schematic showing effect of nitrogen in steel weld in mechanical properties

inert gas like argon, helium or their mixtures. Inert gases are considered to be more effective to protect the weld pool than inactive gases. Thermal decomposition of inactive gases like carbon dioxide at high temperature in arc environment can provide oxygen which results in the development of gaseous pores and oxide inclusion and so reducing the protection of the weld pool (Fig. 3.9). However, inert gases are costlier

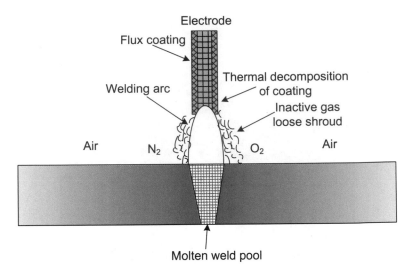

Fig. 3.9 Schematic approach of loose inactive gas shroud around the welding arc

than inactive gases from economy point of view. Further, there are two ways of forming shroud/cover of inactive gas around the weld pool (a) jet of inactive/inert gas coming out from nozzle co-axially like in gas metal arc gas welding and gas tungsten arc welding and (b) inactive gases released by thermal decomposition of hydro-carbons present in coating like in SMAW and FCAW processes. Former approach is found to be more effective to protect the weld pool than later one as reflecting from oxygen and nitrogen content in the weld metal.

Factors affecting the effectiveness of shielding of the weld pool in case of thermal decomposition of the coating approach for providing inactive gases to protect the weld include (a) thickness of flux or coating, (b) traverse speed of welding arc, (c) type of weld bead: stringer or weaver and (d) velocity of ambient gases across the arc. In general, thin coating, high arc travel speed, weaver bead and high velocity of ambient gases result in poor protection of the weld pool from atmospheric gases.

Factors affecting the effectiveness of weld pool shielding using a jet of protecting shielding gas (carbondioxide, argon and helium) include flow rate of shielding gas (5–50 l/min) though the nozzle, type of shielding gas (Ar, He), speed of welding arc, location and position of welding (flat, vertical, overhead), design of nozzle especially in respect of size of nozzle w.r.t. electrode diameter, design of nozzle orifice opening (Fig. 3.10). It has been found that too high/low flow rate and too large/close gap between nozzle orifice and electrode leads to poor protection of the weld pool. Moreover, the effect of weld location on projection of the weld pool depends on type of shielding gas. Ar offers better protection than He in flat and vertical welding, while reverse is true for overhead welding primarily due to density difference (with respect to ambient air) governing the tendency of shielding gases to move up/down during the welding.

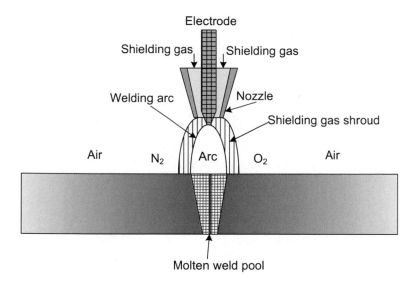

Fig. 3.10 Schematic showing approach of the firm inert gas shroud around the welding arc like in GTAW, GMAW

3.4.2 Covering the Weld Pool Using Molten Flux

Effectiveness of the molten flux cover approach to protect the weld pool (SAW, ESW) depends on basicity of flux, thickness of granular flux cover, size of granular particles in flux, type of flux (halide flux, oxide flux and halide–oxide flux) as per their ability to clean the weld metal from impurities, forming thick enough layer of molten flux to avoid interaction between molten weld metal and air, and ease of melting and free of oxygen (Fig. 3.11). Choice of the flux dictates the ability to protect the weld pool, therefore, considering a metal to be welded, welding constraint, mechanical properties of the weld joint desired and suitable flux/coating need to be selected.

3.5 Cleanliness of Weld Metals and Welding Processes (Approaches for Protection of the Weld Pool)

Since factors determining the effectiveness of shielding weld pool are function of the shielding approach and its process parameters; therefore, the effectiveness of each method for weld pool protection is found to be different. Therefore, contamination of the weld pool from atmospheric gases (in terms of presence of gases such as oxygen, nitrogen and hydrogen) in the weld produced by different arc welding processes varies significantly as evident from Fig. 3.12.

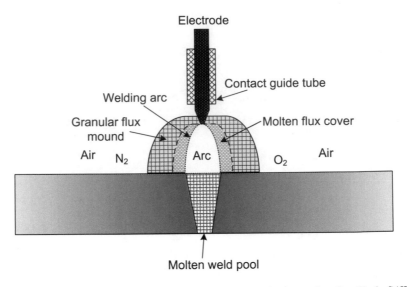

Fig. 3.11 Schematic showing shielding welding and weld pool using molten flux like in SAW

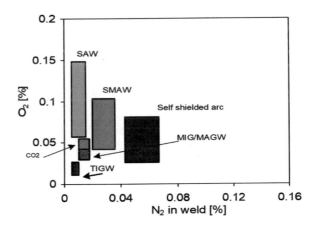

Fig. 3.12 Oxygen and nitrogen content in weld metal produced by different welding processes

3.5.1 Shielded Metal Arc Welding

Protection of the weld in this welding process is based on two approaches (a) developing shroud of inactive carbon dioxide generated by thermal decomposition of flux in the form of coating on the electrode and (b) formation of cover by molten slag layer over the molten weld pool both in liquid and solid states. Shroud of inactive gases formed around the weld pool becomes very weak and incomplete which is easily penetrated even by little ambient air disturbance. Therefore, the protection of the weld pool from atmospheric gases in shielded metal arc welding process is not considered to very effective as weld metal generated contains a lot of oxygen and nitrogen.

Effectiveness of protection depends on how much (thick) and what types of shielding gases are produced apart from welding related parameters such as welding current causing core wire heating, welding speed affecting ambient air relative velocity, arc voltage affecting arc gap so ease of tendency of air entry in the arc zone and type of flux on the core wire namely rutile, basic, cellulosic, acidic and their combinations.

Type and amount of shielding gas are affected by composition of coating material like oxide, halides, or mixture of two. Halide base fluxes offer cleaner weld than oxide-fluxes as halide fluxes becomes free from oxygen. Thickness of coating over the core wire of the electrode is indicated by coating ratio i.e. ratio of diameter of electrode (with coating) and that of core wire. Accordingly, electrode classified as per light, medium and heavy coated electrode for increasing coating ratio.

3.5.2 Submerged Arc Welding

Protection of weld pool in case of SAW is provided by thin layer of molten flux covering the weld pool. The effectiveness of weld pool protection approach of SAW is found to be better than that of SMAW. However, it generally results in higher oxygen concentration in the weld pool due to use of mainly oxide fluxes. Effectiveness of the weld pool protection in SAW is influenced by:

- Thickness of layer of flux cover the weld/arc
- Type of flux: fused, agglomerated, mechanical mix
- Size and size range of flux particles
- Basicity of flux: acidic, neutral and basic flux.

SAW fluxes are generally oxides of different types provide oxygen in the weld metal.

3.5.3 Gas Tungsten Arc Welding

The gas tungsten arc welding process protects the weld pool by forming a shroud using a jet of inert gas around the welding arc. The approach used for protecting the weld pool in GTAW process is very effective due to following reasons (a) stable welding arc, (b) non-consumable W electrode, (c) short arc length and (d) mostly inert gases like Ar, He and their mixtures which are used. Therefore, GTAW results in a cleanest weld among the common arc welding processes. Moreover, effectiveness of protection of the weld pool in this process is determined by

- Type of shielding gas (Ar, He, Ar + He mixture)

- Nozzle and electrodes size
- Flow rate of shielding gas (l/min)
- Speed of welding affecting relative velocity with atmospheric gases
- Welding position especially when He is used.

3.5.4 Gas Metal Arc Welding

Similar to the gas tungsten arc welding, gas metal arc welding process protects the weld pool by forming a cover/shroud using a jet of inert/inactive gas. However, weld metal produced by GMAW is not as clean as that of GTAW due to two reasons.

- Somewhat poor arc stability
- Use of consumable electrode
- Longer arc length
- Use of shielding gas mixtures containing O_2, H_2, N_2.

A variant of gas metal arc welding is flux cored arc welding in which protection of the welding pool from atmospheric gases can be of two types by providing shroud of (a) inactive gases generated by thermal decomposition of the flux in the core of the electrode only and (b) inactive gases generated by thermal decomposition of the flux in the core of the electrode and a jet of inert gas both. The latter option gives cleaner weld than former one.

3.5.5 Electro-Slag and Electro-Gas Welding

Protection of the weld pool in electro-slag welding (ESW) process is achieved by a layer of molten flux covering weld metal (Fig. 3.13), while protection of the weld pool in electro-gas welding (EGW) process is achieved by providing a protective jet of inert gas and flux present in welding arc.

Questions for Self-assessment

- Explain the role of heat in welding.
- What are different ways heat can be generated for welding?
- Why should weld pool be protected from atmospheric gases?
- How should the gas flame torch be adjusted for welding?
- Why does heat generated and peak temperature observed in primary and secondary zones show different trend?
- How does type of flame affect the weld metal properties?
- Describe factors affecting protection of weld pool in case of different welding processes.
- How does heat application affects the characteristics of the weld joints?
- Should the welding position be considered for selection of suitable shielding gas?

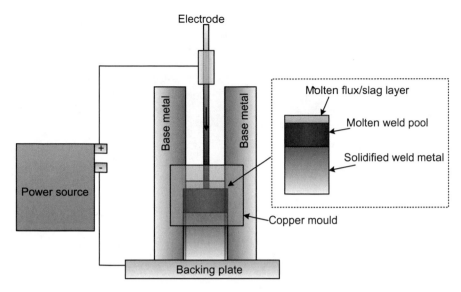

Fig. 3.13 Schematic of shielding approach used in electro-slag welding

• How does the shielding of weld metal in case of FCAW different from SMAW?

Further Reading

American Welding Society (2001) Welding handbook, vol 1, 2 and 3, 9th edn
ASM International Publication (1993) Metals handbook, vol 6
Kou S (2002) Welding metallurgy. Wiley, New York (2002)
Procedure handbook of arc welding, 14th edn. Lincoln Electric Co (2004)

Chapter 4
Power Density and Peak Temperature of Welding Processes

Power density (kJ/mm^2) associated with a welding process plays an important role in developing sound and quality weld joints because it significantly determines the extent of undesirable effects of welding heat on the base metal. Power density of a welding process is largely function of inherent process features such as welding process parameters (current, voltage, shielding gas), diameter of arc/plasma/beam, arc gap and construction of arc, if any. Peak temperature realized by a welding (heat source) process does not depend only on heat generated but also on the area over which generated heat is spread/applied during the welding as it directly affects the rate of heat dissipation to the underlying low-temperature base metal. In general, increase in area (over which heat is applied during welding) increases the heat dissipation rate which in turn reduces peak temperature attained. Therefore, understanding of power density and peak temperature associated with different welding processes will be of significant academic and practical importance as it will be useful in establishing the relation between welding processes and respective weld performance related parameters such as weld area, width of heat affected zone, metallurgical transformations in weld and heat affected zone, hardening and softening of weld and heat affected zone, residual stress and distortion.

4.1 Introduction

Fusion welding processes must be seen with regard to their capability of power density at which these can apply heat for melting the faying surfaces of base metal. Heat required for fusion of faying surfaces of components being welded can be obtained from the different sources like gas, arc, plasma and radiation beam as per type of fusion welding process. Each type of heat source has capability to supply heat at different power densities (kW/mm^2). Different welding processes even for a given arc power [obtained from the product of arc current (I) and arc voltage (V)] deliver heat for melting the faying surfaces at different power densities due to the fact that it

© The Author(s), under exclusive license to Springer Nature Singapore Pte Ltd. 2022   65
D. K. Dwivedi, *Fundamentals of Metal Joining*,
https://doi.org/10.1007/978-981-16-4819-9_4

Table 4.1 Power density and maximum temperature related with different welding processes

Si. no	Welding process	Power density (W/cm^2)	Temperature (°C)
1	Gas welding	10^2–10^3	2500–3500
2	Shielded meta arc welding	10^4	> 6000
	Gas metal arc welding	10^5	8000–10,000
3	Plasma arc welding	10^6	15,000–30,000
4	Electron beam welding	10^7–10^8	20,000–30,000
5	Laser beam welding	> 10^8	> 30,000

is applied over different areas on the faying surfaces by different welding processes. Power density (kW/mm^2) is directly governed by the area over which heat is applied by a particular process for a given set of welding condition. Typical power density and approx. peak temperature generated (in ascending order) by different welding processes from gas welding to arc welding to radiation welding processes are given in Table 4.1.

4.2   Effect of Power Density

Power density associated with a particular welding process directly affects the amount of heat (kJ/mm) required to be supplied for fusion of the faying surfaces. Heat required for melting a metal during welding comprises three components (a) sensible heat, (b) latent heat and c) heat dissipated to base metal. Sensible heat and latent heat for a metal under a given set of ambient condition remain constant, and these must be supplied to ensure fusion of faying surfaces while heat dissipated to the underlying base metal during welding is directly affected by time for which heat is applied (to meet sensible and latent heat requirements) to cause fusion which in turn depends on the power density of the (heat source) welding process. It is important to note that heat required for melting the unit quantity of a given metal is constant and is a property of material. Heat for melting comprises sensible heat and latent heat. Latent heat for steel is 2 kJ/mm^3. An increase in power density decreases the actual heat (input) required for melting faying surfaces to develop a weld joint by lowering the heat dissipated to base metal.

A combination of the power density of fusion-based welding processes and speed of welding heat source affect net heat input, penetration capability which in turn affect the penetration, mode of welding (melt in or keyhole), weld depth-to-width ratio and HAZ width (Fig. 4.1a). An increase in power density with optimized parameters including speed of heat source changes the way fusion weld joint is developed from melt in mode to keyhole mode coupled with increased penetration, weld penetration depth to width and reduced HAZ width (Fig. 4.1b).

A high-power density (heat source) welding process delivers a lot of heat over a small area in short time period. If heat required (sensible and latent heat) for fusion of

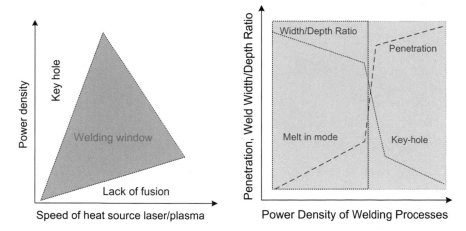

Fig. 4.1 Effect of **a** power density and speed of welding heat source and **b** power density of welding processes on mode of welding and parameters of weld geometry

faying surface is supplied during welding in short time, then melting occurs rapidly and heat dissipated to the underlying low-temperature base metal is reduced. The reduction in time of heat application during welding in turn lowers the amount of heat dissipated away from the faying surfaces to the base metal, so the most of the heat applied on the faying surfaces is used for their fusion only. Basically, heat dissipated to the underlying base metal is reduced with use of high-power density welding process.

Fusion welding processes use localized melting approach using heat energy for development of weld joints. To ensure melting of base metal in short time, it is necessary that power density of welding process is high enough (Fig. 4.2). Time to

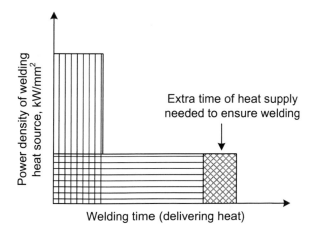

Fig. 4.2 Effect of power density and time on heat input

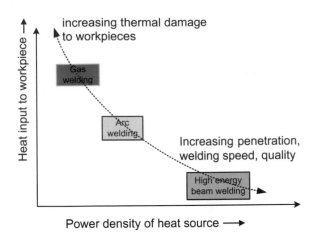

Fig. 4.3 Effect of power density of heat source on heat input required for welding

melt the base metal is found inversely proportional to the power density of heat source which is defined as ratio of power of heat source (arc, flame, plasma, and beam) to the area of workpiece over which it is applied (W/cm^2). Lower the power density of heat source greater will be the time for which heat input will be given for fusion of faying surface welding as a large amount of heat is dissipated to colder base material of workpiece away from the faying surface by thermal conduction (Fig. 4.3).

4.3 Need of Optimum Power Density of Welding Process

As stated, low-power density processes need higher heat input than high-power density processes. Neither too low nor too high heat input is considered to be good for developing a sound weld joint. As low heat input can lead to lack of penetration and poor fusion of faying surfaces during welding, excessive heat input may cause damage to the base metal in terms of distortion, softening of HAZ and reduced mechanical properties (Figs. 4.3 and 4.4).

High heat input in general lowers the tensile strength of many work-hardenable and heat-treatable aluminium alloys due to thermal softening of HAZ and development of undesirable metallurgical properties of the weldment (Fig. 4.5). Moreover, use of high-power density offers many advantages such as deep penetration, high welding speed and improved quality of welding joints. A fusion welding process should have power density approximately 10 W/mm^2. Vaporization of metal takes place at about 10,000 W/mm^2 power density. Processes (electron and laser beam) with such high-power density are used for ablation and controlled removal of metal for shaping of difficult to machine metals. Welding processes with power density in ascending order are shown in Fig. 4.6.

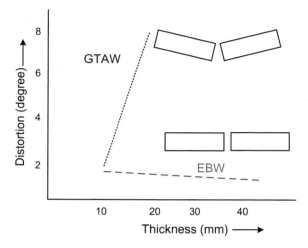

Fig. 4.4 Effect of welding process on angular distortion of weld joint as a function of plate thickness

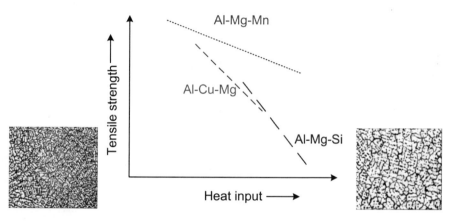

Fig. 4.5 Schematic diagram showing effect of heat input on tensile strength of aluminium alloy weld joints (magnification of micro-graph in figure is 200 X)

Question for Self-assessment

- How the heat generated is used in different welding processes?
- What are approaches used for heat generation in welding?
- How does energy density associated with a particular welding processes affect the weldability and performance of weld joint?
- Compare different welding processes in respect of power density and temperature.
- Does energy density associated with a welding process affect HAZ and distortion tendency?

Fig. 4.6 Power densities of
different welding processes

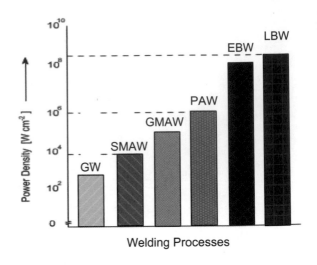

Further Reading

American Welding Society (1983) Welding handbook, 7th edn, vol 1 & 2, USA
Kou S (2003) Welding metallurgy. 2nd edn,. Wiley, USA
Nadkarni SV (2010) Modern arc welding technology. Ador Welding Limited, New Delhi

Part II
Physics of Welding Arc

Chapter 5
Physics of Welding Arc

Electron Emission and Welding Arc

5.1 Introduction

A welding arc is an electric discharge that develops heat primarily due to flow of current from cathode to anode. Flow of current through the gap between electrode and workpiece (called arc gap) needs column of charged particles for having reasonably good electrical conductivity. These charged particles are generated by various mechanisms such as thermal emission, field emission and secondary emission. Density of charged particles in arc gap determines electrical conductivity of gaseous column. In an electric arc, electrons released from cathode (due to electric field or thermo-ionic emission) are accelerated towards the anode because of potential difference between workpiece and electrode. These high-velocity electrons moving from cathode towards anode collide with gaseous molecules (present in gap between workpiece and electrode) and decompose them into charged particles as electrons and ions. These charged particles move towards electrode and workpiece as per their polarity and form a part of welding current (Fig. 5.1). Electrons move towards the anode, while ions move in direction of cathode. Ion current becomes very low as compared to electron current. Ion current is about 1% of electron current because ions are heavier than the electrons so they move slowly. Eventually electrons get merged into anode and generate heat by impact. Arc gap between electrode and workpiece acts as pure resistive load. Heat generated by a welding arc therefore depends on arc voltage and welding current and is given by product of two.

5.2 Emission of Free Electrons

Free electrons and charged particles with enough density are needed in the gap between the electrode and workpiece as mentioned above for initiation and maintenance of welding arc. Ease of emitting electrons by a material is assessed on the basis of two parameters namely work function and ionization potential. Emission of

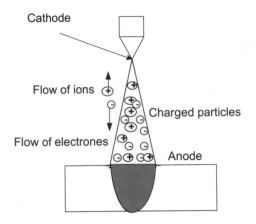

Fig. 5.1 Schematic of flow charged particles in welding arc

electrons from the cathode depends on the work function. The work function is the energy (ev or J) required to get one electron released from the surface of cathode material in consideration. Ionization potential is another measure of ability of a metal to emit the electrons and is defined as energy/unit charge (v) required for removing an electron from an atom. Ionization potential is found to be different for different metals. For example, Ca, K and Na have very low ionization potential (2.1–2.3 eV), while that for Al and Fe is on the higher side with values of 4.2 and 4.5 eV, respectively (Table 5.1). Common mechanisms through which free electrons are emitted during arc welding are described below:

Table 5.1 Ionization potential of common metals

Element	Work function (ev)
Silver	4.2
Aluminium	4.2
Gold	5.1
Caesium	2.1
Copper	4.6
Lithium	2.9
Lead	4.2
Tin	4.4
Chromium	4.6
Molybdenum	4.3
Stainless steel	4.4
Gold	4.8
Tungsten	4.5
Copper	4.5
Nickel	4.6

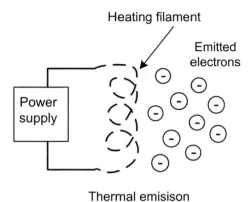

Fig. 5.2 Schematic showing the approach of thermos-ionic emission

5.2.1 Thermo-Ionic Emission

Increase in temperature of the metal increases the kinetic energy of free electrons, and as it goes beyond a certain limit, electrons are ejected from the metal surface (Fig. 5.2). This mechanism of emission of electron due to heating of the metals is called thermo-ionic emission. The temperature at which thermo-ionic emission takes place, the most of the metals melt. Hence, refractory materials like tungsten and carbon having high melting point exhibit good thermo-ionic electron emission tendency.

5.2.2 Field Emission

In this approach, free electrons are pulled out of the metal surface by developing high-strength electromagnetic field (Fig. 5.3). High potential difference (10^7 V/cm) between the workpiece and electrode is established to generate the required electromagnetic field for the emission of the electrons.

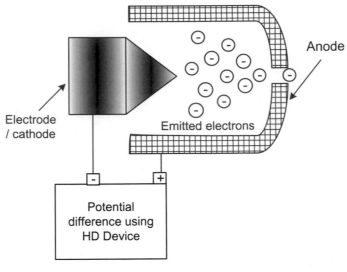

Field emission

Fig. 5.3 Schematic showing the approach of field emission of electrons

5.2.3 Secondary Emission

High-velocity electrons moving from cathode to anode in the arc gap collide with other gaseous molecules. This collision results in decomposition of gaseous molecules into atoms and charged particles like electrons and ions (Fig. 5.4).

5.3 Zones in Arc Gap

On establishing the welding arc, drop in arc voltage is observed across the arc gap. However, rate of drop in arc voltage varies with distance from the electrode tip to the weld pool (Fig. 5.5). Accordingly, five different zones are found in the arc gap namely cathode spot, cathode drop zone, plasma, anode drop zone and anode spot (Fig. 5.6). Three zones namely cathode drop zone, plasma, anode drop zone formed in between cathode and anode can be easily seen in Figs. 5.3 and 5.4, while two zones namely cathode spot and anode spot are formed at the cathode and anode surface, respectively.

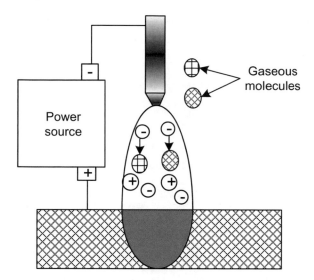

Fig. 5.4 Schematics showing approach of secondary emission of electrons

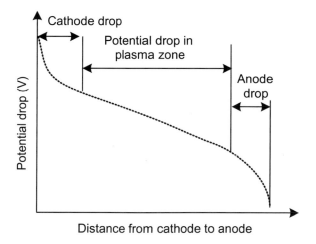

Fig. 5.5 Potential drop as function of distance from the cathode to anode

5.3.1 Cathode Spot

Cathode spot is a region on cathode surface wherefrom electrons are emitted. Three types of cathode spots are generally formed on the surface of cathode namely normal cathode spot, pointed cathode spot and mobile cathode spot (Fig. 5.7). There can be one or more than one cathode spots moving at high speed ranging from 5 to 10 m/s. Mobile cathode spot is usually produced at high current density 100–1000 A/mm^2 and is found during the welding of aluminium and magnesium alloys. The formation of

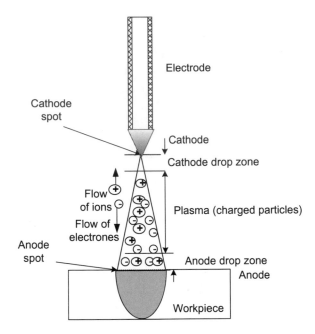

Fig. 5.6 Zones in arc gap of a welding arc

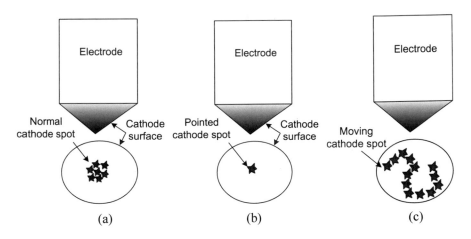

Fig. 5.7 Schematic of **a** normal, **b** point and **c** mobile cathode spot

mobile cathode spot helps in loosening the refractory, coherence and adherent oxide layer formed on surface of reactive metal like aluminium, magnesium and stainless steel during the welding. Therefore, formation of mobile cathode spot is considered to be favourable from cleaning of the weld metal point of view. However, to ensure the formation of mobile cathode spot on the surface of workpiece during the welding, it is required to use reverse polarity, i.e. workpiece is made cathode by connecting negative terminal of the DC power supply to the workpiece. Pointed cathode spot is formed at a point only, e.g. tungsten inert gas welding at about $100A/mm^2$. Pointed tungsten electrode used during tungsten inert gas welding forms the pointed cathode spot. Ball shape tip of coated steel electrode forms the normal cathode spot.

5.3.2 Cathode Drop Region

Cathode drop region is a zone very next (close) to the surface of cathode. A very sharp drop in voltage is observed in this zone due to cooling effect of cathode. Cooling effect of the electrode results in low temperature in cathode drop zone which in turn reduces density of charged particles drastically. Therefore, cathode drop zone experiences a sharp drop in voltage to maintain the flow of current (I). Voltage drop (V_c) in this region directly affects the heat generation near the cathode (V_cI) which in turn governs melting rate of the electrode in case of the consumable arc welding process with straight polarity (electrode is cathode).

5.3.3 Plasma Zone

Plasma zone is a region between electrode and work where mostly uniform flow of charged particles (free electrons and positive ions) takes place. In this region, therefore uniform voltage drop (V_p) takes place as function of distance from the cathode. Heat generated in the plasma region (V_pI) has lesser contribution on melting of the workpiece and electrode than cathode and anode drop zones. A large fraction of the heat generated in plasma zone is dissipated to atmosphere through the arc surfaces.

5.3.4 Anode Drop Region

Like cathode drop region, anode drop region is also formed adjacent to the anode surface, and a very sharp drop in voltage (V_a) takes place in this region due to cooling effect of the anode. Voltage drop in this region affects the heat generation (V_aI) near the anode which in turn determines the melting rate of anode. In case of direct current electrode negative (DCEN) polarity, voltage drop in this zone affects melting of the workpiece and penetration depth.

5.3.5   Anode Spot

Anode spot is a region on the surface of anode where electrons get merged and their impact generates heat for melting of the anode. However, no fixed anode spot is generally noticed on the anode surface like cathode spot.

5.4   Electrical Fundamentals of Welding Arc

The welding arc acts as impedance for the flow of current in an electric conductor. The impedance of arc is usually found a function of temperature and becomes inversely proportional to the density of charge particles and their mobility. Therefore, distribution of charged particles in radial and axial direction in the arc affects (the arc axially and radially) the total impedance of the arc. Three major regions (in addition to cathode and anode spots) are noticed in arc gap that accounts for total potential drop in the arc, i.e. cathode drop region, plasma and anode drop region. Product of potential difference across the arc (V) and welding current (I) gives the power of the arc (W) indicating the heat generation per unit time (kJ/s). Arc voltage (V) considered as sum of potential drop across the cathode drop region (V_c), potential drop across the plasma region (V_p) and potential drop across the anode drop region (V_a) as shown in Fig. 5.8.

$$\text{Power of the arc (P)} = (V_c + V_p + V_a)I \tag{5.1}$$

Potential drop in different zones is expressed in terms of volt (V), welding current in ampere (A) and power of arc P is in watt (W). Equation 5.1 suggests that the distribution of heat in three zones, namely cathode, anode and arc plasma, can be changed if voltage drops in different zones is regulated. Considering rationale of charged particle density governing potential drop, all factors such as arc gap, electrode and workpiece metal, shielding and welding current, influencing the charge particles release/density must be considered for their effect on heat generation in different zones. However, effect of these factors on potential drop in different zones will be different. Variation of arc length mainly affects plasma heat, while shielding gas influences the heat generation in the cathode and anode drop zones also.

Increase in arc gap increases the arc voltage so heat generated in plasma zone due to reduced charged particle density, low arc temperature owing high heat losses and so increased resistance for the flow of current (Fig. 5.9). Increase in arc gap also increases the spread of arc which in turn leads to wider weld bead, shallower penetration as compared to low arc gap.

Use of shielding gas of high ionization potential like He results in high voltage drop so greater heat generation. Addition of low ionization potential materials (such as potassium and sodium) reduces the arc voltage because of increased ionization (so charged particles density) in the arc gap, so increased electrical conductivity in

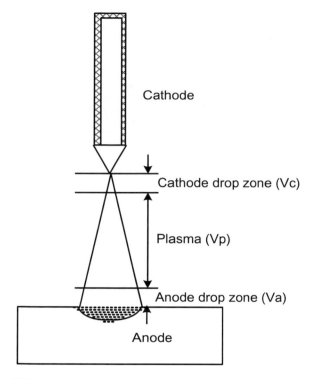

Fig. 5.8 Three different zone in which voltage drop takes place

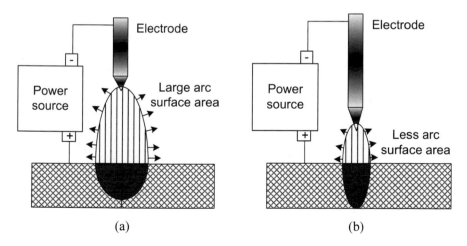

Fig. 5.9 Schematic showing effect of arc length on arc surface area leading varying arc voltage and weld bead geometry

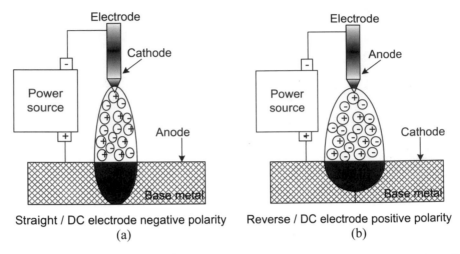

Straight / DC electrode negative polarity   Reverse / DC electrode positive polarity
(a)   (b)

Fig. 5.10 Schematic showing effect of **a** straight and **b** reverse polarity on arc characteristics and weld bead geometry

turn reduces the heat generation in plasma region by reducing the voltage drop for a given current. Heat generation at the anode and cathode drop zones is primarily governed by type of welding process (as per electrode materials, shielding gas) and polarity associated with welding arc (Fig. 5.10).

In case of direct current (DC) welding, when electrode is connected to the negative terminal and workpiece is connected with positive terminal of the power source, then it is termed as direct current electrode negative polarity (DCEN) or straight polarity, and when electrode is connected to the positive terminal of the power source and workpiece is connected with negative terminal, then it is termed as direct current electrode positive polarity (DCEP) or reverse polarity. TIG welding using DCEN with argon as shielding gas shows (3–4) times higher current carrying capacity (without melting) than DCEP. The submerged arc welding with DCEP generates larger amount of heat at anode (electrode) side than cathode (workpiece) side which in turn results in higher melting rate of consumable electrode.

Increase in spacing between the electrode and workpiece generally increases arc voltage because of increased losses of the charge particles by radial migration to cool boundary of the plasma. Increase in the length of the arc column (by bulging) exposes more surface area of arc column to the low-temperature atmospheric gas which in turn imposes the requirement of a greater number of charge particles to maintain the flow of current. Therefore, loss of charged particles must be accommodated by increasing the arc voltage to stabilize the welding arc. The most of the heat generated at anode (workpiece) and cathode (electrode) are used for melting of workpiece and electrode, respectively, in consumable arc welding processes. Therefore, consumable arc welding processes (SAW, GMAW) result in higher thermal efficiencies than non-consumable arc welding processes (GTAW, PAW). This is more evident from the fact that the thermal efficiency of metal arc welding processes is found in range of

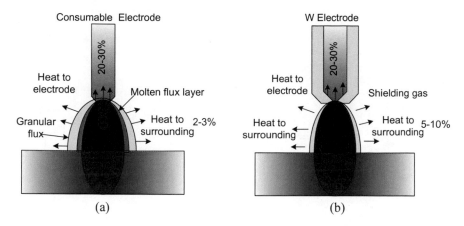

Fig. 5.11 Schematic of tentative dissipation of heat generated by welding arc during **a** consumable arc welding (SAW) and **b** non-consumable arc welding processes (GTAW/PAW)

80–90%, whereas that for non-consumable arc welding processes is found in range of 40–60% (Fig. 5.11).

Questions for Self-assessment

- What is mechanism of development of a welding arc?
- Explain mechanisms of electron emission by (a) thermionic emission method, (b) field emission method and (c) secondary emission.
- Describe different zones in which a welding arc can be divided.
- Schematically show variation in potential drop as a function of distance from cathode to anode and explain why.
- Explain the effect of electrical parameters namely voltage and current on power of the welding arc and heat generation.
- Can polarity affect the melting rate of electrode during consumable arc welding processes?
- Does the choice of shielding gas and electrode material affect the heat generation by welding arc?
- Why does consumable arc welding show thermal efficiency different from of non-consumable arc welding processes?
- How does the selection of polarity affect the cleanliness of the weld metal during welding of stainless steel?
- How does type of cathode spot formation affect development of a sound weld joint?

Further Reading

American Welding Society (1987) Welding handbook, 8th edn, vol 1 & 2, USA
Cary H (1988) Welding technology, 2nd edn. Prentice Hall
Little R (2001) Welding and welding technology, 1st edn. McGraw Hill
Nadkarni SV (2010) Modern arc welding technology. Ador Welding Limited, New Delhi

Chapter 6
Physics of Welding Arc

Arc Initiation and Arc Maintenance

6.1 Arc Initiation

Arc welding processes is based on application of heat generated by an arc for melting the faying surfaces of metals to be joined. Initiation of the welding arc is the first step in direction of developing a fusion weld. There are two most commonly used methods to initiate an electric arc in welding processes, namely touch start and field start. The touch start method is used in case of all common manually controlled arc welding processes, while the later one is preferred for automatic welding operations and also in those arc welding processes where electrode has tendency to form inclusion in the weld metal like in GTAW and PAW processes.

6.1.1 Touch Start

In this method, the electrode is brought in the contact of the workpiece, and then, it is pulled apart to create a very small gap (usually 2–4 mm) between electrode and workpiece. Touching of the electrode to the workpiece resulting closing of electric circuit and so short-circuiting leads to the flow of heavy current through small electrode–workpiece contact region. Flow of heavy current through localized contact zone leads to excessive heat generation by electrical resistive heating. Heat generated causes partial melting and even little evaporation of the metal at the electrode–workpiece contact interface, i.e. electrode tip/base metal. The metal vapour produced at electrode–workpiece contact interface later on ionization produce charged particles (ions, electrons). All these events happen in a very short time of contact usually within few seconds (Fig. 6.1a, b). Heating of electrode tip produces few free electrons by (a) thermal ionization of electrode materials and (b) ionization/dissociation of metal vapours (owing to lower ionization potential of the metal vapours than the atmospheric gases) to produce charged particles, i.e. free electron and positively charged ions. Pulling up of the electrode slightly away from the workpiece starts flow of

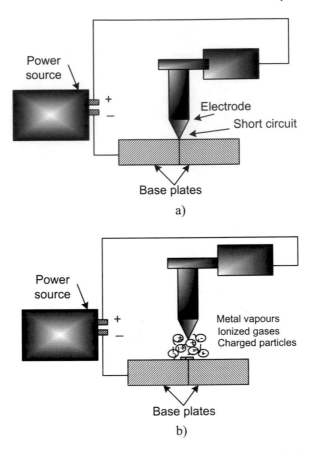

Fig. 6.1 Schematic showing mechanism of arc initiation by touch start method **a** when circuit is closed by touching electrode with workpiece just for a while and **b** emission of electrons on putting them apart

current through gap between electrode and workpiece (having charged particles) if there is enough potential difference. This flow of current in turn results in arc/spark for a moment. To use the heat of electric arc for welding purpose, it is necessary that after initiation of arc, it must be maintained and stabilized to deliver enough heat for consistent penetration (melting up to the required depth), continuous and uniform melting the components to be joined. However, this method suffers with two limitations (a) high non-consumable electrode wear and (b) possibility of inclusion of electrode metal in weld metal. Both these limitations primarily occur due to typical requirement of touch start method to touch the workpiece with electrode and heat generation at electrode–workpiece contact interface.

6.1.2 Field Start

In this method, high-strength electric field is applied between electrode and workpiece for release of electrons from cathode by electromagnetic field emission (Fig. 6.2). High-strength electromagnetic field is developed using high potential difference (10^7 V) either between (a) electrode and work like in GTAW and (b) electrode and nozzle like in PAW. Electromagnetic field facilitates the ejection of electrons from cathode spots. Once the free electrons are available in arc gap, normal potential difference between electrode and workpiece (10–50 V) ensures flow of charged particles to maintain a welding arc. This method of arc initiation is commonly used in case of mechanized arc welding processes such as plasma arc and gas tungsten arc welding process where direct contact between electrode and workpiece is not preferred due to problems associated with (a) W inclusion formation and (b) short electrode life. These are coupled with high rate of electrode degradation due to contamination and thermal damage.

6.2 Maintenance of Arc

Once the electric arc is initiated, maintenance of the arc is necessary to use the heat generated for welding purpose. Maintenance of welding arc helps to deliver given amount of heat to the workpiece during the welding for consistent penetration and uniform welding. For maintaining of the welding arc, two conditions must be fulfilled (a) heat dissipation rate from the welding arc to workpiece, electrode, surrounding air and base metal should be equal to the heat generated by the arc to maintain the temperature of the arc high enough for thermal ionization, so that electrical conductivity of arc gap is maintained (through charged particle density) for the flow of current and (b) number of electrons produced should be equal to that of electrons lost to the workpiece and surroundings from the arc surface.

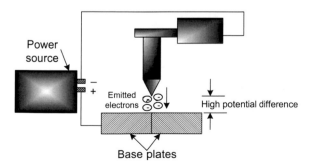

Fig. 6.2 Schematic showing the field start method of arc initiation

An electric arc primarily involves flow of current through the gap between the workpiece and electrode, and hence, there must be sufficient number of charged particles namely electrons and ions. However, some of the electrons are lost from the arc surface, to the weld pool and surroundings, while few electrons reunite with ions (Fig. 6.3). Loss of these electrons must be compensated by generation of enough new free electrons.

In case of direct current (DC), magnitude and direction of current flow do not change with time, and hence, maintaining the flow of electrons becomes easy, and so the arc is maintained without any difficulty. While in case of alternating current (AC), maintenance of welding becomes difficulty due to two reasons (a) both magnitude and direction of current flow change with time and (b) for a moment the flow of current becomes zero (Fig. 6.4). These two conditions make re-ignition of welding

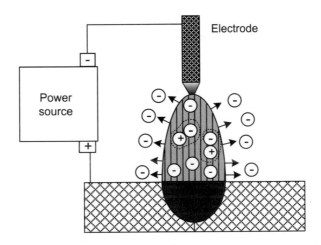

Fig. 6.3 Schematic showing different ways loss of electrons occurs from the welding arc

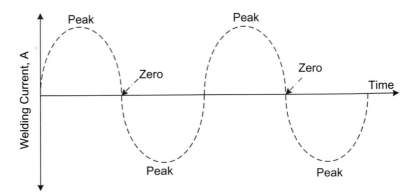

Fig. 6.4 Schematic of AC current wave form showing variation in current as function of time

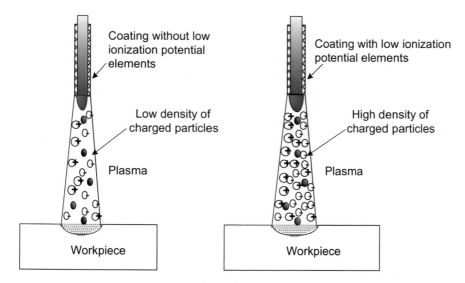

Fig. 6.5 Schematic representation of effect of low ionization potential elements in coating of the electrode on density of charged particles in welding arc **a** without low ionization potential elements and **b** with low ionization potential elements

arc with AC somewhat difficult. Welding with AC, therefore, needs extra precautions and provisions.

There are two commonly used methods for maintaining the arc in AC welding: (a) use of low ionization potential elements in flux/coating and (b) use of low power factor power source.

6.2.1 Low Ionization Potential Elements

In this method, low ionization potential elements such as potassium, calcium and sodium are incorporated in the flux/coating of the electrode. These elements release free electrons under arc welding conditions (even with small potential difference between electrode and workpiece) needed to have reasonably good electrical conductivity for maintaining welding arc (Fig. 6.5).

6.2.2 Low Power Factor

The power factor of an electrical system indicates how effectively the power is being utilized. It is generally preferred to have high power factor of machine or system. Power factor of welding power system is defined as ratio of actual power drawn

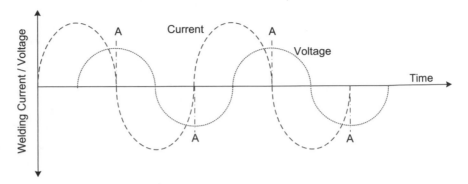

Fig. 6.6 Schematic of current and voltage variation as function of time with proper power factor for maintenance of the arc

(from the power source) to perform the welding and apparent power drawn into the welding circuit line. Welding transformer operates at high power factor (> 0.9). However, in AC welding usually, low power factor is intentionally used to improve the arc stability and better maintenance of welding arc.

In this method, current and voltage wave forms are made out of phase using proper low power factor (~0.3), so that when welding current is zero, full open circuit voltage is available between electrode and workpiece as shown in Fig. 6.6 at location "A". Full open circuit voltage across the electrode and workpiece helps in release of free electrons to the maintain flow of already existing electrons. The charge particle density between electrode and workpiece must be enough for the maintenance of the arc.

6.3 Welding Arc Characteristic

Welding arc characteristic shows a variation in the arc voltage with welding current. A typical welding arc characteristic curve exhibits three different zones namely drooping, flat and rising characteristic zones (Fig. 6.7). Initially at low welding current, when arc is thin, an increase in welding current increases the temperature of arc which in turn increases the number of charged particles in plasma zone of the arc due to thermal ionization of gases present in arc gap and thermo-ionic emission of electrons. As a result, electrical conductivity of arc zone increases (coupled with reduction in electrical resistance) with increase of welding current which in turn decreases arc voltage during initial stage of increase of welding current in drooping characteristic zone. Welding current is very low in this zone. Arc tends to stabilize in drooping characteristic zone. This trend of decrease in arc voltage with increase of current continues up to certain level of current and beyond that increase in welding current increases the diameter/cross-section of arc thereby the surface area of the arc also increases. Increase in surface area of the arc in turn increases loss of heat

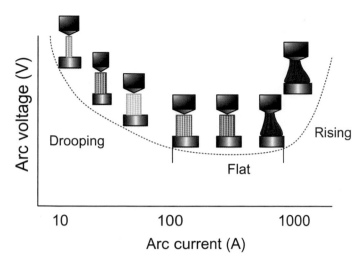

Fig. 6.7 Schematic diagram showing welding arc characteristic curve

and charged particle from the arc surface. Therefore, no significant increase in arc temperature takes place with increase of current in the flat zone; hence, arc voltage is not affected appreciably over a range of current in flat zone of the arc characteristic curve. Further, increase in welding current bulges the arc, which in turn increases the path (so resistance) for flow of current (due to increased losses of charge carriers and heat from arc surface). Therefore, arc voltage starts increasing with increase in welding current in rising characteristic zone. These three zones of arc characteristic curve are called drooping, flat and rising characteristic.

Increase in arc length in general increases arc voltage during welding. However, the extent of increase in arc voltage with increase in arc length varies with process as shown in Fig. 6.7. In general, arc voltage increases almost lineally with increase in arc length (within reasonable limits), and the same is attributed to increase in resistance to the flow of current due to reduction in charged particle density in arc zones with increase in arc length owing to increased loss of heat and charged particles of the surface of arc.

Significant difference in arc voltage vs. arc length relationship for different arc welding processes such as SMAW, GMAW and GTAW is attributed to variation in charged particle densities in respective arc zones (Fig. 6.7). For example, GTAW process due to tungsten electrode (having high electron-emitting capability) results in higher charged particle density in arc region than GMAW and SMAW which in turn leads to lower arc voltage/arc length ratio for GTAW than GMAW and SMAW process as shown by slope of curves in Fig. 6.8.

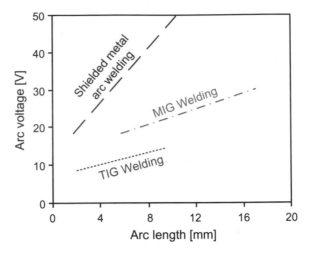

Fig. 6.8 Variation in arc voltage as function of arc length for different arc welding processes

6.4 Temperature of the Arc

In addition to welding parameters such as arc voltage and welding current governing the power of arc, thermal properties (thermal conductivity, specific heat) of electrode and shielding and ionization potential of shielding gases present in arc zone predominantly affect the temperature and its distribution in the arc region (Figs. 6.9, 6.10 and 6.11). In general, increase in arc power, ionization potential and reduction in thermal conductivity of the shielding gases increases the temperature of the arc due to two factors (a) increase in amount of heat generated on increasing the arc power and

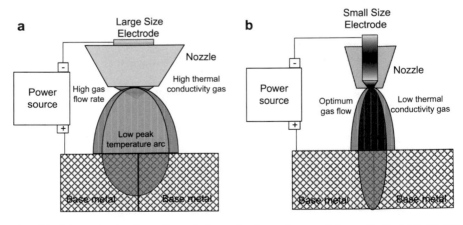

Fig. 6.9 Schematic showing effect of electrode size **a** large and **b** small diameter on arc size/temperature and bead geometry

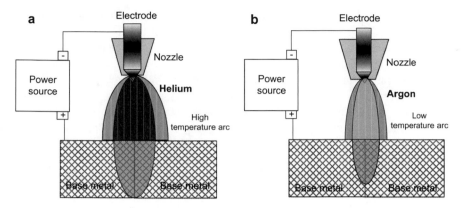

Fig. 6.10 Schematic showing effect of shielding gas **a** helium and **b** argon on arc temperature and bead geometry

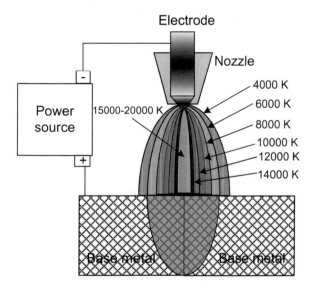

Fig. 6.11 Schematic showing typical temperature distribution in the arc

using high ionization potential shielding gas and (b) reduction in heat loss from the arc surface to the surrounding when using low thermal conductivity shielding gases. Increase in welding current and arc voltage increases the arc power, while use of high ionization potential shielding gases increases the arc voltage which in turn also increases the power of arc. Further, thermal conductivity and diameter of electrode, arc length, flow rate of shielding gases, ionization potential coating/electrode/flux, if any, influencing either heat generation or heat losses also affect temperature of the welding arc (Fig. 6.9). High thermal conductivity, large diameter, high arc length and too high flow rate of shielding gas increase heat dissipation from the arc; therefore, all these factors reduce arc temperature. On the other hand, low ionization potential elements in coating/flux/electrode reduce the heat generation which in lead to reduction in temperature of arc. Thus, all factors affecting heat generation, heat transfer from the arc surface (and so temperature of the arc) are expected to determine the depth of penetration, melting rate, welding speed and above all productivity and economy of the welding.

Thermal conductivity of the most of the gases (He, N, Ar) increases with rise in temperature; however, this increase is not continuous for some of the gases such as helium. Thermal conductivity of base metal and that of shielding gas governs temperature gradient in the arc region (Fig. 6.10). Reduction in thermal conductivity increases the temperature gradient. Therefore, a very sharp decrease in temperature of arc is observed with increase of radial distance from the axis (centre) of the arc towards the arc surface (Fig. 6.11).

Maximum temperature is observed at the core (along the axis of electrode) of the arc, and it decreases rapidly with increase in distance away from the axis. Temperatures in anode and cathode drop zones are generally lower than the plasma zone due to cooling effect of electrode/workpiece. Temperature of arc can vary from 4000 to 30,000 K depending upon the welding process, welding current, arc voltage, shielding gas and plasma gas. For example, in case of SMAW, maximum temperature of arc is about 6000 K while that for TIG/MIG welding arc, it is found in range of 150,000–20,000 K.

Questions for Self-assessment

- What are methods of arc initiation?
- Explain the mechanism of arc initiation by touch start and field start methods.
- What are the conditions necessary for maintenance of welding arc?
- Explain the two approaches used for maintenance of the AC welding arc.
- What is arc characteristic?
- Describe different zones of arc characteristics curve namely drooping, flat and rising zones.
- Why do different arc welding processes show different arc characteristics curves?
- Why does variation in arc voltage vs arc length relationship found different for SMAW and GTAW processes.
- How is arc characteristic affected by arc length?
- Describe the temperature variation in welding arc.
- What are factors affecting temperature in welding arc?
- How does the welding parameters related with a welding process affect the peak temperature of the arc?

Further Reading

American Welding Society (1987) Welding handbook, 8th edn, vol 1 and 2, USA

Cary H (1988) Welding technology, 2nd edn. Prentice Hall
Little R (2001) Welding and welding technology, 1st edn. McGraw Hill
Nadkarni SV (2010) Modern arc welding technology. Ador Welding Limited, New Delhi

Chapter 7
Physics of Welding Arc

Arc Forces, Electrode Polarity and Arc Blow

7.1 Arc Forces and Their Significance on Welding

All forces acting in the arc zone are termed as arc forces. In respect of consumable electrode arc welding processes, influence of arc forces on resisting or facilitating the detachment of molten metal drop hanging at the electrode tip is important in determining the mode of metal transfer, weld metal disposition efficiency and quality of weld metal deposited (Fig. 7.1a–f). Metal transfer is basically detachment and movement of molten metal drops from tip of the consumable electrode to the weld pool in workpiece. Metal transfer is of great practical importance for consumable arc welding processes as it determines (a) flight duration of molten metal drop in arc region (on movement from electrode tip to weld pool) affecting the quality of weld metal and element transfer efficiency, and (b) deposition efficiency. Gravity force, surface tension force and pinch force play more important role in metal transfer than the other arc forces occurring due to impact of charge carrier with molten metal drop, gas eruption and metal vapours. Relevance of these forces have been elaborated in the following section.

7.1.1 Gravity Force

This type of force is caused by gravity effect on molten metal drop hanging at the tip of electrode. Gravitational force depends on the volume of the drop and density of metal. Increase in size and density of molten metal drop hanging at the tip of electrode increases the gravity force. However, impact of gravity force on the metal transfer during welding is dictated by the welding position. In case of down-hand welding (where molten metal drop hanging at the tip of electrode is pointed in downward direction during welding), gravitational force helps in detachment/transfer of molten metal drop from electrode tip (Fig. 7.1). While in case of overhead welding (where molten metal drop is pointed in upward direction during the welding), the gravity

D. K. Dwivedi, *Fundamentals of Metal Joining*,
https://doi.org/10.1007/978-981-16-4819-9_7

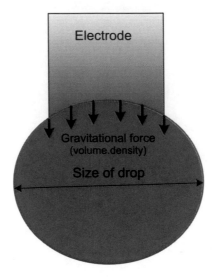

Fig. 7.1 Schematic of gravitational force

force resists the detachment of the drop from the electrode tip.

$$\text{Gravitational force} \left(F_g\right) = \rho V g \tag{7.1}$$

where ρ (kg/m)^3 is the density of metal, V is volume of drop (m^3) and g is gravitational constant (m/s^2).

7.1.2 Surface Tension Force

This force experienced by drop of the molten metal hanging at the tip of electrode is caused by surface tension effect. Magnitude of the surface tension force (Eq. 7.2) is determined by the size of droplet, electrode diameter and surface tension coefficient of liquid metal. The surface tension force always resists the detachment of molten metal drop from electrode tip irrespective of the welding position. However, this effect of surface tension force on molten metal drop detachment is exploited effectively for odd welding position. During flat welding, surface tension force acts against gravitational force, and therefore, it delays the metal transfer until a molten metal drop at the electrode tip grows to large size enough to overcome/counter the surface tension force by gravitation force.

In case of vertical and overhead welding positions, high surface tension force helps in placing the molten weld metal at required place more effectively by reducing tendency of molten weld metal to fall down (Fig. 7.2). Accordingly, flux/electrode coating composition for odd position welding (vertical, horizontal and overhead

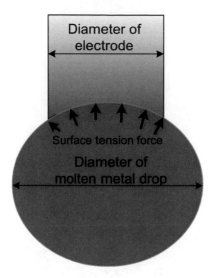

Fig. 7.2 Schematic of surface tension force

welding position) should be designed to have viscous and high surface tension weld metal/slag, so that molten metal from the consumable electrode can be placed effectively in the weld pool/metal.

$$\text{Surface tension force } (F_s) = (2\sigma \times \pi R_e^2)/4R \tag{7.2}$$

where σ is the surface tension coefficient, R is drop radius and R_e is the radius of electrode tip. It is evident from Eq. 7.2 that increase in surface tension coefficient, drop size and decrease in electrode diameter increase the surface tension force. Increase in temperature of the molten metal reduces the surface tension coefficient (σ) which in turn reduces surface tension force. Accordingly, reduction in surface tension force reduces its hindering effect on detachment of the drop from the electrode tip and conversely facilitates the detachment of drop from electrode tip in metal transfer.

7.1.3 Force Due to Impact of Charge Carriers

As per polarity being used during arc welding processes, charged particles namely ions and electrons move towards cathode and anode, respectively; eventually, these charged particles impact with them. This force invariably resists the detachment of the molten metal drop; however, contribution of this force in molten metal transfer is very limited. In case of reverse polarity (DCEP), electrode is connected to the positive terminal and acts as anode, and therefore, fast-moving electrons will be impact with consumable electrode having a molten metal drop at the tip. Force generated due to

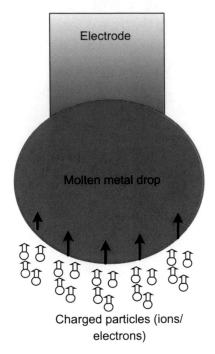

Fig. 7.3 Schematic of force acting on molten metal drop due to impact of charged particles in arc zone

impact of charged particles (electrons) on to the molten metal drop hanging at the tip of electrode tends to hinder the detachment (Fig. 7.3). Ions will be impacting with molten metal drop in case of straight polarity (DCEN). The force due to impact of the charge carriers is given by Eq. 7.3

$$\text{Force due to impact of charged particles } F_m = m(dV/dt) \qquad (7.3)$$

where m is the mass of charge particles, V is the velocity and t is the time.

7.1.4 Force Due to Metal Vapours

Molten metal evaporating from bottom side of drop and weld pool move in upward direction. Assuming metal vapours are lighter than gases present in arc zone. Force generated due to impact caused by upward moving of metal vapours acts (from the bottom side in flat welding position) against the detachment of molten metal drop hanging at the tip of the electrode. Thus, this force tends to hinder the detachment of droplet (Fig. 7.4).

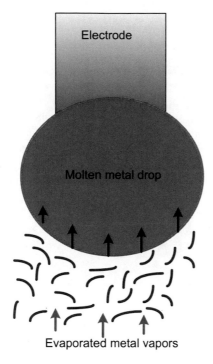

Fig. 7.4 Schematic of force acting on molten metal drop due to impact of evaporated metal vapours

7.1.5   *Force Due to Gas Eruption*

Gases present in the molten metal such as oxygen and hydrogen may react with some of the elements (such as carbon) present in molten metal drop and form gaseous molecules (such as carbon dioxide). The growth of such gas in molten metal drop as a function of time ultimately leads to bursting of metal drops which in turn increases the spattering and reduces the control over handling of molten weld metal (Fig. 7.5).

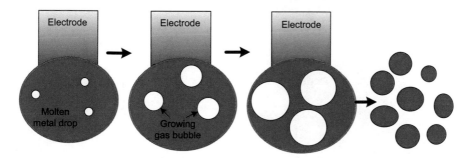

Fig. 7.5 Schematic showing stages of bursting of molten metal drop due to gas eruption

7.1.6   Force Due to Electro Magnetic Field

Flow of welding current through the arc gap develops the electromagnetic field. Charge carriers in arc zone also have their own electromagnetic field, interaction of electromagnetic fields due to (a) flow of welding current with that of (b) charge carriers produces a force which tends to pinch the drop hanging at the tip of the electrode. Therefore, electromagnetic force acting in arc zone is termed pinch force. The pinch force acts at an angle on to the surface of molten metal drop (Fig. 7.6).

The pinch force on resolving gives two components forces (a) horizontal force acting the molten metal drop near the electrode tip and (b) vertical force acting on the drop in the downward direction. The horizontal component of the pinch force reduces the cross-section for molten metal drop near the tip of the electrode and thus helps in detachment of the droplet from the electrode tip. The vertical component of pinch force acting in downward direction tends to pull down the molten metal drop to facilitate the detachment from the electrode tip (Fig. 7.7). The pinch force is given by:

$$\text{Pinch force}(F_p) = (\mu \times I^2)/8\pi \tag{7.4}$$

where μ is the magnetic permeability of metal and I is the welding current flowing through the arc gap.

7.2   Effect of Electrode Polarity

In case of DC welding, polarity depends on the way electrode is connected to the power source, i.e. whether electrode is connected to positive or negative terminal of

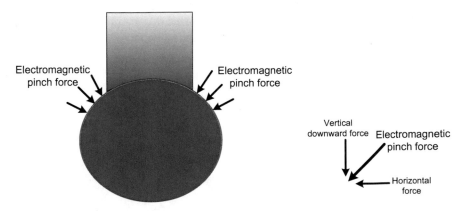

Fig. 7.6 Schematic electromagnetic pinch force **a** acting on molten metal drop near electrode and **b** resolved components of pinch force

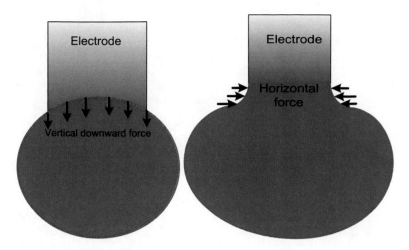

Fig. 7.7 Schematic electromagnetic pinch force acting on molten metal drop **a** vertical downward component trying detach drop from electrode and **b** horizontal component trying reduce cross-section of drop

the power source. If electrode is connected to negative terminal of the power source, then it is called direct current electrode negative (DCEN) or straight polarity, and if electrode is connected to positive terminal of the power source, then it is called direct current electrode positive (DCEP) or reverse polarity. Polarity in case of AC welding does not remain constant as it changes in every half cycle of current. Selection of appropriate polarity is important for successful welding as it affects (Table 7.1):

1. Distribution of heat generated by welding arc at anode and cathode
2. Arc initiation and stability
3. Cleanliness of weld metal.

Table 7.1 Comparison of AC and DC welding power sources

S. No	Parameter	AC	DC
1	Arc stability	Poor	Good
2	Distribution of arc heat	Uniform	Provide better control of heat distribution
3	Efficiency	High	Low
4	Power factor	Low	High
5	Cleaning action	Good	Depends on polarity
6	Maintenance	Less	More
7	Cost	Less	More

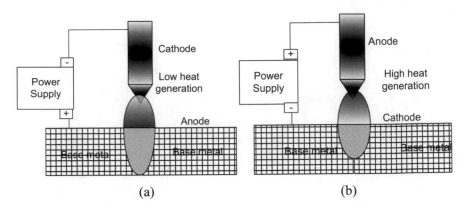

Fig. 7.8 Schematic showing effect of polarity on heat generation and weld bead geometry **a** straight polarity (DCEN) and **b** reverse polarity (DCEP)

7.2.1 Heat Generation

In general, more heat is generated at the anode than the cathode. Of total DC welding arc heat, about two-third of heat is generated at the anode and one-third at the cathode. The differential heat generation at the anode and cathode is due to the fact that impact of high-velocity electrons with anode generates more heat than that of comparatively low-velocity ions with cathode as electrons possess higher kinetic energy than the ions. Ions being heavier than electrons do not get accelerated enough, so these move at low velocity in the arc zone. In case of non-consumable electrode welding processes such as GTAW and PAW, it is always desired that arc is established (between the tungsten electrode and workpiece) with minimum thermal damage/degradation to the electrode so as to achieve longer life of the electrode. Further, DCEN results in larger weld area and deeper penetration due to higher heat generation on base metal side than the DCEP (Fig. 7.8). Therefore, DCEN polarity is commonly used with non-consumable electrode welding processes so as to reduce the thermal degradation of the electrodes. Moreover, DCEP polarity facilitates higher melting rate deposition rate in case of consumable electrode welding process such as SAW, FCAW and GMAW.

7.2.2 Arc Initiation and Stability

Ease of striking and maintenance of the welding arc is decided by fact that how easily high charged particle density is achieved (between the electrode and workpiece) through emission of free electrons by thermal emission and ionization of metal vapours (Fig. 7.9). Free electrons for striking/maintenance of the arc are mainly released by cathode. Cathode can be either electrode or workpiece as per polarity in

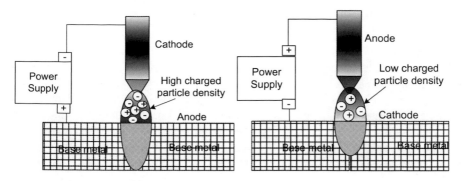

Fig. 7.9 Schematic showing effect of polarity on charge particle density (so arc stability) and weld bead geometry **a** straight polarity (DCEN) and **b** reverse polarity (DCEP)

case of DC welding. Choice of polarity therefore becomes important for all those arc welding processes (such as SMAW, PAW and GTAW) in which mainly electrode is expected to emit free electrons for easy arc initiation and its stability. Electrode in all such processes need to be made cathode using DCEN polarity. Further, poor arc stability and difficult arc striking behaviour are observed if electrode is made anode through DCEP or polarity changes alternately like in AC welding.

Shielded metal arc welding process using electrode with coating having low ionization potential elements provide better stable arc stability with DCEN than DCEP and AC. However, SMA welding with DCEP gives smoother metal transfer. Similarly, in case of GTAW welding, tungsten electrode is expected to emit electrons to provide stable arc, and therefore, DCEN is commonly used except when clearing action is receded in case of reactive metals, e.g. Al, Mg, Ti.

7.2.3 Cleaning Action

Three types of cathode spots are observed on the cathode during welding namely pointed, normal and mobile. Mobile cathode spot helps in loosening the refractory oxide film/layer if present on the surface of cathode. This aspect is utilized to develop a cleaner weld by making workpiece as cathode using DCEP polarity in case of all those metals forming mobile cathode spot during the welding. Mobile cathode spot loosens the tenacious refractory oxide layer during GTAW/GMAW welding of aluminium and magnesium using DCEP polarity. However, DCEP polarity due to high heat generation at the electrode results in lower electrode life than DCEN. Therefore, a compromise is made between the electrode life and cleaning action by selecting the AC for GTAW of aluminium, magnesium and stainless steels.

7.3   Arc Blow

Arc blow is basically a deflection of a welding arc from its intended path, i.e. axis of the electrode. Deflection of arc during welding reduces the control over the handling of molten metal by making it difficult to apply the molten metal at desired place. A severe arc blow increases the spattering which in turn decreases the metal deposition efficiency of the welding process. As per direction of deflection of arc with respect to welding direction, an arc blow can be termed as forward or backward arc blow. Deflection of arc in direction of the welding (ahead of the weld pool) is called forward arc blow and that in reverse direction is known backward arc blow (Fig. 7.10a–c).

7.3.1   Causes of Arc Blow

Arc blow is mainly encountered during DC arc welding due to interaction between different electromagnetic fields in and around the welding arc. Interaction between electromagnetic fields mainly occurs in areas where these fields are localized. There are two common situations of interaction between electromagnetic fields that can lead to arc blow:

Interaction between electromagnetic field due to (a) flow of current through the arc gap and that due to (b) flow of current through plates being welded. Electromagnetic field is generated around the arc in arc gap and interaction of this field with other electromagnetic fields causes to deflection of the arc from its intended path (Fig. 7.11a).

The lines of electromagnetic fields are localized near the edges of the plates as these cannot flow easily in air. Flow of lines of electromagnetic fields through the metal is easier than the air, and therefore, distribution of lines of electromagnetic forces does not remain symmetric around the arc. Rather, these get segregated/concentrated near the edge of the plate (Fig. 7.11b).

7.3.2   Mechanism of Arc Blow

Electromagnetic field is generated in a plane perpendicular to the direction of current flow through a wire. Intensity of self-induced magnetic field ($H = i/2\pi r$) due to flow of current depends upon the distance of point of interest from centre of wire (r) and magnitude of current (i). In general, increase in current and decrease in the distance from the wire increase the intensity of electromagnetic field. Depending upon the direction of current flow through two wires, there can be two types of polarities namely like and unlike polarity, accordingly electromagnetic fields due to current flow interacts with each other. In case of like polarity, the direction of flow of current

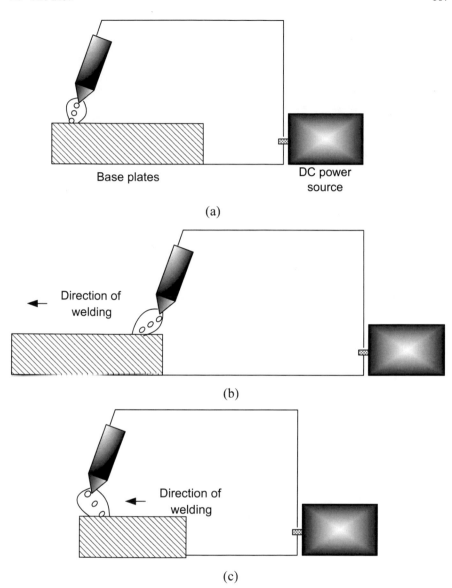

Fig. 7.10 Schematic diagram showing welding **a** without arc blow, **b** forward arc blow and **c** backward arc blow

is same in two conductors. Electromagnetic fields in case of like polarities repel each other while those of unlike polarities attract each other.

The welding arc tends to deflect away from area where electromagnetic flux concentration exit. In practice, such kind of localization of electromagnetic fields and so deflection of arc depends on the position of ground connection as it affects the

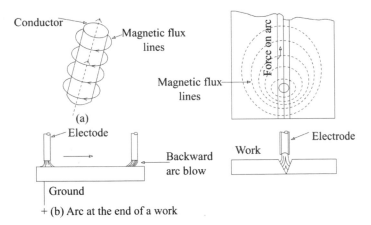

Fig. 7.11 Schematic showing generation of electromagnetic force around the welding arc and electrode causing arc blow

direction of current flow through the plate and so related electromagnetic field. Arc can blow towards or away from the earthing point depending upon the orientation of electromagnetic field around the welding arc. Effect of ground connection on arc blow is called ground effect. Ground effect may add or reduce the arc blow, depending upon the position of arc and ground connection. In general, ground effect causes the deflection of arc in the direction opposite to the ground connection from the arc position.

Arc blow occurring due to interaction between electromagnetic field around the arc and electromagnetic field localized near the edge of the plates, always tends to deflect the arc away from the edges of the plate. The ground connection in opposite side of the edge plate experiencing deflection can help to reduce the arc blow.

7.3.3   Controlling the Arc Blow

Arc blow can be controlled by:

- Reduction of the arc length so as to reduce the possibility and extent of misplacement of molten metal
- Adjust the ground connection as per position of arc so as to use ground effect in favourable manner
- Shifting from DC to AC if possible so as to neutralize the arc blow occurring in each half
- Directing the tip of the electrode in direction opposite to that of the arc blow.

Questions for Self-assessment

- What is the role of arc forces in development of sound weld joint?
- Describe various forces acting in arc zone.
- What are the factors affecting the forces in arc region?
- How does surface tension force affect the detachment of molten metal drop from a consumable electrode tip in flat, and overhead welding positions?
- Describe role of polarity on heat generation during DC welding.
- What is straight (DCEN) and reverse (DCEP) polarity in DC welding?
- How does polarity affect the arc stability and cleaning action during welding?
- What is arc blow? How does arc blow affect the welding?
- Explain the mechanism of arc bow.
- What are causes of arc blow? What steps can be taken to control the arc blow?
- Should metal of workpiece and electrode be considered for selection of suitable polarity for DC welding?

Further Reading

American Welding Society (1987) Welding handbook, 8th edn, vol 1 and 2, USA
Cary H (1988) Welding technology, 2nd edn. Prentice Hall
Little R (2001) Welding and welding technology, 1st edn. McGraw Hill
Nadkarni SV (2010) Modern arc welding technology. Ador Welding Limited, New Delhi

Chapter 8
Physics of Welding Arc

Arc Efficiency, Metal Transfer and Melting Rate

8.1 Arc Efficiency

Arc welding basically involves melting of faying surfaces of base metal using heat generated by arc under a given set of welding conditions, i.e. welding current and arc voltage. However, only a part of heat generated by the arc is used for melting purpose to produce fusion weld joint, and remaining heat is lost in various ways such as heat conduction to base metal, heat convection and radiation to surrounding (Fig. 8.1). Moreover, the heat generation on the workpiece side depends on the polarity in case of DC welding, while heat generated is equally distributed between workpiece and electrode in case of AC welding. Further, it can be recalled that heat generated (in unit time) by arc is dictated by the power of the arc (VI) where V is arc voltage, i.e. mainly sum of voltage drop in cathode drop (V_C), plasma (V_p) and anode drop regions (V_p) apart from of work function related factor, and I is welding current. Product of welding current (I) and voltage drop in particular zone governs the heat generated in that zone, e.g. anode, cathode and plasma region. In case of DCEN polarity, high heat generation at workpiece facilitates melting of base metal to develop a weld joint of thick plates, while DCEP polarity produces high heat at the electrode which in turn results in high melting rate/deposition rate.

8.1.1 Arc Efficiency of Different Welding Processes

On the basis of electrode, arc welding processes can be grouped in two categories (a) consumable electrode (b) non-consumable electrode welding process. As name suggests in case of consumable electrode arc welding processes, electrode is melted using heat generated by the welding arc and is consumed to produce the weld metal, while in case of non-consumable electrode arc welding processes, electrode is not melted, but it gets heated only using heat generated by the welding arc. Rest of the heat generated by welding arc is (a) used to melt the faying surfaces of workpiece to

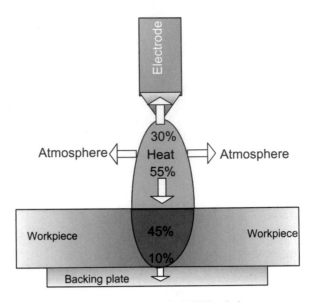

Fig. 8.1 Distribution of heat from the welding arc in DCEN polarity

produce weld metal and (b) dissipated to surrounding by convection and radiation. Heat utilized for either melting of electrode or workpiece during welding effectively contributes to arc efficiency. Heat is generated by arc at both electrode and workpiece sides as per the polarity. However, irrespective of heat used for melting of faying surfaces of base metal, for a given polarity, heat generated at electrode side in non-consumable arc welding processes is not utilized for melting of electrode, while in case of consumable arc welding processes heat generated at electrode side utilized for melting which in turn contributes in development of weld metal. This aspect results in significant difference in arc efficiency of non-consumable arc welding processes and consumable arc welding processes.

Under simplified conditions (with DCEN polarity), ratio of the heat generated at anode and total heat generated by the arc is defined as arc efficiency. However, this ratio shows the arc efficiency of non-consumable arc welding processes such as GTAW, PAW, laser and electron beam welding processes only where filler metal is not commonly used. However, this definition does not reflect true arc efficiency for consumable arc welding processes as it does not include use of heat generated in plasma region and cathode side for melting of electrode/filler metal and base metal (Fig. 8.2). Therefore, arc efficiency equation for consumable arc welding processes must include heat used for melting of both workpiece and electrode.

Since consumable arc welding processes (SMAW, SAW and GMAW) use heat generated both at cathode and anode for melting of filler and base metal, while in case of non-consumable arc welding processes (GTAW, PAW) heat generated at the anode only is used for melting of the base metal; therefore, in general, consumable arc welding processes offer higher arc efficiency than non-consumable arc welding

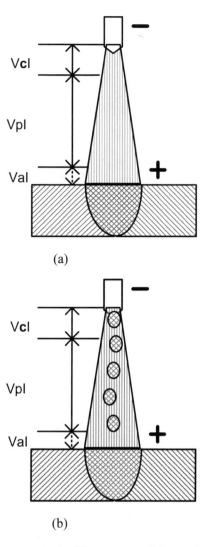

(a)

(b)

Fig. 8.2 Schematic of heat generation in different zones of the arc of **a** non-consumable arc and **b** consumable arc welding processes

processes. Additionally, in case of consumable arc welding processes (SMAW, SAW) heat generated is more effectively used because of reduced heat losses to surrounding as weld pool is covered by molten flux and slag. Welding processes in ascending order of arc efficiency are GTA, GMA, SMA and SAW. GTAW offers lower arc efficiency (21–48%) than SMAW/GMAW (66–85%) and SA welding (90–99%).

8.1.2   Calculations of Arc Efficiency

Non-consumable arc welding processes

Heat generated at the anode is found from sum of heat generated due to electron emission and that from anode drop zone.

$$Q_a = [\phi + V_a] I \tag{8.1}$$

where Q_a is the heat generated at anode.
    ϕ is work function of base metal at temperature T

$$\phi = (\phi_0 + 1.5\,kT) \tag{8.2}$$

ϕ_0 is work function of base metal at temperature 0 K.
k is the Boltzmann constant.
T is temperature in Kelvin.
V_a is anode voltage drop.
I is welding current.
Heat generated in plasma region is given by

$$Q_p = V_p I \tag{8.3}$$

Say fraction m (%) of the heat generated in plasma region goes to anode/workpiece for melting

$$= m\,(V_p I) \tag{8.4}$$

So arc efficiency = total heat used/total heat generated in arc

$$= [Q_a + m\,(V_p I)]/V I \tag{8.5}$$

where V is arc voltage $= V_a + V_p + V_c$.
    Another way is that [{total heat generated in arc − (heat with plasma region + heat of cathode drop zone)}/total heat generated in arc}].
    So arc efficiency

$$[\{VI - Q_c + (1 - m)\,(V_p I)\}/VI\}][\text{ or }[\{VI - [V_c I + (1 - m)\,(V_p I)\}/VI\}] \tag{8.6}$$

where q_c is the heat generated in cathode drop zone.

Arc efficiency for consumable arc welding processes

Arc efficiency for consumable arc welding processes can be obtained from the ratio of sum of heat generated/used for melting of electrode in cathode and heat used for melting of workpiece at anode (comprising heat generated at the anode and fraction of heat generated in plasma zone for melting of workpiece) to total heat generated by the arc. This can be expressed as following:

$$\text{Arc efficiency: } (Q_c + Q_a + m Q_p)/Q = (IV_c + IV_a + mIV_p)/IV$$
$$= (V_c + V_a + mV_p)/V$$

8.2 Metal Transfer

Metal transfer refers to the transfer of molten metal from the tip of a consumable electrode to the weld pool. It is of academic and practical importance for consumable electrode welding processes as it directly affects the control over the handling of molten metal, slag and spattering and quality of the weld metal. However, metal transfer is considered to be more of academic importance for GMA and SA welding than practical need due to more technological advance nature of the phenomenon. Shielding gas, composition, diameter and extension of the electrode and welding current are some of the arc welding-related parameters, which affect the mode of metal transfer for a given power setting. Four common modes of metal transfer, namely short circuiting, globular, spray and rotational metal transfer, are generally observed during consumable arc welding depending upon the welding process and welding conditions. Further, short circuiting, globular, spray and rotational metal transfer occur in the increasing order of welding current. During transfer of the molten metal from the consumable electrode tip to the weld pool takes places through three stages such as (a) melting of electrode tip leading formation of molten drop, (b) growth of molten metal until its detachment to get transferred to the weld pool and (c) detachment of molten metal drop from the electrode tip under the influences of various forces acting in the arc zone. The growth of the molten metal drop reduces the effective arc gap for the flow current needed to maintain the arc which in turn affects the resistance for the flow of current, and so welding current gets adjusted accordingly. The variation in arc gap as per the mode of metal transfer affects the welding current (Fig. 8.3). Mechanism and welding condition for different types of metal transfers have been described in the following sections.

8.2.1 Short Circuit Transfer and Dip Transfer

This kind of metal transfer takes place in the presence of two welding conditions such as (a) welding current is very low but high enough to have stable arc and (b) narrow arc gap. Under these welding conditions, molten metal droplet grows slowly (due to low welding current) at the tip of the electrode, and then as soon as drop touches weld

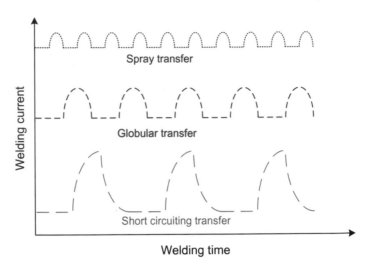

Fig. 8.3 Schematic of welding current variation during different modes of metal transfer

pool, short circuiting takes place. Further, due to narrow arc gap, molten drop does not attain a size big enough to fall down on its own (by weight) under gravitational force, but before that short circuiting takes place. On occurrence of short circuit, welding current flowing through the droplet to the weld pool increases abruptly which in turn results in excessive heat generation. High heat generation makes the molten metal of droplet thinner (due to reduction in surface tension and viscocity) which in turn reduces the surface tension force trying to resist the detachment of the droplet at the electrode tip. Touching of the molten metal drop to weld pool leads to transfer of molten metal into weld pool due to pull effect of welding pool caused by surface tension (Fig. 8.4a). Once molten metal is transferred to the weld pool, an arc gap is established which in turn increases arc voltage abruptly. This increase in arc voltage (due to creation of the arc gap on transfer of molten metal drop) reignites arc, and flow of current starts. This whole process is repeated rapidly at a rate ranging from 20 to more than 200 times per second during the welding. Schematically variation in welding current and arc voltage for short circuit metal transfer is shown in Fig. 8.4c.

Dip type of metal transfer is observed when welding current is very low and feed rate is high. Under these welding conditions, electrode is short-circuited with weld pool, which leads to the melting of electrode and transfer of molten drop (Fig. 8.4b). Approach-wise dip transfer is similar to that of short circuit metal transfer, and many times two are used interchangeably. However, these two differ in respect of welding conditions especially arc gap and feed rate that lead to these two types of metal transfers. Low welding current and narrow arc gap (at normal feed rate) result in short circuit mode of metal transfer, while the dip transfer is primarily caused by abnormally high feed rate even when working with recommended range of welding current and arc gap.

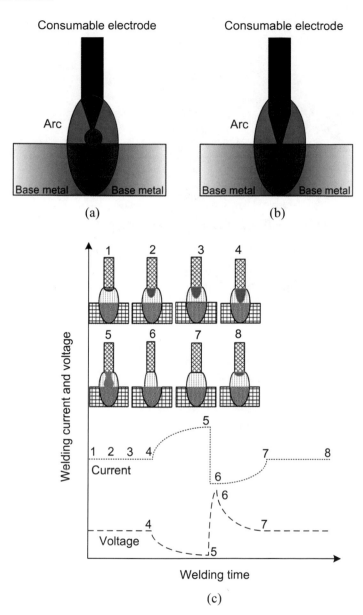

Fig. 8.4 Schematic of **a** short circuiting metal transfer, **b** dip transfer and **c** current and voltage variation during short circuiting transfer

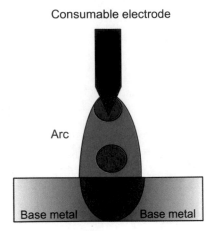

Consumable electrode

Arc

Base metal   Base metal

Fig. 8.5   Schematic of globular metal transfer

8.2.2   Globular Transfer

Globular metal transfer takes place in the presence of two welding conditions (a) welding current is low (but higher than that is needed for short circuit metal transfer) and (b) arc gap is large enough, so molten metal droplets can grow slowly (at the tip of the electrode) with melting of the electrode tip (Fig. 8.5). In this mode of the metal transfer, the droplet continues to grow until gravitational force acting on droplet (due to its own weight) exceeds the surface tension force and other forces if any trying to hold the drop at the tip of electrode. As soon as drop attains a large size enough and so gravitational force becomes more than drop holding forces (such as surface tension force), the drop detaches from the electrode tip and is transferred to the weld pool. The transfer of molten metal drop normally occurs when it attains size larger than the electrode diameter. No short circuit takes place in this mode of metal transfer.

8.2.3   Spray Transfer

This kind of metal transfer takes place when welding current is higher than that is required for globular transfer. High welding current results in high melting rate and greater pinch force as both melting rate and pinch force are directly related with welding current. Melting rate and pinch force vary in proportion to square of welding current. Therefore, at high welding current, droplets are formed rapidly and are pinched off from the tip of electrode quickly by high pinch force even when they are of very small in size (Fig. 8.6a). Another reason for detachment of small droplets is that high welding current results in high arc power, high heat generation, so high temperature of arc zone which in turn reduces the surface tension force. Reduction

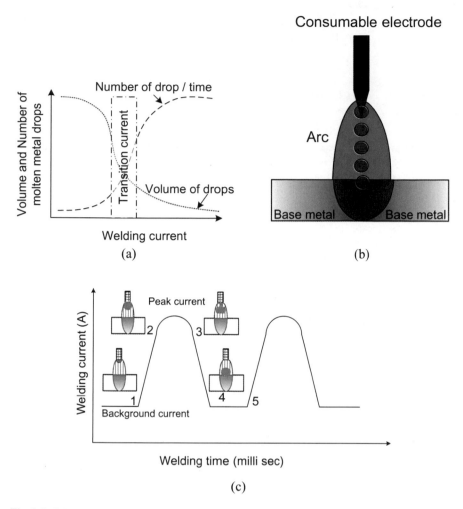

Fig. 8.6 Schematic of **a** volume/number drop transfer as function of welding current, **b** spray metal transfer and **c** variation in welding current in case of spray transfer during pulse GMAW process

in surface tension force decreases the resistance to detachment of drops from the electrode tip (drop holding forces) which in turn facilities detachment of drops from the tip of the electrode even when they are of small size. Since the transfer of molten metal from electrode tip appears similar to that of spray in line of axis of the electrode, therefore, it is called spray transfer (Fig. 8.6b). Spray like feature of this mode of metal transfer helps to direct the molten metal in proper place where it is required especially in difficult to access areas. However, high temperature of molten metal makes control over the weld pool difficult especially in odd welding positions as it increases the tendency of the molten metal to fall down due to low surface tension.

A typical variation in welding current in case of spray transfer during pulse GMAW process is schematically shown in Fig. 8.6c.

8.2.4   Rotational Transfer

The rotational transfer is observed when two welding condition prevail (a) very high welding current (greater than the spray transfer) and (b) thin wire of low stiffness and high electrical resistivity metal. These welding conditions lead to excessive electrical resistance heat generation is leading to significant thermal softening of electrode wire. A softened electrode wire under influence of the heavy electromagnetic force generated by high welding current results in rotation of electrode which is melting at high rate by arc heat during welding (Fig. 8.7). A combination of melting and rotation of electrode wire produces rotational metal transfer. The control over the rotational metal transfer becomes poor due to two reasons (a) difficult placement of molten metal at the desired place and (b) increased spattering.

8.3   Melting Rate

In consumable arc welding processes, weld metal deposition rate is governed by the rate at which electrode is melted during the welding. Melting of the electrode needs sensible and latent heat, which is supplied by arc through the electrical reactions, i.e. heat generated at anode $(I \cdot V_a)$, cathode $(I \cdot V_c)$ and plasma zone $(I \cdot V_p)$. In case of DCEN polarity, heat generated in anode drop region and plasma region does not influence melting of electrode tip appreciably as electrode (cathode) in case of straight polarity (DCEN) gets very negligible heat from these two regions (anode and plasma). Hence, in case of straight polarity (DCEN), melting rate of electrode is primarily determined by the heat generated due to (a) cathode reaction and (b) electrical resistance heating. Accordingly, melting rate of electrode for consumable arc welding processes is given by the following equation:

$$\text{Melting Rate} = a \times I + b \times L \times I^2 \qquad (8.7)$$

where a and b are constants {(independent of electrode extension (L) and welding current (I)}.

Value of constant "a" depends on ionization potential of electrode material (ability to emit the charge carriers), polarity, composition of electrode and anode/cathode voltage drops, while another constant "b" accounts for electrical resistance of electrode (which in turn depends on electrode diameters and resistivity of electrode metal).

Melting rate equation suggests that first factor $(a \times I)$ accounts for electrode melting due to heat generated by anode/cathode reaction and second factor $(b \times L \times$

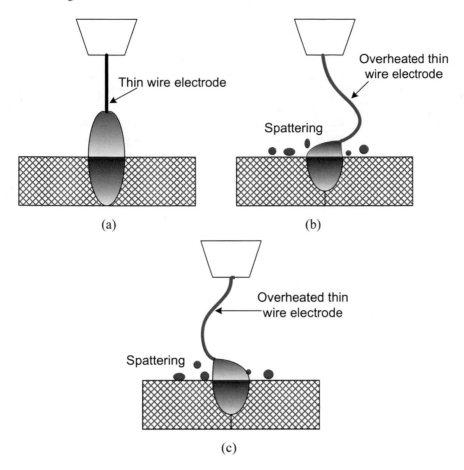

Fig. 8.7 Schematics showing rotational metal transfer and its effect on metal deposition and bead geometry: **a** low current, no rotation and **b** and **c** high current beyond current carrying capacity of the electrode

I^2) considers the melting rate due to heat generated by electrical resistance heating. Relative contribution of these two factors in melting of the electrode is primarily governed by welding current and electrode (material and its extension). Melting rate is mainly governed by the first factor when welding current is low, electrode diameter is large, extension is small, and electrical resistivity of the electrode metal is low; whereas second factor significantly determines the melting rate of electrode predominantly when welding current in high, electrode diameter is small, extension is large, and electrical resistivity of electrode metal is high.

8.3.1 Factors Limiting the Melting Rate

Difference in values of constant a and b and welding parameters results in variation of melting rate of the electrode in case of different welding processes. For a given welding set (electrode and its extension), welding current is main factor controlling the melting rate. To increase the melting rate, welding current for a specific welding process can be increased up to a limit. The upper limit of welding current is influenced by two factors: a) extent of overheating of electrode caused by electrical resistance heating and so related thermal degradation of the electrode coating (if any) and b) required mode of metal transfer for smooth deposition of weld metal with minimum spatter. For example, in semi-automatic welding process such MIG/SAW, minimum welding current is determined by the current level at which short circuit metal transfer starts and upper level of current is limited by appearance of rotational spray transfer. For a given electrode material and diameter, upper limit of current in case of SMAW is dictated by thermal composition of the electrode coating and that in case of GTAW is determined by thermal damage to tungsten electrode (Fig. 8.8). Lower level of current in general determined is by arc stability (the current at which stable arc is developed) or current level below which arc is not stable. In additional to arc stability, lower limit of welding current is also dictated by other requirement of welding such as penetration, proper placement of the weld metal and control over the weld pool especially in case of vertical and overhead welding positions and those related with poor accessibility. Depending upon these factors, higher and lower limits of welding current melting rate are decided.

Example A TIG welding process uses DCEN polarity, arc voltage of 30 V and welding current of 120 A for welding of 2 mm thin plate. Assuming (a) the voltage

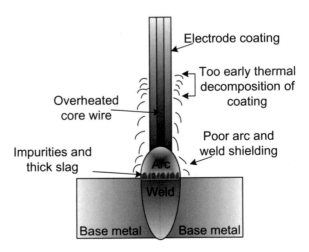

Fig. 8.8 Schematic showing premature thermal decomposition of the electrode coating in SMAW due to overheating of the core wire caused by high welding current beyond its current carrying capacity leading to the poor protection of the weld pool

drop in anode, cathode and plasma regions is 16 V, 10 V, 4 V, respectively, and (b) 20% of heat generated in plasma zone is used for melting of base metal and (c) all heat generated in anode drop zone is used for welding. Neglecting the voltage drop on account of work function of metal during welding, calculate the arc efficiency.

Solution

Arc efficiency: (Heat generated in anode drop zone + heat generated in plasma used welding)/all heat produced by welding arc

$: V_a \times I + m(V_p \times I)/VI \sim (V_a + mV_p)/V$
$(16 + 0.2 \times 4)/30 \sim 16.8/30$
Arc efficiency: $0.56 \sim 56\%$

Questions for Self-assessment

- Define arc efficiency. What are the factors affecting the arc efficiency?
- Why do consumable arc welding processes offer higher arc efficiency than non-consumable arc welding processes?
- Electroslag welding of 50-mm-thick steel plates was performed using current 480A and voltage 34 V. The heat losses to the water-cooled copper shoes and by radiation from the surface of the slag pool were 1275 and 375 cal/s, respectively. Calculate the heat source efficiency.
- Establish the equation of arc efficiency for consumable and non-consumable arc welding processes.
- What is metal transfer? How does metal transfer affect the development of sound weld joints?
- Explain the mechanism of different modes of metal transfer during consumable arc welding processes?
- Describe the factors affecting the mode of metal transfer.
- Enlist the welding conditions and positions where different modes of metal transfer are preferred.
- How can we control the melting rate during consumable arc welding processes?
- Explain the factors limiting the melting rate in different consumable arc welding processes.

Further Reading

Little R (2001) Welding and welding technology, 1st edn. McGraw Hill
Cary H (1988) Welding technology, 2nd edn. Prentice Hall
Nadkarni SV (2010) Modern arc welding technology. Ador Welding Limited, New Delhi
American Welding Society (1987) Welding handbook, 8th edn., vols. 1 & 2. American Welding Society, USA

Part III
Arc Welding Power Source

Chapter 9
Arc Welding Power Source

Basic Characteristics and Self-regulating Arc

9.1 Introduction

One of the main requirements of a welding power source is to deliver the controllable current at desired voltage according to the fluctuating demand of the welding process. Each welding process has distinct features (in respect of controls, current/voltage variation with changing arc lengths during welding) from other processes for smooth and stable welding arc. Therefore, application of correct type of arc welding power source plays a very important role in successful welding. The conventional welding power sources are:

Power Source	Supply
(i) Welding Transformer	AC
(ii) Welding Rectifier	DC
(iii) Welding Generators	AC/DC
(iv) Inverter type welding power source	DC.

Welding transformers, rectifiers and DC generators are used in manufacturing industries and fabrication shops while engine coupled DC and AC generators are used at the sites where domestic line supply is not available. Rectifiers and transformers are usually preferred over generators because of lower noise, higher efficiency and lower maintenance.

The inverter type welding power source first transforms the AC into DC. The DC power is then fed into a step-down transformer to produce the desired welding voltage/current. The pulse of high voltage and high frequency DC is then fed to the main step-down transformer and there it is transformed into low voltage and high frequency DC suitable for welding. Finally, low voltage and high frequency DC is passed through filters for rectification. The switching on and off is performed by solid state switches at frequencies above 10,000 Hz. The high switching frequency reduces the size/volume of the step-down transformer. The inverter type of power source provides better features for power control and overload protection. These

D. K. Dwivedi, *Fundamentals of Metal Joining*,
https://doi.org/10.1007/978-981-16-4819-9_9

systems are found to be more efficient and better in respect of control of welding parameters than other type of welding power system.

The invertor type welding power source with microcontrollers allow changes in electrical characteristics of the welding power using software in real time. This can be done even on a cycle by cycle basis so as to provide features such as pulsing the welding current, variable ratios of peak/base current and current densities, stepped variable frequencies.

Selection of a power source with requisite features (open circuit voltage, duty cycle, and power factor, static and dynamic characteristics) mainly depends on the welding process and welding consumables to be used for arc welding. An understanding of the features and characteristics of welding power source is useful for selection of suitable welding power source for a welding process as per welding conditions. The open circuit voltage normally ranges between 50 and 90 V in case of welding transformers while that in case of rectifiers varies from 20 to 50 V. Moreover, arc voltage becomes lower than open circuit voltage of the power source. Welding power sources can be classified based on different various technical features/parameters as under:

- Type of current: A.C., D.C. or both.
- Cooling medium: Air, water, oil cooled.
- Cooling system: Forced or natural cooling.
- Static characteristics: Constant current, constant voltage, rising characteristics.

9.2 Characteristics of Power Source

A welding power source possess a set of characteristics indicating the capability and quality of the power source. These characteristics help in selection of suitable welding power source for a given welding conditions e.g. welding process, base metal, welding current, polarity, duty cycle, welding consumables, weld pool shielding approach. Basic characteristics of a welding power source are given below:

- Open circuit voltage (OCV)
- Power factor (pf)
- Static characteristics
- Dynamic characteristics
- Current rating and duty cycle
- Class of insulation.

9.2.1 Open Circuit Voltage (OCV)

The open circuit voltage of a power source shows the potential difference between the two terminals of the power source when there is no load. Setting up of correct open circuit voltage in a welding power source is important for stability of welding arc especially when AC is used. The selection of an optimum OCV (30–100V) depends on the type of base metal, composition of electrode coating, type of welding current and polarity, type of welding process etc. Base metal of low ionization potential (indicating capability of easy emission free of electrons) needs lower OCV than that of high ionization potential metal. Presence of low ionization potential elements such as K, Na and Ca in electrode coating/flux in optimum amount reduces OCV required for welding. AC welding needs higher OCV compared with DC primarily due to poor arc stability in case of AC in which welding current continuously changes its direction and magnitude; while in case DC magnitude and direction of current flow remains constant. In the same line, GTAW needs lower OCV than GMAW and other welding processes like SMAW and SAW because GTAW uses tungsten electrode which has good free electron emitting capability by thermal and field emission mechanism while in case of other welding processes like SMAW and SAW electrode metal usually does not exhibit electron emission behavior similar to tungsten. Abundance of free electron in GTAW under welding conditions therefore lowers the OCV required for stable welding arc.

Too high OCV may cause electric shock to the operator, therefore, if possible, it is preferred to use low OCV. Arc voltage is found to be lower than OCV depending up on the electric resistance offered by the different connectors in the circuit namely cable, connectors, electrode itself for the flow of the welding current. Greater is the resistance offered by the cables and connectors for the flow of current, higher is difference in OCV and arc voltage. Arc voltage is a potential difference between the electrode tip and work piece surface when there is a flow of current. Any fluctuation in arc length affects the resistance to flow of current through plasma and hence arc voltage is also affected. Increase in arc length or electrode extension are therefore known to increase the arc voltage. Electrical resistance heating of electrode increases with electrode extension for a given set of welding parameters which in turn increases the melting rate of the electrode.

9.2.2 Power Factor (pf)

Power factor of a power source is defined as a ratio of actual power (kW) used to produce the rated load (which is registered on the power meter) and apparent power drawn from the supply line (kVA) during the welding. It is always desired to have high power factor (pf) of the power source. Low power factor of the power source indicates unnecessary wastage of power and less efficient utilization of power during welding. Welding transformers usually offer higher power factor than other types of power sources. However, sometimes low power factor is intentionally used in case of welding transformers to increase the stability of arc for AC welding. The basic principle of using low power factor for better arc stability has been explained and presented in Sect. 6.2.2. In brief, a low power factor welding power source provides maximum open circuit voltage (OCV) when current is zero during AC welding. Application of a welding power source with high power factor offers many advantages such as:

- Reduction of the reactive power in a system, which in turn reduces the power consumption and so drop in cost of power for welding
- Economical operations at an electrical installation (more effective power for the same apparent power)
- Improved voltage quality and fewer voltage drops
- Use of welding cable of smaller cross-section
- Low transmission losses.

9.2.3 Static Characteristic of Power Source

The static characteristic of a welding power source exhibits the trend of variation in arc voltage with welding current when power source is connected to a pure resistive load. The welding power sources are of three types as far as variation in arc voltage with welding current is concerned namely constant current (CC), constant voltage (CV), and rising voltage (RV) characteristic power source.

9.2.3.1 Constant Current Power Source

In constant current type of the power source, variation in arc voltage (due to fluctuating arc length) leads to a minor change in welding current. A plot showing relationship between arc voltage and current exhibits a singnificant negative slope. Therefore, it is termed as 'dropper'. Since a change in arc voltage leads to very little variation in welding current, therefore, melting rate electrode and penetration in base metal during consumable arc welding processes remain fairly constant despite of minor fluctuations in arc length (Fig. 9.1). These power sources are used for welding processes where (a) arc is primarily controlled manually, (b) use of relatively thicker consumable electrodes, (c) possibility for electrode getting stuck with work piece high and (d) touching of electrode with base metal for starting of arc may damage the electrode (like in GTAW) and (e) under welding conditions when the short-circuiting current need to be limited to provide safety to power source and the electrode.

In constant current power source, variation in welding current with arc voltage (due to fluctuations in arc length) is very small therefore welding current remains more or less constant despite of fluctuations in arc voltage/length. Hence, this type of power source is also found suitable for all those welding processes where large fluctuation in arc length is likely to take place e.g., MMA, PAW and GTAW.

9.2.3.2 Constant Voltage Power Source

In constant voltage (CV) type of the power sources, arc voltage remains almost constant during welding despite of fluctuations in arc length. In CV power source, a small variation in arc voltage (due to fluctuations in arc length) causes significant

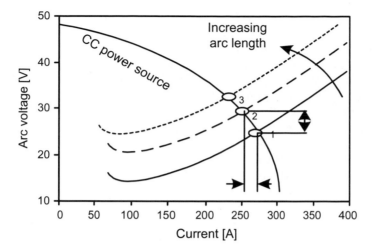

Fig. 9.1 Static characteristics of constant current welding power source

change in welding current. However, the constant voltage power source does not provide a perfectly constant voltage output as suggested by their voltage vs. current relationship curve. The curve in fact shows a little negative slope. This negative slope is attributed to internal electrical resistance and inductance in the welding circuit that causes a minor drop in the output volt-ampere characteristics of the power source (Fig. 9.2). This type of power source is found more suitable for all those welding processes where fluctuations in arc length during welding is limited like in semiautomatic welding process MIG, SAW and FCAW having mechanized electrode

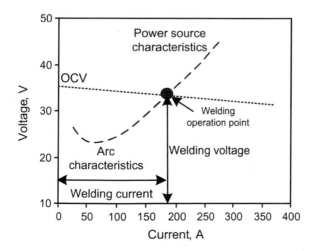

Fig. 9.2 Static characteristics of constant voltage welding power source

wire feeding system. The power source supplies the required welding current to have desired melt rate of the electrode so as to maintain the preset arc voltage/arc length.

A mechanized electrode feed drive system is used to control feed rate of the electrode with these power sources. A little change in arc length or melting rate leads to variation in arc voltage which in turn results in a significant change in welding current. This change in welding current affects the melting rate of the electrode in a big way. For example, a slight reduction in arc length decreases the arc voltage which increases the welding current significantly. Increase in welding current increases the melting rate which in turn increases the arc gap for given feed rate of the electrode. Thus, CV power source helps to maintain the arc gap. The use of such power source in conjunction with a constant speed electrode wire feed drive results in a self regulating or self adjusting arc system.

Self-regulating arc

Semiautomatic arc welding processes (GMAW, SAW) using a constant voltage power source, automatic constant speed wire feed system for provide small diameter consumable electrode, *arc length is maintained by self-regulating arc*. Self-regulating arc is one, which adjusts the melting/burn off rate of the electrode (by changing the current) so that feed rate becomes equal to melting rate for maintaining the arc length. For example, increase in arc length due to any reason shifts the operating point from 2 to 3 thus increases the arc voltage (Fig. 9.3). Operating point is the point of inter- section of power source characteristics with arc characteristics. In CV power source, rise in arc voltage decreases the welding current significantly. Decrease in welding current reduces the melting rate (see melting rate equation) of the electrode which in

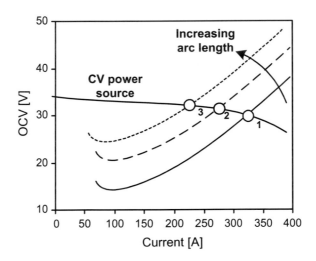

Fig. 9.3 Static characteristics of constant voltage welding power showing operating points with increasing arc length

turn decreases the arc gap if electrode is fed at a constant speed. Reverse phenomenon happens if arc length decreases (shifting the operating point from 2 to 1).

Questions for Self-assessment

- How are welding power-sources different from conventional domestic supply power sources?
- Describe the common welding power sources namely welding transformer, welding generator and rectifier.
- How can welding power sources be classified?
- What are basic characteristics of welding power sources?
- Describe the following characteristics of welding power sources along with their significance in welding

  - Open circuit voltage
  - Power factor
  - Dynamic characteristics
  - Static characteristics

- What is operating point in arc characteristic curve for given welding power sources?
- How is the operating point affected by arc length?
- What is self regulating arc and how can it be achieved in SAW/GMAW processes?
- Explain the self-regulating arc using suitable example for changing operating points during welding.
- How does a suitable selection of power source for manually controlled arc welding process help in obtaining consistent and uniform penetration?
- How can different welding related factors affect the operating point?

Further Reading

Little R (2001) Welding and welding technology, 1st edn. McGraw Hill
Cary H (1988) Welding technology, 2nd edn. Prentice Hall
Nadkarni SV (2010) Modern arc welding technology. Ador Welding Limited, New Delhi
American Welding Society (1987) Welding handbook, 8th edn., vols. 1 & 2. American Welding Society, USA
Parmar RS, Welding process and technology. Khanna Publisher, New Delhi

Chapter 10
Arc Welding Power Source

Basic Characteristics, and Duty Cycle

10.1 Rising Characteristics

Power sources with rising characteristics show increase in arc voltage with increase of welding current (Fig. 10.1). In automatic welding processes where strictly constant voltage power is required, power sources with rising characteristics are used.

10.2 Dynamic Characteristic

A welding arc is generally subjected to severe and rapid fluctuations in arc voltage (due to continuous minor changes in arc length) and welding current. These fluctuations have been exhibited schematically in Fig. 10.2. In the figure, numbers from 1 to 4 in figure indicates different stages of welding arc during pulse GTAW welding. The behavior of welding arc suggests that it is never in a steady state. Welding arc shows transients during arc striking, extinction and re-ignition stages after each half cycle in A.C. welding. To cope up with these fluctuating conditions of welding arc, the welding power source should have good dynamic characteristic to obtain stable and smooth arc. Dynamic characteristic of the power source indicates how instantaneous and fast variation in arc voltage with change in welding current over an extremely short period of welding takes place. A power source with good dynamic characteristic results in an immediate change in arc voltage and welding current corresponding to the change in welding conditions so as to give smooth and stable arc.

10.3 Duty Cycle

During the welding, heavy current is drawn from the power source. Flow of heavy current through the transformer coil and connecting cables causes electrical resistance

© The Author(s), under exclusive license to Springer Nature Singapore Pte Ltd. 2022
D. K. Dwivedi, *Fundamentals of Metal Joining*,
https://doi.org/10.1007/978-981-16-4819-9_10

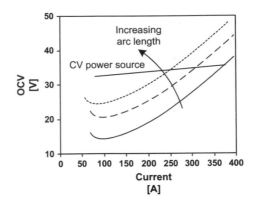

Fig. 10.1 Static characteristics of 'rising-voltage characteristic' welding power source showing operating points for different arc lengths

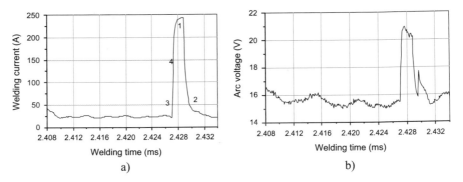

Fig. 10.2 Dynamic characteristics of a power source showing **a** current versus time and **b** voltage versus time relationship

heating. Continuous heating during welding for long time may damage coils and cables. Therefore, welding operation need to be stopped for some time depending upon the level of welding current being drawn from the power source. The total weld cycle is taken as sum of actual welding time and rest time (when welding current is not being drawn). Duty cycle refers to the percentage of welding time of total welding cycle i.e. welding time divided by welding time plus and rest time. Total welding cycle of 5 min is normally taken in India as per European standard. For example, welding for 3 min and followed by rest of 2 min in total welding cycle time of 5 min corresponds to 60% duty cycle. Duty cycle and associated welding current are important. As selection of proper welding current corresponding to required duty cycle ensures that power source is safe and windings are not damaged due to increase in temperature caused by electrical resistance heating beyond specified limit. Moreover, the maximum current which can be drawn from a power source at given a duty cycle depends upon size of winding wire, type of insulation and cooling system

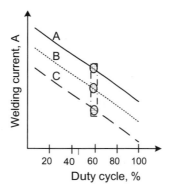

Fig. 10.3 Schematic showing relationship between welding current and duty cycle for welding power sources of the different qualities (A, B, C)

of the power source. In general, large diameter of cable wire, coils and winding, high temperature resistant insulation and forced cooling system allow drawing of high welding current from the welding source at a given duty cycle.

Duty cycle of a welding power source indicates its capability to deliver a welding current for specified period during the welding. It is basically linked with ability of the power source (windings, coils, cables, connectors) to sustain heat generated during supply of welding current for a particular period. In general, supply of low welding current from a power sources for the welding purpose is possible for longer period or even continuously and vice versa (Fig. 10.3).

Duty cycle is defined as ratio of arcing time to the weld cycle time multiplied by 100. Welding cycle time is either 5 min as per European standards or 10 min as per American standard and accordingly power sources are designed.

If welding current is supplied by a power source continuously for 5 min then as per European standard it is considered as 100% duty cycle and for the same case the duty cycle will be 50% duty cycle as per American standard. In case of continuous welding (more than 5 min or 10 min as per relevant standards), power source works at 100% duty cycle. Minimum current is drawn from the welding power source at 100% duty cycle. Welding power source operating at low duty cycle allows supply of high welding current. The welding current which can be drawn at a given duty cycle can be evaluated from the following equation:

$$D_R \times I_R^2 = I_{100}^2 \times D_{100} \qquad (10.1)$$

where

I Welding current at 100% duty cycle
D_{100} 100% duty cycle
I_R Welding current at required duty cycle
D_R Required duty cycle.

10.4   Class of Insulation

The duty cycle of a power source for a given current setting is primarily governed by the maximum allowable temperature of various components (primary and secondary coils, cables, connectors etc.), which in turn depends on the quality and type of insulation and materials of coils/cables used for manufacturing of power source. The insulation is classified as A, E, B, F and G class in increase order of their maximum allowable temperature 60, 75, 80, 100 and 125 °C respectively. Application of high-class insulation (like F and G) increases the duty cycle for a given current setting.

10.5   High Frequency Unit

Some of the power sources are designed with high frequency (HF) unit to start the arc like in GTAW and PAW. The role of high frequency unit is limited to initial stage of the welding arc initiation. Filters are used between the regular welding power circuit and HF unit to avoid damage of control circuit, because HF unit provides very high voltage at high frequency. High frequency unit supplies pulses of high voltage (of the order of few kV) at low current and high frequency (of few kHz). The high voltage pulses supplied by HF unit ionize the gaseous medium between electrode and work piece/nozzle to produce starting pilot arc which eventually leads to the ignition of the main arc. Although, high voltage can be fatal for operator but at high frequencies current passes through the skin and does not enter the body. This is called skin effect i.e. current passes through the skin without any damage to the operator.

10.6   Feed Drives for Constant Arc Length

Two types of electrode wire feed systems are generally used for maintaining the arc length (a) constant speed wire feed drive and (b) variable speed wire feed drive (Fig. 10.4a, b). In constant speed wire feed drives, feed rollers are rotated at constant speed for pushing/pulling electrode wire to feed into the welding arc zone so as to maintain the constant arc length during welding (Fig. 10.4b). This type drive is normally used with constant voltage power sources in combination with small diameter electrode where self regulating arc helps to attain the constancy in arc length.

   In case of variable speed wire feed drives, feed rollers used for feeding electrode wire (in consumable arc welding processes like SAW and GMAW) are rotated at varying speed as per need to maintain the arc length during welding. Fluctuation in arc length due to any reason is compensated by increasing or decreasing the electrode feed rate. The electrode feed rate is controlled by regulating the rotational speed of

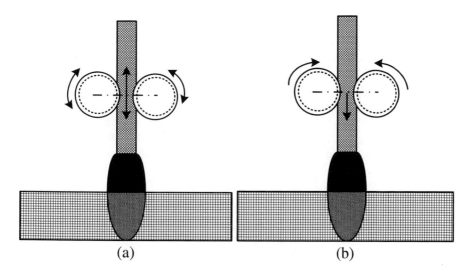

(a)   (b)

Fig. 10.4 Schematics show electrode feed drives for controlling arc length: **a** variable speed feed drive and **b** constant speed feed drive

feed rollers powered by electric motor (Fig. 10.4a). Input power to the variable speed motor is regulated with help of sensor which takes inputs from fluctuations in the arc gap. For example, an increase in arc gap sensed by sensor increases the input power to the variable speed motor to increase the rotational speed of roller which in turn increases feed rate of electrode so as to maintain arc gap.

Example Current rating for a welding power source at 60% duty cycle is 400 A. Determine the welding current for automatic continuous welding i.e. 100% duty cycle.

Solution:

  Rated current: 400 A
  Rated duty cycle: 60%
  Desired duty cycle: 100%
  Desired current?

$$\textbf{Desired duty cycle} = \frac{(\text{rated current})^2 \times \text{rated duty cycle}}{(\text{desired current})^2}$$

$$100 = \frac{(400)^2 \times 6}{(\text{desired current})^2}$$

  Answer :   Desired current : 310 A

Questions for Self-assessment

- Describe methods used for maintaining the arc length.
- Why is dynamic characteristic of a welding power source important for arc welding?
- Why does a welding power source designed to supply low welding current for high duty cycle?
- What are the factors affecting duty cycle for a power source at a welding current?
- How can a welding power sources of high duty cycle be designed?
- What is high frequency unit and how is does work?
- Which type of insulation should be used for high duty cycle power sources?
- Describe static characteristics of welding power sources that are commonly used for consumable and non-consumable arc welding processes.

Further Reading

Little R (2001) Welding and welding technology, 1st edn. McGraw Hill
Cary H (1988) Welding technology, 2nd edn. Prentice Hall
Nadkarni SV (2010) Modern arc welding technology. Ador Welding Limited, New Delhi
American Welding Society (1987) Welding handbook, 8th edn., vols. 1 & 2. American Welding
    Society, USA
Parmar RS, Welding process and technology. Khanna Publisher, New Delhi

Part IV
Arc Welding Processes

Chapter 11
Arc Welding Processes: Shielded Metal Arc Welding—Principle, Electrode and Parameters

11.1 Arc Welding Process

All arc welding processes apply heat generated by an electric arc for melting the faying surfaces of the base metal to develop a weld joint (Fig. 11.1). Common arc welding processes are manual metal arc or shielded metal arc welding (MMA or SMA), gas metal arc (GMA), gas tungsten arc (GTA), submerged arc (SA), plasma arc (PA), carbon arc (CA) welding, etc. However, each arc welding process uses different approaches of protecting the molten weld metal from the atmospheric gases. These may also differ in terms of electrode, type of power source, preferred polarity, power density and peak temperature produced during welding. Therefore, performance in terms of mechanical and corrosion properties, microstructure, quality of the weld metal, heat affected zone structure and properties of the weld joint produced by each process differs from those produced by other arc welding processes.

11.2 Shielded Metal Arc Welding (SMAW)

In this process, the heat is generated by an electric arc established between base metal and a consumable electrode. Electrode movement is controlled manually; hence, this process is also termed as manual metal arc welding. This process is extensively used for depositing weld metal for welding and weld surfacing. This process is very popular in fabrication industries due to its simplicity, ease to deposit the molten weld metal at the right place at the desired location, the consumable electrode provides filler metal, no separate shielding requirement, simple welding power source, semi-skilled welder requirement and above all favorable economics. This process is commonly used for developing somewhat less critical and non-critical weld joints of the metals which are comparatively less sensitive to the atmospheric gases, conversely those metals having limited affinity with gases like oxygen, nitrogen, etc.

D. K. Dwivedi, *Fundamentals of Metal Joining*, https://doi.org/10.1007/978-981-16-4819-9_11

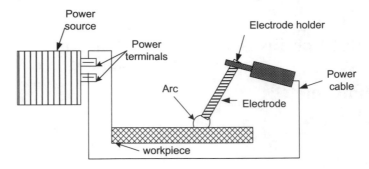

Fig. 11.1 Schematic showing various elements of SMA welding system

Shielded metal arc welding process can work on both AC and DC. The constant current DC power source is invariably used with all types of electrodes (basic, rutile and cellulosic) irrespective of base metal (ferrous and non-ferrous). However, AC can be unsuitable for certain types of electrodes and base metals. Therefore, AC should be used in the light of manufacturer's recommendations for the electrode application. In case of DC welding, heat liberated at anode is greater than that of the arc column and cathode side. The amount of heat generated at the anode and cathode may also differ appreciably depending upon the flux composition of coating, polarity and the nature of arc plasma. In case of DC welding, polarity determines the distribution of the heat generated at the cathode and anode and accordingly the melting rate of electrode and penetration into the base metal by the welding arc are affected with change in polarity.

Heat generated by a welding arc (J) = Arc voltage (V)

$$\times \text{ Arc current } (A) \times \text{ Welding time } (s)$$

$$(11.1)$$

If welding arc is moving at speed S (mm/min), then net heat input can be calculated as:

$$H_{net} = VI(60)/(S \times 1000) \text{ kJ/mm} \tag{11.2}$$

11.3   Shielding in SMA Welding

To avoid contamination of the molten weld metal from the atmospheric gases present in and around the welding arc, protective environment must be provided. In different arc welding processes, the protection of the weld pool is provided using different approaches (Table 11.1). In case of shielded metal arc welding, the protection

Table 11.1 Role of common constituents added in flux of SMAW electrode

Constituent in flux	Role on welding arc features
Quartz (SiO_2)	Increases current carrying capacity
Rutile (TiO_2)	Increases slag viscosity and good restriking
Magnetite (Fe_3O_4)	Refines transfer of droplets through the arc
Calcareous spar ($CaCO_3$)	Reduces arc voltage and produces inactive shielding gas and slag formation
Fluorspar (CaF_2)	Increases slag viscosity of basic electrodes and decreases ionization
Calcareous fluorspar ($K_2O\ Al_2O_3\ 6SiO_2$)	Improves arc stability by easy ionization
Ferro-manganese and ferro-silicon	Acts as deoxidant
Cellulose	Produces inactive shielding gas
Potassium sodium silicate (K_2SiO_3/Na_2SiO_3)	Acts as a bonding agent

to the weld pool is provided by covering of (a) slag formed over the surface of weld pool/metal and (b) inactive gases generated through thermal decomposition of flux/coating material on the electrode (Fig. 11.2). However, relative effect of above two aspects/ways of the protection to the weld metal depends on type of flux coating. Few fluxes (like cellulosic) provide large amount of inactive gases for shielding of weld pool, while other fluxes form slag in ample amount to cover/protect the weld pool from the atmospheric gases. Shielding of the weld pool by inactive gases in SMAW is not found to be very effective due to two reasons: (a) gases generated by thermal decomposition of electrode coating do not necessarily form a proper firm cover around the arc and welding pool, and (b) continuous movement of arc and fluctuating arc gap during welding further decreases the effectiveness of generated inactive shielding gas to protect the weld pool. Therefore, SMAW weld joints are often contaminated with atmospheric gases (in the form of porosity) and inclusions of oxides and nitrides, and hence, these welds are not clean enough for developing critical joints. In view of above, this welding process is not usually recommended for

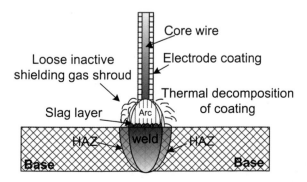

Fig. 11.2 Schematic showing different elements related to SMAW electrode and weld zone

developing weld joints of reactive metals like Al, Mg, Ti, Cr and stainless steel. These reactive metal systems are therefore commonly welded using welding processes like GTAW, GMAW, etc., that provide more effective shielding of inert or inactive gases to the weld pool from atmospheric contamination.

11.4 Coating on SMAW Electrode

The welding electrodes used in shielded metal arc welding process are called by different names like stick electrode, covered electrode and coated electrode. Coating or cover on the electrode core wire is composed of various hydrocarbons, compound and elements to perform specific roles such as low ionization potential element, produce inactive gases, binders, etc.(Fig. 11.3). Na and K silicates are invariably used as binders in all kinds of electrode coatings. Summarized role of different constituents of SMAW electrode coating is given in Table 11.1. Coating on the electrode for SMAW is provided to perform some of the following objectives:

- To increase the arc stability with the help of low ionization potential elements like Na and K.
- To provide protective shielding gas cover to the arc zone and weld pool with the help of inactive gases (like carbon dioxide) generated by thermal decomposition of constituents present in coatings such as hydrocarbon, cellulose, charcoal, cotton, starch and wood flour.
- To remove impurities from the weld pool by forming slag using constituents in coatings such as titania, fluorspar and china clay which react with impurities and oxides, if any, in present weld pool to form slag. Slag being lighter than molten weld metal floats over the surface of weld pool which is removed after solidification of weld through chipping.
- Controlled alloying of the weld metal (to achieve specific properties) can be done by incorporating required alloying elements in electrode coatings, and during welding, these elements get transferred from coating to the weld pool. However,

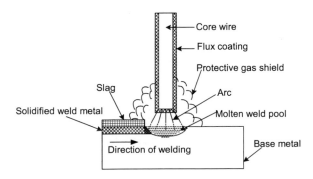

Fig. 11.3 Schematic showing constituents of SMAW

element transfer efficiency showing ability to transfer the alloying element from coating to weld pool is influenced by the welding parameters (current and type of coating) and welding process itself especially in respect of shielding of molten weld pool and alloying element evaporation tendency in the arc environment.

- To deoxidization and cleaning of the weld metal: Elements oxidized in the weld pool may act as inclusions and deteriorate the mechanical and corrosion performance of the weld joint. Therefore, metal oxides and other impurities present in weld metal are removed by deoxidation and slag formation. For this purpose, deoxidizers like ferro-Mn, ferro-Si and silicates of Mg and Al are frequently incorporated in the coating material.

- To increase viscosity/surface tension of the slag/molten metal so as to reduce their flowability in order to decrease tendency of falling down of weld metal during welding in horizontal, overhead and vertical welding position. This is done by adding constituents like TiO_2 and CaF_2 in the coating material. These constituents increase the viscosity of the slag.

11.5 Types of SMAW Electrodes

The steel electrode of a given composition is made available with different types of flux coating in order to make them suitable for different arc characteristics, welding position, welding speed, deposition rate, weld metal recovery, weld metal properties and variety of other weld quality requirements. The selection of correct type of electrode coating results in weld metal with desired quality characteristics at low cost. In general, welding electrode is selected in such a way that characteristics of weld metal are similar or better than the base metal while keeping in mind the welding position and weld joint design as these aspects significantly affect the properties of the weld metal.

11.5.1 Rutile Electrode

These electrodes predominantly contain rutile (TiO_2) besides other constituents and are known to offer almost 100% weld metal recovery, easy arc striking and restriking. Rutile electrode is found suitable for (a) fillet welds, (b) welding of sheet metal, (c) good gap bridging capability, (d) free from spatter losses and (e) all position welding. These electrodes are recommended for welding of low strength steel (< 440 MPa). For welding of high strength steel (> 440 MPa), generally weld metal should have low hydrogen level and therefore such weld joints are developed using basic, basic rutile and zircon-based electrode.

11.5.2   Cellulosic Electrodes

Cellulosic electrodes are composed of large amount of hydrocarbon compounds and calcium carbonates besides other constituents and are found suitable for (a) all welding positions especially for vertical and overhead welding position and (b) realizing good mechanical properties in a weld metal of radiographic quality. These electrodes are preferred for vertical downward welding. However, these produce high hydrogen content in weld metal besides deep penetration.

11.5.3   Acidic Electrode

Acidic electrodes offer (a) easier arc striking than basic electrodes but poorer arc striking than rutile electrodes, (b) moderate welding speed, (c) smooth weld bead, (d) good slag detachability. However, acidic electrodes are usually replaced by rutile electrode and basic electrode for flat and odd position welding, respectively. The ductility and toughness of the weld metal developed by acidic electrode are better than those developed with rutile electrodes; however, yield and ultimate tensile strength are found inferior. This type of electrode results in minimal penetration which is good for welding of thin sheet, but these are sensitive to moisture pickup. Moisture pickup can enhance problems related to hydrogen-induced cracking in transformation hardenable steel weld joints.

11.5.4   Basic Electrode

These electrodes have basic (alkali) coatings containing calcium carbonate/calcium fluoride. The basic electrodes are preferred over other types of electrode for developing weld joints of high-strength steel (480–550 MPa) in high joint-restraint condition for producing a weld metal having (a) low hydrogen, (b) good low-temperature toughness, (c) resistance to hot and cold cracking. However, these electrodes suffer from comparatively poor slag detachability. The welding speed and deposition rate offered by the basic electrodes especially in vertical welding position are much higher than those offered by rutile and acidic electrode. Basic electrodes can sustain higher welding current even in vertical welding position.

11.5.5   Basic Rutile Electrode

This type of electrode combines positives of both basic and rutile electrodes and therefore recommended for horizontal–vertical fillet welds of high-strength steels.

11.6   Welding Parameters for SMAW

Shielded metal arc welding process normally uses constant current type of power source with welding current in a range from 50 to 600 A and voltage from 20 to 80 V at 60% duty cycle. Welding transformer (AC welding) and generator or rectifiers (DC welding) are commonly used as welding power sources. In the case of AC welding open-circuit voltage (OCV) is usually kept 10–20% higher than DC welding so as to overcome the poor arc stability problem related to AC welding. Because in the case AC both magnitude and direction of current change in every half-cycle while those remain constant in DC welding. OCV requirement is primarily determined by factors like type of welding current and electrode composition affecting the arc stability. The presence of low ionization potential elements (Ca, K) in coating reduces the OCV required for the stable arc.

Importance of Welding Current

Selection of welding current required for developing a sound weld joints is primarily determined by the section thickness of base metal to be welded. In general, increase in section thickness to be welded increases the requirement of heat input to ensure proper melting, penetration, welding cross-sectional area and deposition rate (Fig. 11.4). This increased requirement of heat input can be fulfilled using (a) higher welding current and (b) lower welding speed. However, for given set of welding conditions, there is an optimum welding speed which results in maximum penetration depth, while arc speed other than optimum welding speed (either high or low) reduces the penetration depth (Fig. 11.5). Therefore, heat input requirement is primarily fulfilled by adjusting the welding current.

Need of high welding current for high heat input dictates use of large diameter electrode. SMAW electrode is commercially available in different sizes (core wire diameters) and is generally found in core wire diameter ranging from 1 to 12.5 mm in steps like 1.25, 1.6, 2, 2.5, 3.15, 4, 5, 6.3, 8, 10 and 12.5 mm.

Upper and lower limits of welding current for SMAW are determined by tendency of thermal decomposition of electrode coating material and arc stability, respectively. Welding current (A) is generally selected in a range of 40–60 times of electrode core wire diameter (mm). Too high welding current creates problem of thermal decomposition of the electrode coating material due to excessive electrical resistance heating of the core wire besides turbulence in the arc. Turbulence in the arc zone can lead to spatter and entrainment of atmospheric gases in arc. On other hand, low current results in the unstable welding arc, poor weld penetration and reduce the fluidity of molten weld metal. All these tend to develop discontinuities in weld joints.

In shielded metal arc welding process, lower limit of current is decided on the basis of requirement for stable arc, smooth metal transfer and penetration, whereas higher limit of current is decided on the basis of extent of overheating of core wire that an electrode coating can bear without any thermal damage (Fig. 11.6). High current coupled with long electrode extension causes overheating of core wire of

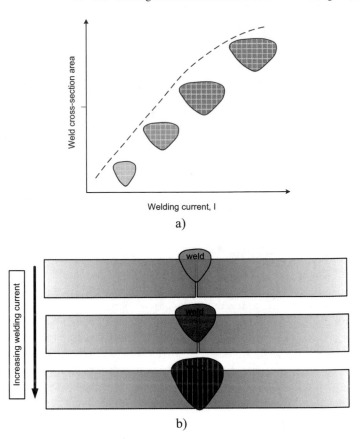

Fig. 11.4 Schematic showing effect of welding current on weld bead geometry, e.g. penetration, width and weld area: **a** plot and **b** weld joint

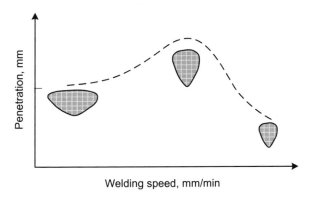

Fig. 11.5 Schematic showing effect of welding speed on weld bead geometry, e.g. penetration, width and weld area

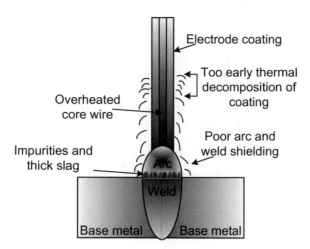

Fig. 11.6 Schematic showing premature thermal decomposition of electrode coating due to overheating of core wire

electrode due to electrical resistive heating. Excessive heating may cause the combustion/decomposition of electrode coating much earlier than when it is required to provide inactive shielding gases for protecting the weld pool and arc. Therefore, large diameter electrodes are selected for welding of thick sections as they can work with high welding current. Large diameter electrodes allow high current setting without any adverse effect on electrode coating materials because increased cross-scctional area of electrode reduces resistance to the flow of current which in turn reduces the electrical resistance heating tendency of the core wire of the electrodes.

Questions for Self-assessment

- Define arc welding processes and enlist common arc welding process.
- Explain the principle of shielded metal arc welding process with suitable schematic diagram.
- How is weld metal protected in SMAW process?
- What is the role of welding parameters on development of sound weld joint by SMAW?
- What is the role of coating on SMAW electrode?
- Write common constituents used in coating of electrode along with their role?
- How do we select the welding electrode of correct diameter?
- Explain the factors affecting the choice of coated electrodes.
- Elaborate factors determine the upper and lower level for welding current for SMAW.
- Elaborate factors need to be considered for selection of proper OCV for SMAW.

Further Reading

Cary H (1988) Welding technology, 2nd edn. Prentice Hall

Little R (2001) Welding and welding technology, 1st edn. McGraw Hill

Metals Handbook (1993) Welding, brazing and soldering, 10th edn, vol 6. American Society for Metals, USA

Nadkarni SV (2010) Modern arc welding technology. Ador Welding Limited, New Delhi

Parmar RS Welding process and technology. Khanna Publisher, New Delhi

Parmar RS (2002) Welding engineering and technology, 2nd edn. Khanna Publisher, New Delhi

Technical document (2005) MMAW, Aachen. ISF, Germany

Welding Handbook (1987) 8th edn, vols 1 and 2. American Welding Society, USA

Chapter 12
Arc Welding Processes: Shielded Metal Arc Welding: Welding Current and Metal Transfer

12.1 Selection of Type of Welding Current

It is important to consider various aspects while selecting suitable type of welding current for developing weld joints in a given situation. Some of the points need careful considerations for selection of welding current are given below.

1. Thickness of plate/sheet to be welded: DC is preferred for welding of thin sheet and thickness plates in order to have better control over heat input as per need to apply less or more heat, respectively.
2. Length of cable required: AC for situations where long cables are required during welding (e.g. ship weld fabrication) as AC results in lesser voltage drop, i.e. loading on power source, under such welding conditions.
3. Ease of arc initiation and maintenance needed even with low current: DC preferred over AC.
4. Arc blow: AC helps to overcome the arc blow as it is primarily observed with DC only.
5. Odd position welding: DC is preferred over AC for odd position welding (horizontal, vertical and overhead) due to better control over heat input and so fluidity of molten weld metal.
6. Polarity selection for achieving the desired melting rate, penetration and weld deposition rate: DC preferred over AC as it helps in generating more/less heat at electrode/workpiece side through suitable selection of polarity. DCEP is for higher electrode melting and deposition rate.
7. AC offers the penetration and electrode melting rate somewhat in between that is obtained by DCEN and DCEP.

DC offers the advantage of polarity selection (DCEN and DCEP) which helps in controlling the melting rate, penetration and required deposition rate of the weld metal (Fig. 12.1). DCEN results in more heat at workpiece, thus producing high welding speed with deeper penetration. DCEN polarity is generally used for welding of all types of steel. DCEP is commonly used for welding of non-ferrous metals like

D. K. Dwivedi, *Fundamentals of Metal Joining*,
https://doi.org/10.1007/978-981-16-4819-9_12

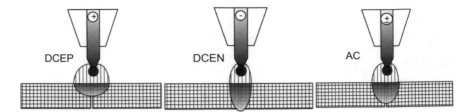

Fig. 12.1 Schematic showing effect of welding current and polarity

Al, Mg besides other metal systems. AC gives the penetration and electrode melting rate somewhat in between of that is offered by DCEN and DCEP.

12.2   Electrode Size and Coating Factor

Diameter of the core wire of an electrode refers to electrode diameter (d). Diameter of electrode with coating (D) as compared to that of core wire (d) is used to characterize the coating thickness factor (Fig. 12.2). The ratio of electrode diameter with coating and core diameter of the electrode is called coating factor (D/d). Coating factor usually ranges from 1.2 to 2.2. According to the coating factor, coated electrodes are grouped into three categories, namely light coated (with coating factor from 1.2 to 1.35), medium coated (with coating factor from 1.4 to 1.7) and heavy coated (with coating factor from 1.8 to 2.2). Stick electrodes are generally manufactured of

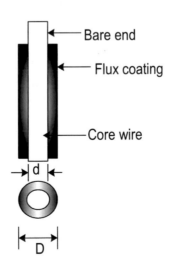

Fig. 12.2 Schematic of electrode showing electrode section size with (D) and without coating (d) and its different components

length ranging from 250 to 400 mm. During the welding, length of the electrode is determined by (a) welder's convenience to strike the arc, (b) electrode diameter and (c) current carrying capacity of electrode without causing excessive electric resistive heating of coating material due to flow of current through the core wire. Bare end (uncoated portion) of electrode is used to make electrical connection with power source using suitable connectors/electrode holder.

12.3   Weld Beads

Two types of weld beads are generally produced during welding, namely stringer bead and weaver bead. Deposition of the weld metal largely in straight line is called stringer bead (Fig. 12.3a). In case of weaver bead, weld metal is deposited in different paths during the welding, i.e. zigzag, irregular, curved, etc. (Fig. 12.3b). Therefore, stringer weld bead offers higher speed of welding arc (in the welding direction) than weaver bead. Hence, weaver bead helps to apply more heat input per unit length during welding than stringer bead. Therefore, weaver beads are commonly used to avoid problems associated with welding of thin plates and molten metal handling issues in odd position (vertical and overhead) welding in order to avoid melt through and weld metal falling tendency, respectively.

a)   b)

Fig. 12.3 Schematic diagram showing weld bead **a** stringer bead and **b** weaver bead

12.4 Metal Transfer in SMAW

Metal transfer refers to the transfer of molten metal droplets from the electrode tip to the weld pool in consumable arc welding processes. Metal transfer in SMA welding is primarily affected by surface tension of molten metal at the electrode tip. Presence of impurities and foreign elements in molten metal generally lowers the surface tension which in turn facilitates easy detachment of molten metal drops from the electrode tip. For details of different types of metal transfer modes, please refer to the Sect. 17.5. Since the purity and quality of the molten metal in SMAW are influenced of electrode coatings and weld pool protection, therefore, type and amount of coating on electrode and their effectiveness for shielding of arc zone from the atmospheric gases appreciably affect the mode of metal transfer.

Acidic, cellulosic and oxide types of electrodes produce molten metal with large amount of oxygen and hydrogen. Presence of these gases in the molten weld metal lowers the surface tension and produces spray like metal transfer by facilitating detachment of drop of molten metal from the electrode tip even when those are small in size primarily due to reduced drop holding surface tension force (Fig. 12.4a). Rutile electrodes are primarily composed of TiO_2 due to which molten metal droplet hanging at tip of electrode is not much oxidized, and therefore, surface tension of the molten weld metal is not reduced appreciably. Hence, rutile electrodes produce more globular and less spray like transfer (Fig. 12.4b). Basic electrodes contain de-oxidizers, and at the same time moisture is completely driven off to render low hydrogen electrodes. Therefore, melt droplets at the tip of the basic electrode are of killed steel type (very–very low oxygen in weld metal) which in turn results in high surface tension. Since high surface tension of molten metal resists the detachment of drops from the electrode tip, the size of molten metal drop at tip of electrode grows until it is detached under the effect of gravitational and electromagnetic pinch forces. These conditions in turn produce a globular metal transfer with basic electrode (Fig. 12.4c). Moreover, the metal transfer from basic electrode can occur by short circuiting mode if molten metal drop touches the weld pool and melt is transferred to weld pool by surface tension effect.

In case of light-coated electrodes incomplete de-oxidation (due to lack of enough flux), CO gas bubble is formed. This bubble remains with molten weld metal droplet until it grows to about half of electrode diameter. Eventually, molten metal drop with bubble of CO bursts due to internal pressure of increasing CO gas volume. Bursting of gas bubble in molten metal drop in turn results in metal transfer in form of fine drops coupled with spatter (Fig. 12.5).

Questions for Self-assessment

- What is coating factor? How does it affect the welding?
- Explain the factors affecting the selection of type of welding current for SMAW.
- How does DC in SMAW help in thick plate/thin sheets?
- Which type of power sources is commonly used with SMAW process and why?
- What are common modes of metal transfer in SMAW?

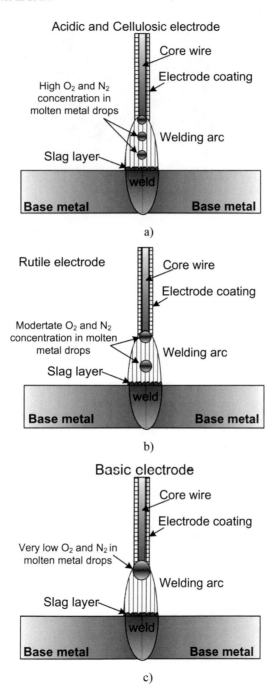

Fig. 12.4 Schematic showing effect of type of electrode on mode of metal transfer in SMAW using **a** cellulosic/acidic, **b** rutile and **c** basic electrodes

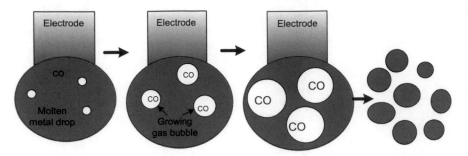

Fig. 12.5 Schematic showing possibilities related to use of light-coated electrode leading to bursting of molten metal drops during welding

- How does quality of the weld metal affect metal transfer in SMAW?
- What are factors determining the selection of welding current and OCV in SMAW?
- Schematically show different types of weld beads along with their application in specific welding condition.
- How does type of electrode affect the metal transfer?
- Why does basic electrode generally produce globular metal transfer?

Further Reading

Cary H (1988) Welding technology, 2nd edn. Prentice Hall

Little R (2001) Welding and welding technology, 1st edn. McGraw Hill

Metals Handbook (1993) Welding, brazing and soldering, 10th edn, vol 6. American Society for Metals, USA

Nadkarni SV (2010) Modern arc welding technology. Ador Welding Limited, New Delhi

Parmar RS Welding process and technology. Khanna Publisher, New Delhi

Welding Handbook (1987) 8th edn, vols 1 and 2. American Welding Society, USA

Chapter 13
Arc Welding Processes: Submerged Arc Welding: Principle, Parameters and Applications

13.1 Introduction

Submerged arc welding (SAW) process uses heat generated by an electric arc established between a bare consumable electrode wire and the workpiece. Since in this process, welding arc and the weld pool both are completely submerged under mound of granular fusible and molten flux; therefore, it is called submerged arc welding. During welding, granular flux is melted from the heat generated by arc and forms a shield/cover of molten flux layer (Fig. 13.1).

The arc length/arc voltage determines surface area of arc exposed to the surrounding granular flux cover experiencing the heat of the arc. Greater the surface area of the welding arc, higher will be melting/consumption rate of the flux during welding. The molten flux layer enclosing arc and pool reduces (a) access of atmospheric gases to the arc zone and the weld pool, (b) spatter tendency and (c) heat transfer from the arc and pool to atmosphere. Further, the molten flux reacts with the impurities, if any, present in the molten weld metal to form slag. Slag (being lighter than the weld metal) floats over the surface of the weld pool. Layer of the slag over the molten weld metal results in the following:

- Increased protection of weld metal from atmospheric gases thus improves the soundness and mechanical properties of weld joint
- Reduced cooling rate of weld metal and HAZ owing to shielding of the weld pool by granular and molten flux and solidified slag in turn leads to (a) a smoother weld bead and (b) reduced the cracking tendency of transformation hardenable steels.

13.2 Components of SAW System

SAW is known to be a high current (sometimes even greater 1000 A) welding process and is mostly used for joining of heavy sections and thick plates as it offers deep

D. K. Dwivedi, *Fundamentals of Metal Joining*,
https://doi.org/10.1007/978-981-16-4819-9_13

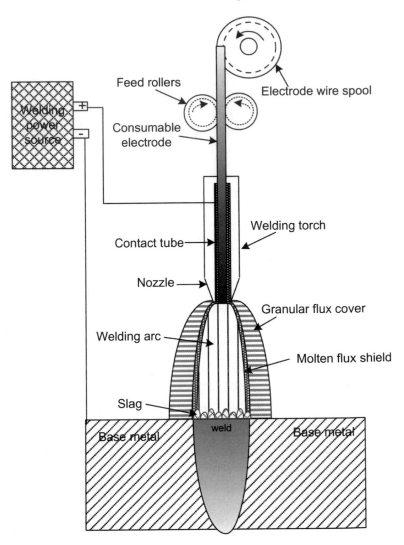

Fig. 13.1 Schematic showing main elements/components of SAW process

penetration with high deposition rate and high welding speed. High welding current can be applied in this process due to three reasons (a) absence of spatter, (b) reduced possibility of air entrainment in arc zone as molten flux and slag form shield to the weld metal and (c) use of large diameter copper-coated electrodes. Continuous feeding of granular flux around the weld arc from the flux hopper provides shielding to the weld pool from the atmospheric gases and also helps in adjusting/controlling the weld metal composition through addition of alloying element incorporated in the flux. Complete cover of the molten flux around electrode tip and the welding pool during the actual welding operation produces weld joint without spatter and smoke.

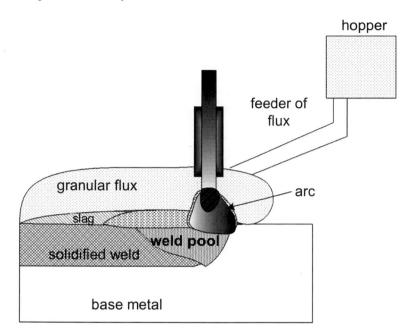

Fig. 13.2 Schematic of cross section of weld during submerged arc welding

In the following section, important components of SAW system and their role have been presented (Fig. 13.2).

13.2.1 Power Source

Generally, submerged arc welding process uses the power source at 100% duty cycle for welding long plates and sections; which means that the welding is done continuously for minimum 5 min or more without a break. Depending upon the electrode diameter, type of flux and electrical resistivity, the submerged arc welding can use either AC or DC. Alternating current and DCEN polarity are generally used with large diameter electrode (> 4 mm). DC with constant voltage power source provides good control over bead shape, penetration and welding speed. However, DC can cause problem of arc blow under few welding conditions. Polarity should be selected carefully as it affects weld bead geometry, penetration and deposition rate. DCEP offers advantage of self-regulating arc in case of small diameter electrodes (< 2.4 mm) and high deposition rate.

13.2.2   Welding Electrode

The diameter of electrodes used in submerged arc welding generally ranges from 1 to 5 mm. The electrode wire is fed from the spool through a contact tube connected to the power source, and workpiece is also made a part of welding circuit. Electrode wire of steel is generally copper coated for two reasons (a) to provide protection to the steel from atmospheric corrosion and (b) to increase the current carrying capacity. However, stainless steel wires being corrosion resistant on its own are therefore not coated with copper.

13.2.3   SAW Flux

Primary role of fluxes in SAW is largely similar to that of coating in stick electrodes of SMAW, i.e. protection of the weld pool from inactive shielding gases generated by thermal decomposition of coating material. SAW fluxes can influence the weld metal composition appreciably in the form of addition or depletion of alloying elements through gas metal and slag metal reactions. Such compositional variations must be taken care of to avoid any adverse effect on mechanical, corrosion performance of the weld joints. Few hygroscopic fluxes are baked (at 250–300 °C for 1–2 hours) to remove the moisture. There are four types of common SAW fluxes, namely fused flux, agglomerated flux, bonded flux and mechanical fluxes. Manufacturing steps of these fluxes are given below.

- Fused fluxes: raw constituents-mixed-melted-quenched-crushed-screened-graded
- Bonded fluxes: raw constituents-powdered-dry mixed-bonded using K/Na silicates-wet mixed-pelletized-crushed-screened
- Agglomerated fluxes: made in similar way to bonded fluxes but ceramic binder replaces silicate binder
- Mechanically mixed fluxes: mix any two or three types of above fluxes in desired ratio to achieve the desired properties in flux.

Specific Characteristics of Each Type of Flux

Fused Fluxes

- Fused fluxes are considered to have many good characteristics

  - Uniformity of the chemical composition
  - No effect of segregation or removal of fine particles on flux composition
  - Non-hygroscopic: easy handling and storage
  - Easy recycling without much change in particle size and composition.

- Limitation of fused fluxes is related with difficulty in

  - Incorporating de-oxidizers and ferro-alloys

– Melting issues due to need of high temperature.

Bonded Fluxes

- Bonded fluxes have many desired characteristics such as

  – Easy to add de-oxidizers and alloying elements
  – Allow thicker layer of flux during welding.

- However, these suffer from the following limitations

  – Hygroscopic
  – Gas evolution tendency
  – Possibility of change in flux composition due to segregation/removal of fine particles from the flux.

Agglomerated Fluxes

These fluxes are similar to that of bonded fluxes except that these use ceramic binders.

Mechanical Fluxes

- Positives of the mechanical fluxes include

  – Several commercial fluxes can be easily mixed and made to suit applications to get desired results

- Major limitations of mechanical fluxes

  – Segregation tendency of various constituents during storage/handling in feeder and recovery system
  – Inconsistency in flux from mix to mix.

13.3 Composition of the SAW Fluxes

The fused and agglomerated type of fluxes usually consist of different types of halides and oxides such as MnO, SiO_2, CaO, MgO, Al_2O_3, TiO_2, FeO and CaF_2 and sodium/potassium silicate. Halide fluxes are used for high-quality weld joints of high strength steel for critical applications, while oxide fluxes are used for developing weld joints of non-critical applications. Some of oxides such as CaO, MgO, BaO, CaF_2, Na_2O, K_2O and MnO are basic in nature (donors of oxygen), and few other oxides such as SiO_2, TiO_2 and Al_2O_3 are acidic (acceptors of oxygen). The relative amount of the acidic fluxes and basic fluxes in combination with neutral fluxes determines the basicity index of a given flux mixture. The basicity index of a flux indicates ratio of sum of (wt. %) all basic oxides to all non-basic oxides. All non-basic oxides include all acidic and neutral oxides.

Basicity of flux is important as it affects the slag detachability, bead geometry, mechanical properties and current carrying capacity. Low basicity fluxes offer high current carrying capacity, good slag detachability, good bead appearance and poor mechanical properties and poor crack resistance of the weld metal, while high basicity fluxes produce opposite effects on above characteristics of the weld.

13.4   Fluxes for SAW and Recycling of Slag

The protection to the weld pool in submerged arc welding process is provided by both molten flux layer and unfused layer of granular flux covering to the weld pool. Neutral fluxes are mostly free from de-oxidizers (like Si, Mn); therefore, loss of alloying elements from weld metal with neutral fluxes becomes negligible. Hence, chemical composition of the weld metal is not appreciably affected by the application of neutral fluxes. Active fluxes contain small amount of de-oxidizer such as manganese, silicon singly or in combination (Table 13.1). The de-oxidizers enhance resistance to porosity and weld cracking tendency.

The fluxes during submerged arc welding produce a lot of slag which is generally disposed of as a waste. The disposal of slag, however, imposes many issues related to storage and environmental pollution. A better disposal route for a slag is to use it for recycling of SAW flux.

The recycling of the used flux can reduce flux production cost appreciably without compromise on the quality of the weld. However, recycling needs extensive experimentation to optimize the composition of recycled fluxes so as to achieve the desired operational characteristics and the performance of the weld joints.

The recycling of fluxes basically involves the mixing of slag with fresh flux. The slag developed from SAW process is crushed and mixed with new flux. This process is different from recycling of unfused flux which is collected from the clean surface of the workpiece and reused. Slag produced during submerged arc welding while using a specific kind/brand of the flux is crushed and then used as flux or used after mixing with original unused flux to ensure better control over the properties of the weld metal. Blending of the slag with unused flux modifies the characteristics of original unused flux; therefore, the blending ratio must be optimized for achieving the desired quality of weld joints.

Table 13.1   Typical metal oxides

Type of oxide	For agglomerated SAW fluxes						
Acidic	SiO_2	TiO_2	P_2O_5	V_2O_5			
Basic	K_2O	Na_2O	CaO	MgO	BaO	MnO	FeO
Neutral	Al_2O_3	Fe_2O_3	Cr_2O_3	V_2O_3	ZnO		

13.5 Welding Parameters

Welding parameters, namely electrode size, welding voltage, welding current and welding speed are four most important parameters (apart from flux) that play a major role in determining the soundness and performance of the SAW weld joint; therefore, these must be selected carefully before welding.

13.5.1 Welding Current

Welding current is the most influential process parameter for SAW because it determines the melting rate of electrode, penetration depth, weld bead geometry and heat input and the cooling rate of the weld joint after welding. However, too high welding current for a given section thickness may lead to burn through owing to deep penetration apart from excessive reinforcement, increased residual stresses and high heat input-related problems like weld distortion and wide heat affected zone. On the other hand, selection of very low current is known to cause lack of penetration, lack of fusion and unstable arc. Selection of welding current is primarily determined by thickness of plates to be welded, and accordingly electrode of proper diameter is chosen so that it can withstand/sustain the welding current for developing sound weld with requisite deposition rate and penetration (Fig. 13.3a).

Diameter (mm)	Welding current (A)
1.6	150–300
2.0	200–400
2.5	250–600
3.15	300–700
4.0	400–800
6.0	700–1200

13.5.2 Welding Voltage

Welding voltage has marginal effect on the melting rate of the electrode. Welding voltage commonly used in SAW ranges from 20 to 35 V. Selection of too high welding voltage (greater arc length) leads to flatter and wider weld bead, higher flux consumption and increased gap bridging capability under poor fit-up conditions while low welding voltage results in narrow and peaked bead and poor slag detachability (Fig. 13.3b).

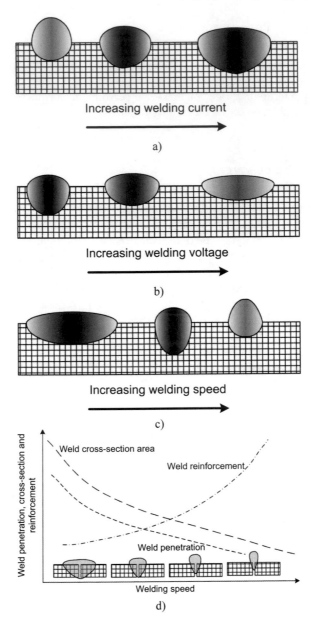

Fig. 13.3 Influence of welding parameters on weld bead geometry **a** welding current, **b** arc voltage, **c**, **d** welding speed

13.5.3 Welding Speed

Required bead geometry and penetration in a weld joint are obtained only with an optimum speed of welding arc during SAW. The selection of a welding speed higher than optimum one reduces heat input per unit length which in turn results in low deposition rate of weld metal, decreased weld reinforcement and shallow penetration (Fig. 13.3c, d). Further, application of too high welding speed increases tendency for (a) undercut in weld owing to reduced heat input, (b) arc blow due to higher relative movement of arc with respect to ambient gases and (c) porosity as air pocket is entrapped due to rapid solidification of the weld metal. On other hand, low welding speed increases heat input per unit length which in turn may lead to increased tendency of melt through and reduction in tendency for development of porosity and slag inclusion.

13.6 Bead Geometry and Effect of Welding Parameters

Bead geometry including depth of penetration is important characteristic of the weld bead which is influenced by size of the electrode for a given welding current setting. In general, an increase in size of the electrode for a given welding current decreases the depth of penetration and increases width of weld bead (Fig. 13.4). Large diameter electrodes are primarily selected to take two advantages: (a) high deposition rate owing to their high current carrying capacity and (b) good gap bridging capability under poor fit-up conditions of the plates to be welded as it helps in developing a wider weld bead.

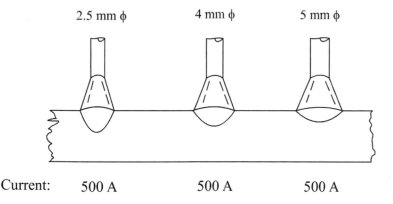

Fig. 13.4 Influence of electrode diameter on weld bead geometry

13.7   Advantage

Two unique features of SAW, namely welding arc submerged under flux and use of high welding current compared with other welding process, make it suitable for various applications with following advantages:

- High productivity due to high deposition rate of the weld metal and capability weld continuously without interruptions as electrode is fed from spool, and the process works under 100% duty cycle.
- High depth of penetration allows welding of thick sections.
- Smooth weld bead is produced without stresses raisers as it allows welding under molten flux cover, at low cooling rate, without sparks, smoke and spatter.

13.8   Limitations

There are three main limitations of SAW including (a) invisibility of welding arc during welding, (b) difficulty in maintaining mound of the flux cover around the arc during odd welding positions and welding of cylindrical components of small diameter and (c) increased tendency of melt through when welding thin sheets. Invisibility of welding arc submerged under unmelted and melted flux cover in SAW makes it difficult to ensure the location where weld metal is being deposited during the welding. Therefore, it becomes mandatory to use an automatic device (like welding tractors) for accurate and guided movement of the welding arc along the weld centre-line so that the weld metal is deposited correctly along weld line only. Applications of SAW process are mainly limited to flat position only as developing a mound of flux in odd position to cover the welding arc becomes difficult which is a requisite for SAW. Similarly, circumferential welds are difficult to develop on small diameter cylindrical components as flux tends to fall down away from weld zone. Plates of thickness less than 5 mm are generally not welded due to risk of burn/melt through.

Further, SAW process is known as high heat input welding process. High heat input, however, is not considered good for welding of many steels as it leads to significant grain growth in weld and HAZ due to low cooling rate experienced under flux/slag cover during welding. Low cooling rate increases the effective transformation temperature (from austinite to other soft phases during steel welding) which in turn lowers nucleation rate and increases the growth rate during solid state transformation. A combination of low nucleation rate and high the growth rate results in coarse grain structure in the weld as well as HAZ. Coarse grain structure deteriorates the mechanical properties of the weld joint specifically toughness and ductility. Therefore, SAW weld joints are sometimes normalized to refine the grain structure and enhance/restore the mechanical properties so as to reduce the adverse effect of high heat input of SAW process on mechanical properties of the weld joints.

13.9 Applications

Submerged arc welding is used for welding of different grades of steels in many sectors such as shipbuilding, offshore, structural and pressure vessel industries for fabrication of pipes, penstocks, LPG cylinders and bridge girders. Apart from the welding, SAW is also used for the surfacing of large surface area of worn out parts for different purposes such as reclamation, hard facing and cladding. The typical application of submerged arc welding for weld surfacing includes surfacing of roller barrels and wear plates. Submerged arc welding is widely used for cladding of carbon and alloy steels with stainless steel and nickel alloy deposits for improved corrosion resistant and high-temperature thermal stability.

Questions for Self-assessment

a. Explain the principle of SAW process and its application.
b. What is the role of flux in SAW?
c. What are common types of fluxes for SAW? Write limitations and advantages of each type of the SAW flux.
d. Write factors affecting the selection of welding power sources and polarity in SAW process with justification.
e. What is the role of welding parameters on development of sound weld joint by SAW?
f. How does increase in arc voltage/arc length increase the flux consumption?
g. Write advantages, limitations and applications of SAW process.
h. Why does SAW process suit for high volume production where extensive welding is needed on heavy sections?
i. How does increase in electrode diameter (for a given setting of voltage and current) affect the weld joint characteristic?
j. How can HAZ properties of SAW joints of steel can be compared with SMAW?
k. Why does it require to perform post-weld heat treatment of SAW joints of steel?

Further Reading

Cary H (1988) Welding technology, 2nd edn. Prentice Hall
Little R (2001) Welding and welding technology, 1st edn. McGraw Hill
Metals Handbook (1993) Welding, brazing and soldering, 10th edn, vol 6. American Society for Metals, USA
Nadkarni SV (2010) Modern arc welding technology. Ador Welding Limited, New Delhi
Parmar RS Welding process and technology. Khanna Publisher, New Delhi
Welding Handbook (1987) 8th edn, vols 1 and 2. American Welding Society, USA

Chapter 14
Arc Welding Processes: Gas Tungsten Arc Welding: Principle and System Components

14.1 Introduction

Tungsten inert gas welding process also called as gas tungsten arc welding is named so because it uses (a) an electrode primarily made of tungsten and (b) inert gas for shielding the weld pool. To prevent the contamination of weld pool from atmospheric gases, a very effective shielding is needed especially when welding highly reactive metals and alloys such as stainless steel, aluminium and magnesium alloys, to develop high-quality weld joints for critical applications like nuclear reactors, aircraft, etc. Invention of this process in middle of twentieth century gave a big boost to fabricators of these reactive metals as none of the processes (SMAW and gas welding) prevailing at that time were not able to weld the reactive metals successfully primarily due to two limitations (a) contamination of weld from atmospheric gases and (b) poor control over the heat input leading to inconsistent penetration of base metals (Fig. 14.1). Moreover, welding of aluminium and its alloys with shielded metal arc welding process can be realized using halide flux-coated electrodes. The halide electrodes reduce problems associated with Al_2O_3 formation; however, halides fluxes are considered to be very corrosive. Therefore, welding of aluminium is preferably carried out using inert shielding environment with the help of processes like GTAW and GMAW. Despite so many developments in the field of welding, GTAW process is invariably recommended for joining of thin aluminium sheets of thickness less than 1 mm.

14.2 TIG Welding System

There are four main components (Fig. 14.2) of TIG welding system, namely (a) DC/AC power source to deliver the welding current as per need, (b) welding torch (air/water cooled) with tungsten electrode and gas nozzle, (c) inert shielding gas (He, Ar or their mixture) for protecting the molten weld pool and electrode from

© The Author(s), under exclusive license to Springer Nature Singapore Pte Ltd. 2022   171
D. K. Dwivedi, *Fundamentals of Metal Joining*,
https://doi.org/10.1007/978-981-16-4819-9_14

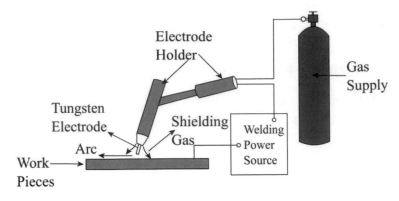

Fig. 14.1 Schematic of tungsten inert gas welding process

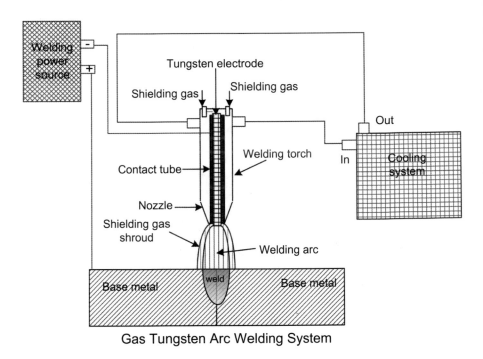

Fig. 14.2 Details of components of GTAW system

contamination by atmospheric gases and (d) control system for moving the welding torch as per mode of operation (manual, semi-automatic and automatic). This process uses the heat generated by an electric arc established between the non-consumable tungsten electrode and the workpiece (mostly reactive metals like stainless steel, Al, Mg, etc.) for melting of faying surfaces, and inert gas is used for shielding the arc zone and weld pool from the atmospheric gases.

14.2.1 Power Source

TIG welding is normally performed using constant current type of power source with welding current ranging from 3 to 200 A or 5–300 A or higher and arc voltage ranging from 10 to 35 V at 60% duty cycle. Pure tungsten electrode having a ball shape tip with DCEN polarity provides good arc stability. Moreover, thorium, zirconium and lanthanum-coated/modified tungsten electrodes can be used with AC and DCEP as coating of these elements on pure tungsten electrodes improves the electron emission capability which in turn increases the arc stability. TIG welding with DCEP is preferred for welding of reactive metals like aluminium to take advantage of cleaning action by formation of mobile cathode spots in workpiece side during welding. The mobile cathode spot loosens the tenacious refractory alumina oxide layer and thus helps in removing the same from the molten weld metal for developing the cleaner weld. DCEN polarity is used for welding of metal such as carbon steel and alloys that do not form such oxides and mobile cathode spot.

14.2.2 Welding Torch

TIG welding torch comprises three main components, namely non-consumable tungsten electrode, collets and nozzle. A collet is primarily used to hold a tungsten electrode of varying diameters in position. Nozzle helps to form a firm jet of inert gas around the arc, weld pool and the tungsten electrode. The diameter of the nozzle must be selected in the light of expected size of weld pool so that proper shielding of the weld pool can be obtained by forming a cover of inert gas. The nozzle needs to be replaced at regular intervals as the size and shape of the nozzle are thermally damaged by wear and tear due to exposure to the intense heat of the welding arc. Damaged nozzle does not form uniform and firm jet of inert gas around the weld pool for protection from the atmospheric gases. Typical flow rate of shielding inert gas may vary from 5 to 50 L/min.

TIG welding torch is generally rated on the basis of their current carrying capacity as it directly affects the welding speed and so the deposition rate. Depending upon the current carrying capacity, the welding torch can be either water or air cooled. Air-cooled welding torch is generally used for lower range of welding current (< 150 A) than water-cooled torches. Air-cooled torches can sustain the limited heat generation of the arc while effective cooling associated with water-cooled torches allows higher welding current than the air-cooled torches. Water cooling reduces operating temperature of nozzle, and so the rate of thermal damage to nozzle is also reduced which in turn allows higher welding current for the same life or longer life for the same welding current.

14.2.3   Filler Wire

Filler metal is generally not used for welding of thin sheet by GTAW. Welding of thick steel plates by GTAW process to produce high-quality welds for critical applications requires addition of filler metal to fill the groove. The filler wire can be fed either manually or with the help of some wire feed mechanism. For feeding small diameter filler wires (0.8–2.4 mm), usually push type wire feed mechanism with speed control device is used. Selection of filler metal is very critical for successful welding because in some cases even use of filler metal similar to that base metal causes cracking of weld metal especially when their solidification temperature range of base/weld metal is every wide (> 50 °C). Therefore, the selection of the filler metal should be done carefully after giving full consideration to the following aspects such as mechanical property requirement, metallurgical compatibility, solidification temperature range, cracking tendency of base metal under welding conditions and fabrication conditions.

For welding of aluminium alloys, Al-(5–12wt.%) Si filler is used as general-purpose filler metal. Al-5%Mg filler is also used for welding of few aluminium alloys. Welding of dissimilar steels, namely stainless steel with carbon or alloy steels for high-temperature applications, needs development of buttering layer before welding for reducing carbon migration and residual stress development-related problems. Similarly, welding of austenitic stainless steel and filler metal is selected in such a way that 3–5% ferrite in matrix of austenite is formed to reduce the solidification cracking tendency.

14.2.4   Shielding Gas

Helium, Argon and their mixtures are commonly used as shielding gas for protecting the weld pool depending upon the metal to be welded, criticality of application and economics. Helium, argon with 1–2 vol.% of hydrogen is sometimes used for specific purposes such as increasing the arc voltage and arc stability which in turn helps to increase the heat of arc. The selection of inert gases to be used as a shielding gas during GTAW and GMAW process in general depends upon the type of metal to be welded and criticality of the application of the weld joints. Carbon dioxide is not used with GTAW process, because at high temperature in arc environment, the thermal decomposition of the carbon dioxide produces CO and O_2. Generation of these gases adversely affects the quality and soundness of the weld joint and reduces the life of tungsten electrode.

14.3 Effect of Shielding Gases on GTAW Characteristics

Argon and helium are the mostly commonly used shielding gases for developing high-quality weld joints of reactive metals and ferrous metals. Small amount of hydrogen or helium is often added in argon to increase the penetration capability and welding speed. These two inert gases as shielding gas exhibit different features during welding. Some of these features are described in following section. Sometime, hydrogen and oxygen (1–3%) are also added with Ar to take benefits similar to that of helium for increasing depth of penetration.

14.3.1 Heat of Welding Arc

The ionization potential of He (25 eV) is higher than Ar (16 eV). Therefore, application of argon as shielding gas results in higher arc voltage than helium. Hence, choice of the shielding gas affects voltage–current (VI) characteristics of arc. In general, arc voltage generated by helium for a given arc length during welding is found higher than argon. Higher arc voltage for a given current setting results in hotter helium arc than argon arc (Fig. 14.3). Hence, helium is preferred for the welding of thick plates at high speed especially the metals having high thermal conductivity and high melting point.

14.3.2 Arc Efficiency

Helium offers higher thermal conductivity than argon. Hence, He effectively transfers the heat from arc to the base metal which in turn helps in increasing the welding speed and arc efficiency under identical welding conditions.

14.3.3 Arc Stability

Helium is considered to cause more problems related to arc stability and arc initiation than Ar as a shielding gas. This behaviour of helium is primarily attributed to its higher ionization potential than Ar. High ionization potential of helium means it will produce fewer charged particles between electrode and workpiece required for initiation and maintenance of welding arc under a given set of welding parameters (Fig. 14.4).

Therefore, arc characteristics are found to be different for Ar and He. A minima arc voltage is found in VI characteristic curve of an arc when both the gases are used as shielding gas. Moreover, minima arc voltage occurs at different level of

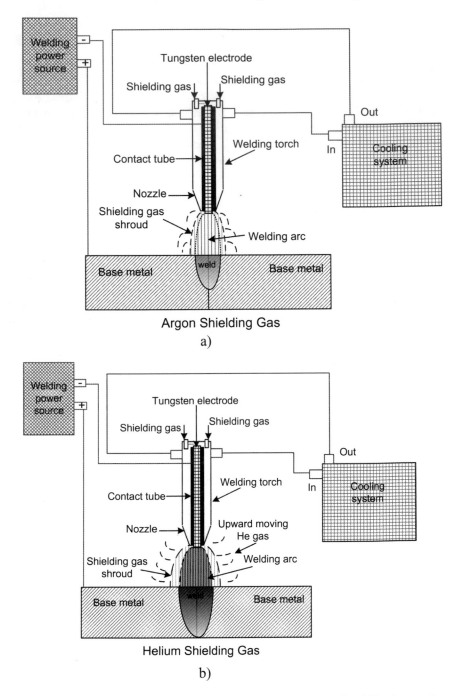

Fig. 14.3 Schematic showing effect of **a** Ar and **b** He on arc heat and weld bead geometry of tungsten inert gas welding process

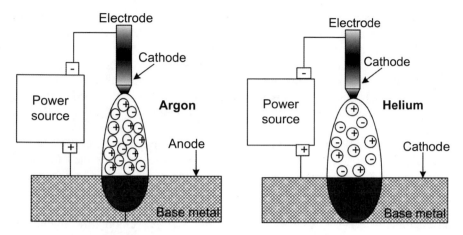

Fig. 14.4 Schematic showing effect of **a** Ar and **b** He on charge particle density in arc zone affecting arc stability and weld bead geometry of tungsten inert gas welding process

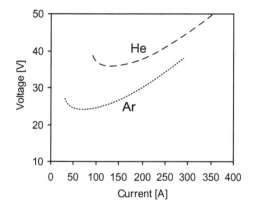

Fig. 14.5 Influence of shielding gas on VI characteristics of GTAW process

welding currents. With argon as shielding gas, the welding current corresponding to the lowest arc voltage is found around 50 A while that for helium occurs at around 150 A (Fig. 14.5). Reduction in welding current below this critical level (up to certain range) increases the arc voltage; which permits some flexibility in arc length to control the welding operation.

14.3.4 Flow Rate of Shielding Gas

Argon (density 1.783 g/L) is about 1.33 and 10 times heavier than the air and the helium, respectively. The difference in density of the air with the shielding gases determines the way shielding gas will behave after coming of nozzle which in turn affects flow rate of particular shielding gas required to form a firm cover/blanket over the weld pool and arc zone to provide protection against the atmospheric gases. Helium being lighter than air tends to rise up immediately in turbulent manner away from the weld pool after coming out of the nozzle; while argon tends to settle down around the welding arc/pool as it is heavier than air. Therefore, for effective shielding of the arc zone, flow rate of helium (12–22 L/min) must be 2–3 times higher than the argon (5–12 L/min).

Flow rate of shielding gas to be supplied for effective protection of weld pool is also determined by the size of molten weld pool, sizes of electrode and nozzle, distance between the electrode and workpiece, extent of turbulence being created by ambient air movement (problematic above 8–10 km/h). For a given combination of welding parameters and welding torch, the flow rate of the shielding gas should be such that it produces a jet of shielding gas so as to overcome the ambient air turbulence and provide perfect cover around the weld pool. Unnecessarily, high flow rate of the shielding gas leads to poor arc stability and weld pool contamination from atmospheric gases due to suction effect.

14.3.5   Mixture of Shielding Gases

Small addition of hydrogen in argon increases arc voltage which burns the arc hotter. Therefore, a combination of high temperature and more heat in turn results in increasing the weld penetration and welding speed. To take the advantage of good characteristics of both helium (high thermal conductivity, high arc temperature) and Ar (good arc initiation and arc stability), a mixture of two gases Ar-(25–75%) He is also used for welding. Increasing proportion of He in Ar-He mixture increases the welding speed and depth of penetration of weld. Addition of oxygen in argon also helps to increase the penetration capability of GTAW process owing to increased arc temperature and plasma velocity (Fig. 14.6).

14.3.6   Advantages of Ar Over He as Shielding Gas

For general-purpose quality weld, argon offers many advantages over helium (a) easy arc initiation, (b) cost effective and good availability, (c) good cleaning action (with AC/DCEP in case of aluminium and magnesium welding) and (d) shallow penetration required for thin sheet welding of aluminium and magnesium alloys.

Questions for Self-study

- Explain the principle of GTAW process and write its applications.
- Why does GTAW offer cleanest weld among other common arc welding processes?

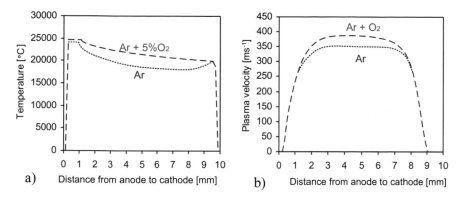

Fig. 14.6 Influence of oxygen addition in Ar on **a** arc temperature and **b** plasma velocity of GTAW process

- Write role of important components of GTAW process.
- Describe factors affecting the selection of power sources and polarity in GTAW process with justification.
- What are shielded gases used in GTAW process? Write factors affecting the effectiveness of protection of the weld pool in GTAW process.
- What is the role of welding parameters, namely welding current, arc voltage and welding speed on development of sound weld joint by GTAW?
- Write advantages and limitations of GTAW process.
- Explain the factors determining the current carrying capacity of electrode in GTAW process?
- Why do oxygen and hydrogen addition in argon are found to be useful in welding by GTAW?
- Explain the factors affecting the welding current carrying capacity of the tungsten electrode in GTAW.
- What are the factors needed to be considered for the selection of filler metal in GTAW?

Further Reading

Cary H (1988) Welding technology, 2nd edn. Prentice Hall
Little R (2001) Welding and welding technology, 1st edn. McGraw Hill
Metals Handbook (1993) Welding, brazing and soldering, 10th edn, vol 6. American Society for Metals, USA
Nadkarni SV (2010) Modern arc welding technology. Ador Welding Limited, New Delhi
Parmar RS Welding process and technology. Khanna Publisher, New Delhi
Parmar RS (2002) Welding engineering and technology, 2nd edn. Khanna Publisher, New Delhi
Welding Handbook (1987) 8th edn, vols 1 and 2. American Welding Society, USA

Chapter 15
Arc Welding Processes: Gas Tungsten Arc Welding: Electrode, Polarity and Pulse Variant

15.1 Electrode for TIG Torch

The electrode for tungsten inert gas welding process can be pure (uncoated) or coated with Zr, La or Th. However, pure tungsten electrode offers shorter life than coated electrodes because of rapid wear and tear owing to thermal damage caused by their comparatively lower current carrying capacity than coated electrodes. The damage to electrode primarily occurs due to formation of tungsten carbide during steel welding by reaction between W and C. The tungsten carbide has lower melting point than pure tungsten, therefore contaminated tip of the electrode degrades and wears out rapidly. Particles generated from pure tungsten electrode due to thermal damage further cause contamination of the weld metal in the form of tungsten particle inclusion. Therefore, pure tungsten electrodes are generally not preferred for critical welding applications.

Pure tungsten electrodes are frequently coated oxides of Th, Zr, La and Ce. These oxides perform two functions (a) improve the arc stability and (b) increasing the current carrying capacity of the electrode.

Increase in arc stability of tungsten electrode in the presence of the oxides of thorium, cerium, zirconium and lanthanum is primarily attributed to lower work function of these oxides than the pure tungsten. Work function of pure tungsten electrode is 4.4 eV while that of Zr, Th, La and Ce is 4.2, 3.4, 3.3 and 2.6 eV, respectively. Lower the work function of the electrode material, easier will be emission of electrons in the gap between electrode and workpiece. The ease of emission of electrons improves the arc stability even at low arc voltage and welding current.

Addition of the oxides of thorium, cerium, zirconium and lanthanum helps to increase the current carrying capacity of pure tungsten electrode up to tenfolds. Size of tungsten electrode is specified on the basis of its diameter as it largely determines the current carrying capacity of a given electrode material. The current carrying capacity of an electrode to be used with a given GTAW set-up is also influenced by cooling arrangement in a welding torch (air/water cooled), type of power source (DC, AC) and polarity (DCEP/DCEN), electrode extension beyond collets, nozzle diameter and shielding gas. Water cooling, DC power source with straight polarity

© The Author(s), under exclusive license to Springer Nature Singapore Pte Ltd. 2022 181
D. K. Dwivedi, *Fundamentals of Metal Joining*,
https://doi.org/10.1007/978-981-16-4819-9_15

(DCEN), short electrode extension and argon as shielding gas increase the current carrying capacity (CCC) of a given electrode while air cooling, DC reverse polarity (DCEP), long electrode extension and He as a shielding gas lower the current carrying capacity of the electrode (Fig. 15.1). Typical electrodes for GTA welding and suitable type of current are given below:

- 2% cerium-coated electrodes: good for both AC and DC welding
- 1.5–2% lanthanum-coated electrode: gives excellent low current starts for AC and DC welding
- 2% thorium-coated electrode: commonly used for DC welding and is not preferred for AC.

15.2 Welding Torch

Air-cooled welding torch offers lower current carrying capacity than water cooled due to the fact that water cooling reduces overheating of the electrode and nozzle during welding by extracting the heat effectively from the electrode. Therefore, high current GTAW welding systems are usually water cooled while GTAW torch for welding of thin sheets can be air cooled as these have very low current rating.

15.2.1 Type of Welding Current and Polarity

Current carrying capacity of an electrode with DCEN polarity is found to be higher than DCEP and AC because DCEN generates lesser (30% of arc power) heat in the electrode side as compared to the DCEP and AC. Therefore, electrodes with DCEN polarity offer longer life for the same level of welding current conversely higher current capacity for the same life. Size of the welding electrode for DCEP (for the same welding current and electrode life) should be larger than that for DCEN so that it can sustain the higher heat generation at anode than cathode for the same welding current. Current carrying capacity of electrode for AC welding is generally found in between that of DCEP and DCEN. Continuous change in polarity during the AC welding allows the somewhat less heating coupled with possibility of cooling of the electrode when (a) welding current is passing through zero or low current value and (b) electrode is negative for one half of the cycle so experiencing lesser heat exposure.

The selection of polarity for GTAW is primarily determined by the type of metal and section thickness of plates to be welded. The DCEN polarity is preferred for welding of steel and nickel alloys and other metals when mobile cathode spot is not formed and this polarity does not contribute in cleaning the weld pool appreciably for developing a sound weld joint. The application of DCEP polarity is not common, and it is preferred mainly for shallow penetration welding applications like thin sheet welding. AC is commonly used for welding of aluminium and magnesium to get advantage of cleaning action and avoiding overheating of tungsten electrode.

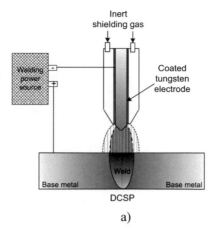

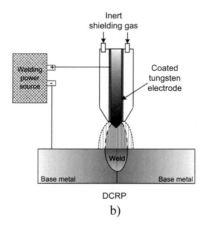

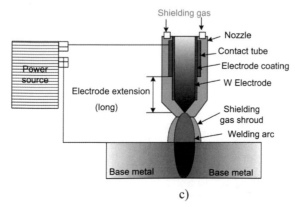

Fig. 15.1 Schematic showing effect of polarity **a** DCSP and **b** DCRP and electrode extension **c** long extension and **d** short extension on arc heat generation on electrode side and weld bead geometry

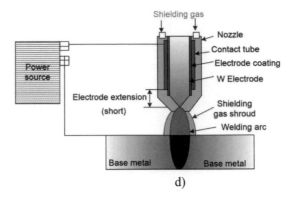

Fig. 15.1 (continued)

15.2.2   Electrode Diameter and Welding Current

The size (diameter) of tungsten electrode for the GTAW is usually found in a range of 0.3–8 mm and length varies from 75 to 610 mm. The selection of electrode material and diameter is governed by the section thickness of the material to be welded. Thick plates demand greater heat input so high welding current which in turn dictates the choice for large diameter electrodes. Small diameter electrodes experience more overheating than large diameter electrode under identical welding conditions of current, electrode extension, etc., primarily due to difference in heat conduction and electrical resistive heating (Fig. 15.2). Excessive welding current causes erosion of electrodes and tungsten particle inclusion in the weld metal due to thermal damage. Erosion of electrode reduces the electrode life.

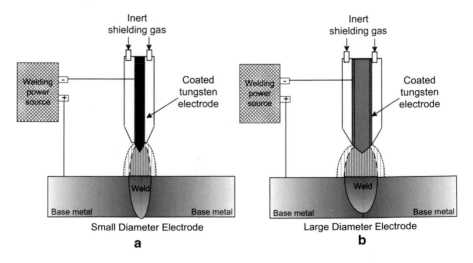

Fig. 15.2   Schematic showing effect of electrode diameter **a** small and **b** large on arc heat generation on electrode side and weld bead geometry

Electrode Taper Angle

Low welding current results in erratic wandering of welding arc over the tip of electrode, which reduces the arc stability. However, wandering of the arc at low current can be corrected by tapering the electrode tip from included angle 30–120° (Fig. 15.3). Low included taper angle is used to correct the arc wandering at low welding current. Taper angle affects the penetration and weld bead width. Low taper angle results in deeper the penetration and narrower weld bead than high angle taper.

15.3 TIG Arc Initiation

The touch start method of initiating GTAW arc is not considered good because it generally leads to many undesirable effects such as (a) contamination of tungsten electrode, (b) partial melting of electrode tip (due to short circuiting), (c) increased erosion of the electrode (d) reduction in life of the electrode, (e) formation of tungsten inclusions in the weld metal and (f) deterioration of the mechanical performance of weld joint. Therefore, alternative methods of GTAW arc initiation have been developed over the years so as to avoid undesirable effects of touch start method for striking the arc. Three methods are commonly used for initiating GTAW arc using (a) carbon block as scrap material, (b) high-frequency high-voltage unit and (c) low current pilot arc.

15.3.1 Carbon Block Method

This method is based on the principle similar to that of touch start method. In this method, the tungsten electrode is brought in contact of a scrap material or carbon block placed in area close to the region where arc is to be initiated and applied for heat for welding (Fig. 15.4). However, this method does not necessarily prevent electrode melting, contamination and erosion, but it reduces tendency for the tungsten inclusions in the weld metal.

15.3.2 Field Start Method Using High-Frequency Unit

This method is based on field emission principle for starting the arc by applying pulses of high frequency (100–2000 kHz) and high voltage (3000–10,000 V) to initiate the welding arc (Fig. 15.5). The high-voltage pulse facilitates the (a) availability of free electrons in arc gap by field emission and (b) ionization of gases between the electrode and the workpiece. Presence of the free electrons and charged particles in the gap between the workpiece and electrode with requisite voltage initiates the welding arc.

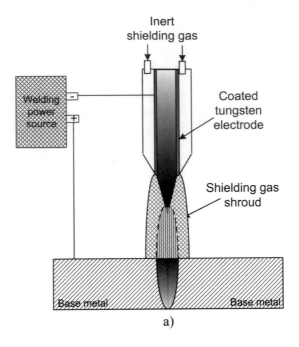

a)

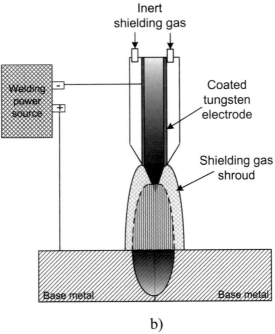

b)

Fig. 15.3 Schematic showing effect of electrode tip taper angle **a** low, **b** medium and **c** high on electrode heating and weld bead geometry

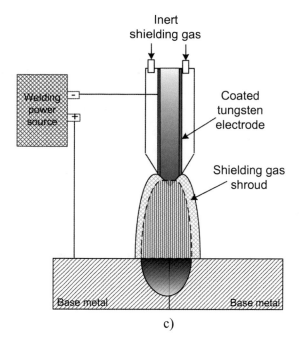

c)

Fig. 15.3 (continued)

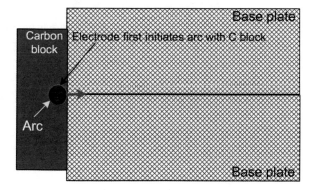

Fig. 15.4 Schematic of carbon block method to initiate welding arc

Once the arc is initiated, HF unit is taken off the welding circuit. This method is mainly used in automatic GTAW process. Absence of contact between electrode and workpiece avoids melting of electrode tip, reduces the electrode contamination and tungsten particle inclusions in the weld metal. These factors in turn increase life of the electrode.

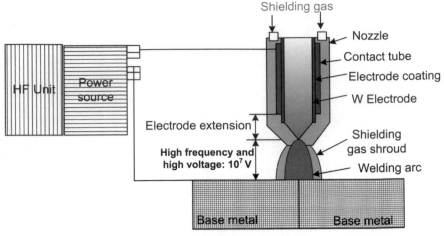

Fig. 15.5   Schematic of field start method to initiate a welding arc

15.3.3   Pilot Arc Method

Pilot arc method is based on the principle of using low current for initiating the arc so as to reduce the adverse effects of high heat generation in form of electrode contamination and electrode melting during the arc initiation (Fig. 15.6). For this purpose, an additional power source is used to strike the pilot arc between the tungsten electrode and auxiliary anode (fitted in nozzle) using low current. This pilot arc is then brought close to base metal to be welded so as to ignite the main arc between electrode and workpiece. Once the main arc is established, auxiliary power source is taken off the welding circuit.

15.4   Maintenance of TIG Welding Arc

Arc maintenance in GTA welding with DC power supply does not create any problem. However, in case of AC welding, to have smooth and stable welding arc, other methods like use of high OCV, imposing the high-frequency and high-voltage pulse at the moment when current is zero can be used so that arc does not extinguish.

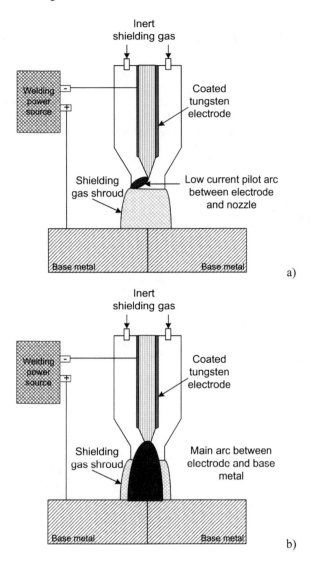

Fig. 15.6 Schematic showing the **a** pilot arc initiation using auxiliary power source and **b** normal welding established using regular welding power source

15.5 Pulse GTA Welding

Pulse GTAW is a variant of gas tungsten arc welding. In this process, welding current is varied in controlled way between a high and a low level at regular time intervals. This variation in welding current between high and low level is called pulsation of welding current (Fig. 15.7). High-level current is termed as peak current and is primarily used for melting of faying surfaces of the base metal while low current

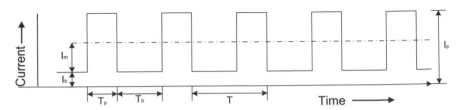

Fig. 15.7 Schematic showing parameters related to the pulse current and time where I_p, I_b and I_m are peak current, base current and mean current respectively, while T_p, T_b and T show pulse current duration, base current duration and total cycle time for one pulse, i.e. sum of pulse and base current period (in ms)

is generally called background or base current and it performs two functions: (1) maintenance of the welding arc while generating minimum possible heat and (2) allows time for solidification of the weld pool by dissipating the heat to the base metal. This feature of current pulsation associated with pulse GTAW effectively reduces net heat input required for welding of the base metal. The reduction in net heat input for welding facilitates (a) easy welding of thin sheets, (b) refinement of grain structure of the weld and (c) reduced effect of high heat input in the form of reduction in residual stress and distortion, reducing width of HAZ. Conversely, reduction in net heat input using arc pulsation decreases undesirable effects of comparatively high heat input of conventional TIG welding such as melt through, wrapping/buckling and fit-up.

15.5.1   Process Parameters of Pulse TIG Welding

Important variables in this variant of GTAW are peak current, background current, duration of peak current (pulse duration) and that of background current. Peak and background current and corresponding durations can be controlled independently depending upon the characteristics of the base metal to be welded such as thickness and materials.

Questions for Self-assessment

- Explain principle of pulse GTAW process and mention its suitable application.
- Describe different types of electrodes used in GTAW process.
- Explain the effect of electrode tip angle on shape and power density distribution for GTAW process.
- Describe the principle of initiating the GTAW arc.
- How does pulse GTAW process help in developing sound weld joints of thin sheet?
- How does pilot arc methods offer better electrode life than when touch start method is used?
- What is the role of pulse and background current in pulse GTAW?

- How does pulse GTAW offer different characteristics of weld joints that of conventional GTAW?
- How does touch start method promote tungsten inclusions in the weld metal?
- What are the factors to be considered when selecting suitable electrode diameters?

Further Reading

Cary H (1988) Welding technology, 2nd edn. Prentice Hall
Little R (2001) Welding and welding technology, 1st edn. McGraw Hill
Metals Handbook (1993) Welding, brazing and soldering, 10th edn, vol 6. American Society for Metals, USA
Nadkarni SV (2010) Modern arc welding technology. Ador Welding Limited, New Delhi
Parmar RS Welding process and technology. Khanna Publisher, New Delhi
Welding Handbook (1987) 8th edn, vols 1 and 2. American Welding Society, USA

Chapter 16
Arc Welding Processes: Gas Tungsten Arc Welding: Pulse Current, Hot Wire and Activated Flux-Assisted GTAW: Plasma Arc Welding: Principle, System, Application

16.1 Selection of Pulse Parameters for Pulse GTAW

High peak current setting is required for welding of thick sections and metals having high thermal conductivity to facilitate through thickness penetration. Background current or low level of current must be high enough to maintain the stable arc with the lowest possible heat generation so that solidification of the molten weld can take place without any heat build-up. Duration of the pulse and background currents determines the pulse frequency. The frequency of the pulses and so duration of pulse/base current are selected in such a way that required heat input is provided while achieving desired degree of control over the weld pool. In pulsed GTA welding, the weld bead is composed of a series of overlapping weld spots/pools, which can be observed especially when welding is done at low frequency (Fig. 16.1).

Due to continuous variation in current in P-GTAW, calculations of heat input for pulse GTAW are different from conventional GTAW. It required to calculate the average welding current during pulse welding for calculation of heat input. Average current can be obtained by using following equation:

$$Im = [(I_p \times t_p) + (I_b \times t_b)]/(t_p + t_b) \tag{16.1}$$

where I_p is peak current (A), T_p is peak pulse current duration (ms), I_b is background current (A), T_b is background current duration (ms), and I_m is average current (A).

16.1.1 Pulse Current

Generally, background current varies from 10 to 25% of peak current depending upon the thickness base metal and cooling conditions desired. The peak current is set at about 150–200% of steady (constant) current corresponding to the conventional GTA welding for the same thickness of base metal. Greater the thickness, higher

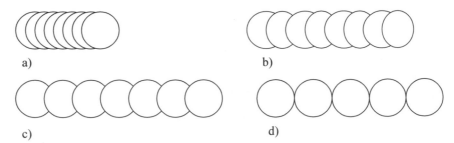

a)

b)

c)

d)

Fig. 16.1 Relationship between the overlapping of weld spot and pulse frequency in reducing order (for a given welding speed)

is the peak current. Selection of the pulse/peak current duration depends on the weld pool size and penetration required for welding of the workpiece of a particular thickness (Fig. 16.2). Increase in peak current and peak current duration in general increases the average welding current which in turn (for a given voltage setting) increases the heat generated by pulse GTAW arc. Increase in heat input decreases the cooling rate experienced by the weld metal and heat affected zone. The background current duration is determined on the basis of cooling rate required in the weld during solidification to achieve the better control over the weld pool and the microstructure of weld metal and heat affected zone so that desired mechanical performance of the weld joints can be obtained and weld cracking tendency is reduced.

16.1.2   Pulse Frequency

Reduction in pulse frequency (conversely longer background current duration and short peak current duration) during pulse GTA welding reduces heat input available for welding which in turn increases the cooling rate and so solidification rate. Too high solidification rate can increase porosity in the weld primarily because of entrapment of gases due to inadequacy of time available for escaping of gases from the weld pool during the solidification. Moreover, high solidification rate fines grain structure. A fine-grained structure can also be achieved using high pulse frequencies owing to different grain refining mechanisms (Fig. 16.3). Fine microstructure is known to improve the mechanical properties of the weld joint in general except creep resistance. Low pulse frequency (up to 20 Hz) has more effect on the microstructure and mechanical properties than high pulse frequency. Pulse GTA welding is commonly used for root pass welding of tubes, pipe welding and joining of thin sheets to take the advantage of the low heat input.

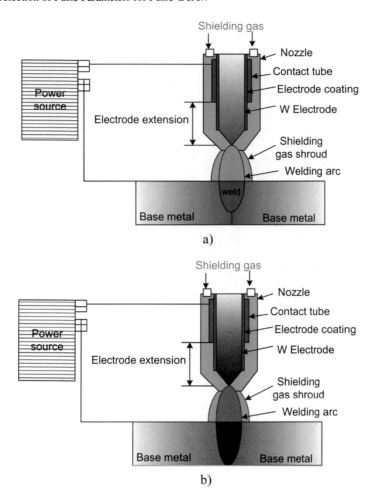

Fig. 16.2 Schematic showing effect of pulse current on heat generation and weld bead geometry **a** low pulse current and **b** high pulse current

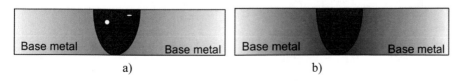

Fig. 16.3 Schematic showing effect of pulse frequency **a** low (5 Hz) and **b** high frequency (30 Hz)

16.2   Hot Wire Gas Tungsten Arc Welding

The hot wire gas tungsten arc welding (HW-GTAW) process is based on the principle of using preheated filler wire in conventional GTAW. This process is primarily designed to reduce heat input to the base metal while realizing higher deposition rate for welding of thick sections (Fig. 16.4). Preheating of the filler wire reduces the heat required from arc for its melting as some of the sensible heat is provided to filler during preheating. Therefore, preheating of filler increases melting rate and so welding speed even when using low heat input. Increased welding speed due to high deposition rate even at low heat input results in high productivity. Preheating of the filler can be done using an external source of heat. AC is commonly used to preheat the filler wire by electrical resistance heating principle (Fig. 16.5). Application of DC for preheating of wire can interfere with welding arc due to interaction of electromagnetic field to cause arc blow. This process can be effectively used for welding of ferrous metals and Ni alloys. Welding of aluminium and copper by this process is somewhat limited mainly due to difficulties associated with preheating of

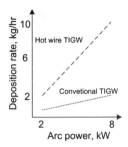

Fig. 16.4 Comparative deposition rates of conventional and hot wire GTAW process

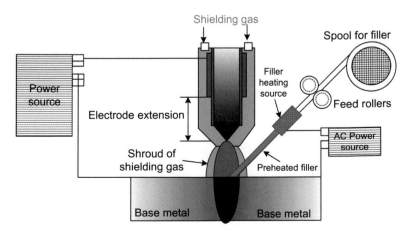

Fig. 16.5 Schematic showing the principle of hot wire GTAW process

Al and Cu fillers by electrical resistance heating as such high (thermal and electrical) conductivity metals need heavy current for electrical resistive heating of filler wire owing to their low electrical resistivity.

16.3 Activated Flux-Assisted Welding Processes

Activated flux-assisted GTA and GMA welding processes are being explored to take advantage of high penetration which can be achieved by these processes. Typically, flux-assisting GTAW is an autogenous welding process. Moreover, filler can be used as per need for joining of dissimilar metals. No special edge preparation in needed for joining base metals. Edges of the base metals need to be made square only. The flux-assisted processes use common fluxes like TiO_2, SiO_2, Cr_2O_3, ZrO_2 and halide fluxes. The flux is usually applied in the form of paste on to the faying surfaces at the top surface of base metal (along the weld centreline) followed by the application of welding arc for melting the base metal (Fig. 16.6). Application of activating flux results in many desirable effects during welding (a) increases the arc voltage compared with conventional GTAW process under identical conditions of arc length, and welding current which in turn burns the arc hotter and increases the depth of penetration and (b) increases the constriction of the arc which in turn facilitates the development of high depth to width ratio weld (Fig. 16.7).

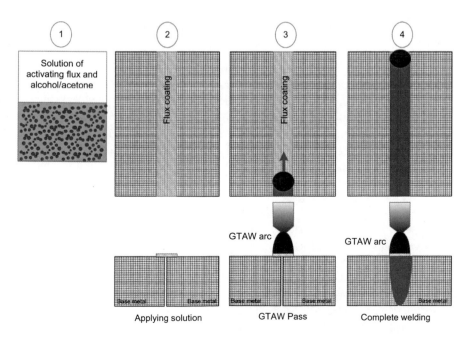

Fig. 16.6 Schematic of steps used in flux-assisted GTAW process

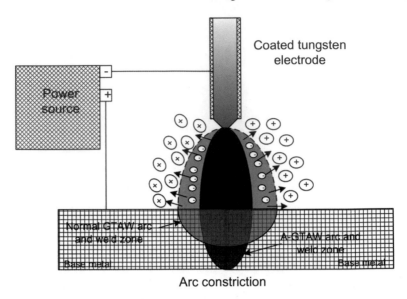

Coated tungsten electrode

Arc constriction

Fig. 16.7 Schematic of A-GTAW showing constriction of normal arc and corresponding change in weld bead geometry

Increase in depth of the penetration in turn increases the rate of lateral heat flow from the weld pool to the base metal. Increased rate of heat flow from the weld pool causes grain refinement owing to the high cooling rate and low solidification time. High depth to width ratio of the weld bead lowers the angular distortion and residual stresses due to smaller area of the weld. High depth to width ratio effect observed in the weld pool by activated fluxes is similar to that of the high energy density processes like laser and electron beam welding. The depth of width ratio of the flux-assisted GTAW weld is influenced by welding current, speed and shielding gas controlling the heat input, type of flux and flux density (g/mm^2). Increase in depth to width ratio of weld is attributed to two basic mechanisms: (a) arc constriction and (b) reversal of Marangoni convection. Arc constriction reduced arc diameter which in turn increases the energy density which in turn increase the penetration and reduces the width. According to Marangoni convection, the flow of molten weld metal occurs in normal fusion weld from the centre of weld (arc) at the highest temperature zone (low surface tension) towards fusion boundary of minimum molten metal temperature zone (high surface tension). This flow pattern facilitates the transfer heat supplied by the arc from the centre to fusion boundary; therefore, in conventional fusion welds wider weld bead with shallow penetration is realized. Application of activating flux causes reversal of Marangoni convection flow (due to reversal in surface tension vs temperature relationship) in the weld pool, so molten metal flows from fusion boundary to weld centre (Fig. 16.8). In such a situation, heat supplied by the welding arc is transferred from top weld pool surface to root of weld pool which in turn results in deeper and deeper penetration to produce high depth to width ratio welds

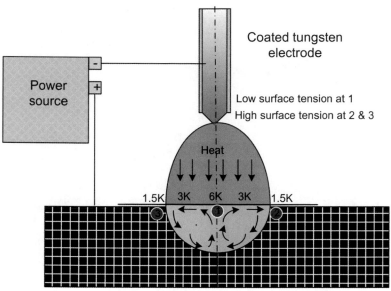

Normal Convection Current Flow Pattern of Molten Metal

a)

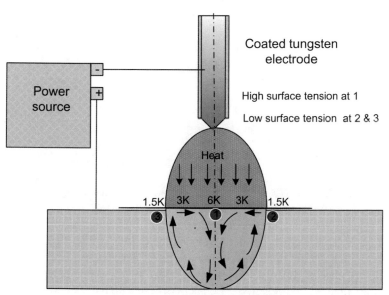

Reversal Flow Pattern of Molten Metal: Reverse
Marangoni Convection

b)

Fig. 16.8 Schematic of showing pattern of flow of molten metal **a** centrifugal in conventional GTAW as per Marangoni convection current and **b** centripetal in A-GTAW as per reverse Marangoni convection current

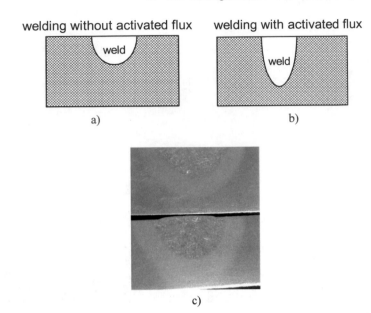

c)

Fig. 16.9 Schematic of activated flux TIG welding: **a** weld bead geometry without activated flux, **b** weld bead geometry with activated flux and **c** photograph of weld bead geometry without activated flux and with activated flux GTAW

(Fig. 16.9). Activated flux-assisted GTA welding processes have been developed for joining of titanium and steel for nuclear and aerospace applications.

Plasma Arc Welding

16.4   Introduction

The plasma arc welding (PAW) is considered as an advanced version of GTA welding. Like GTAW, PAW also uses the tungsten electrode for establishing the welding arc and inert gases for shielding of the molten metal in the weld pool. Welding arc of GTAW differs from PAW primarily with regard to plasma velocity, arc size and peak temperature attained in arc zone. Welding arc of GTA welding is characterized as low velocity plasma, diffused arc and comparatively low peak temperature in while PAW offers high plasma velocity, coherent plasma and high peak temperature. Large surface area of the GTAW is exposed to ambient air, and base metal results in greater heat losses from the arc to the ambient air and base metal than PAW which in turn reduces the power density of GTAW and peak temperature appreciably. Therefore, GTAW arc burns at temperature lower than plasma arc. On the other hand, plasma arc burns hotter and offers higher energy density than GTAW to due to smaller cross

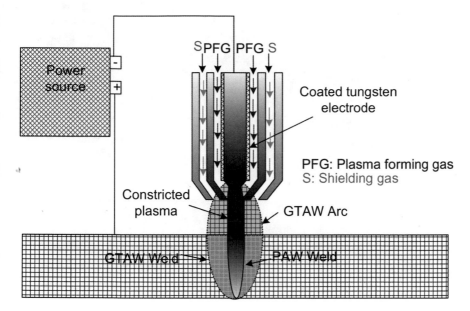

Fig. 16.10 Schematic showing comparative features of GTAW and PAW

section of the plasma arc produced by constriction (Fig. 16.10). Therefore, plasma arc welding produces weld joint at low heat input with deeper penetration, narrow weld bead and smaller width of heat affected zone. In view of above, PAW weld joints generally offer better characteristics than GTAW joint under identical welding condition.

16.5 Principle of PAW

This process uses the heat generated by plasma for fusion of faying surfaces of the base metal to produce a fusion weld joint. To produce plasma, an arc is established between nozzle/auxiliary electrode and tungsten electrode, and then plasma forming gas is passed through the arc. High-temperature exposure of the plasma forming gas passing through the arc results in thermal ionization to form high-temperature plasma. This plasma is passed through a (water cooled) copper nozzle. The nozzle causes the constriction of the arc (Fig. 16.11). The constriction reduces cross section of plasma jet and accelerates it to produce high velocity plasma. Constriction of plasma arc leads to (a) reduction in cross-sectional area of the arc, (b) increase in the energy density, (c) increase in velocity of plasma approaching to the sound velocity and (d) increase in peak temperature to about 25,000 °C. These factors together make PAW, a high energy density and low heat input welding process; therefore, it poses fewer problems associated with weld thermal cycle to the base metal.

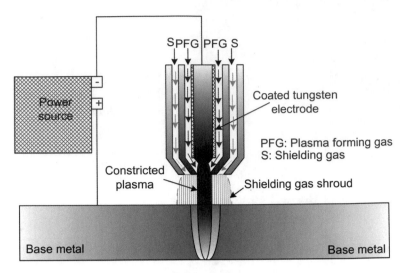

Fig. 16.11 Schematic showing working principle of PAW process

Constriction of arc increases the penetration and reduces the width of weld bead. Energy associated with plasma depends on plasma current, size of nozzle and plasma gas (Fig. 16.12). A coherent, culminated and stiff plasma is formed due to constriction; therefore, it does not get deflected and diffused during the welding. Hence, heat is transferred to the base metal over a very small area which in turns results in high energy density and deep of penetration and small width of the weld pool/keyhole/cut.

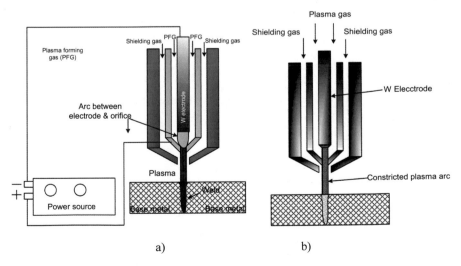

Fig. 16.12 Schematic of **a** plasma arc welding system and **b** arc constriction and so penetration capability

Further, stiff and coherent plasma makes it possible to use stable arc even at very low current levels (< 15 A). This feature has led to the development of micro-plasma system.

Power density and penetration capability of plasma jet are determined by the various input process parameters, namely plasma current, nozzle orifice diameter and shape, plasma forming gas (Air, He, Ar) and flow rate of plasma forming gas. Increase of plasma current, flow rate, thermal conductivity of plasma forming gas and reducing nozzle orifice diameter together result in increase of the power density and penetration capability of plasma jet. In general, the plasma cutting uses high energy density in combination with high plasma velocity and high flow rate of high thermal conductivity plasma forming gas. A combination of such characteristics for plasma cutting is achieved by controlling above process parameters. Further, thermal conductivity of plasma forming gas must be high enough for cutting operation so that heat can be effectively transferred to the base metal rapidly. Plasma welding needs comparatively lower energy density and lower velocity plasma than the plasma cutting to avoid melt through or blowing away tendency of molten metal.

High-power density associated with plasma arc produces a temperature of order of 20,000 to 25,000 °C. This process uses the heat transferred by plasma (produced by a plasma forming gas such as Ar, Ar-H_2 mixture passing through an electric arc) for melting of faying surfaces. An additional and separate arrangement is used to supply inert gas (Ar, He) around the plasma arc to protect the molten weld pool from the atmospheric gases. Charged particles (electrons and ions) formed as a result of ionization of plasma gas tends to reunite after impinging the surface of workpiece. Recombination of charged particles also liberates heat which is used for melting of base metal.

Electric arc for producing plasma can be established either between a non-consumable tungsten electrode and workpiece or non-consumable electrode and nozzle, and accordingly plasma processes is characterized as transferred and non-transferred plasma arc welding. As discussed above, plasma arc welding uses two types of gases: one is called plasma forming gas and other is inert gas. Roles performed by these gases are different. Inert gas is primarily used for shielding the weld pool from the contamination by atmospheric gases. Plasma forming gas is used to develop plasma by passing through the arc zone and transferring the heat to the weld pool.

PAW uses a constant current-type power source with DCEN polarity. The DCEN polarity is invariably used in PAW, because tungsten electrode is used for developing the arc through which plasma forming gas is passed. Tungsten electrode has good electron emitting capability; therefore, it is made cathode. Further, DCEN polarity causes less thermal damage to the electrode during welding as about one-third of total heat is generated at the cathode and balance two-third of arc heat is generated at the anode side, i.e. workpiece. DCEP polarity does not help the process in either way. Current for PAW can vary from 2 to 200 A.

The plasma arc in PAW is not initiated by the conventional touch start method. Moreover, arc initiation in PAW heavily depends on use of high-frequency unit (Fig. 16.12). Plasma is generated using two cycles approach: (a) pilot arc and (b) main arc. Pilot arc is very small high-intensity spark (pilot arc) between the electrode and nozzle within the torch body by imposing pulses of high voltage, high frequency and low current about 50 A (from HF unit) which in turn generates a small pocket of plasma gas. Then as the plasma torch is approached to the workpiece, main current starts flowing between electrode and workpiece leading to the ignition of the main transferred arc. At this stage, pilot arc is extinguished and taken off the welding circuit.

16.6   Types of PAW

Plasma generated by arc between the non-consumable electrode and workpiece is called transferred plasma, whereas that is produced by arc between non-consumable electrode and nozzle is called non-transferred plasma. Non-transferred plasma system to a large extent becomes independent of nozzle to workpiece distance. Workpiece is not a part of welding circuit in case of non-transferred plasma.

Transferred plasma offers higher energy density than non-transferred plasma, and therefore, it is preferred for welding of thick section and cutting of high-speed steel, ceramic, aluminium, etc. Non-transferred plasma is usually applied for welding of somewhat thin section and thermal spray application of steel and other common metals. Depending upon the current, plasma gas flow rate and the orifice diameter, following variants of PAW have been developed:

- Micro-plasma (< 15 A)
- Melt-in mode (15–400 A) plasma arc
- Keyhole mode (> 400 A) plasma arc.

Micro-plasma welding work with very low plasma forming current (generally lower than 15 A) which in turn results in comparatively low-power density and low plasma velocity. These conditions become good enough for welding of thin sheets (Fig. 16.13a).

Plasma for melt-in mode uses somewhat higher current and greater plasma velocity than micro-plasma system for welding applications (Fig. 16.13b). Melt-in mode of PAW is generally used up to 2.4 mm thickness. For thickness of sheet greater than 2.5 mm, normally welding is performed using keyhole technique.

The keyhole technique uses high current and high-pressure plasma gas to ensure the keyhole formation. High-power density and high heat associated with plasma melt the faying surfaces of base metal, and high-pressure plasma jet pushes the molten metal against (unmelted) vertical wall of base metal for developing a keyhole (Fig. 16.13c). Plasma velocity should be such that it does not push molten metal out of the keyhole, while at the same it is able (to form keyhole) by pushing molten metal along the hole created. Keyhole formation needs a good combination of heat for melting and plasma velocity for pushing molten metal along the hole. The keyhole is therefore formed under certain combination of plasma current, orifice gas flow rate and velocity of plasma for a given welding torch, and any disturbance to above

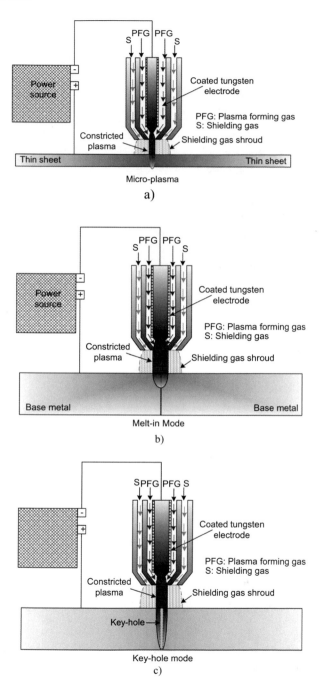

Fig. 16.13 Schematic showing PAW of type **a** micro-plasma, **b** melt-in mode and **c** keyhole mode

parameters will lead to a loss of keyhole. For keyholing, flow rate of the plasma is very crucial, and therefore, it is controlled accurately $\pm$ 0.14 L/min. Nozzles are generally specified with welding current and plasma gas flow rate.

16.7   Advantage of PAW

With regard to power density, PAW stands between GTAW/GMAW and EBW/LBW, and accordingly it can be applied using melt-in mode or keyhole mode. Melt-in mode approach of welding offers greater heat input and higher width to depth of weld ratio than keyhole mode. Higher-power density associated with PAW than GTAW produces narrow heat affected zone and lowers residual stress and distortion-related problems due to smaller weld cross section. Further, high depth to width ratio weld produced by PAW reduces the tendency of angular distortion. PAW generally uses about one-tenth of welding current as compared to GTAW for same thickness; therefore, it can be effectively applied for joining of the thin sheets. Further, non-transferred plasma offers the flexibility of variation in stand-off distance between nozzle and workpiece without extinction of the arc; therefore, it can be effectively used for thermal spray process.

16.8   Limitation of PAW

Infrared and ultraviolet rays generated during the PA welding are found harmful to the operator and human being around. High noise (100 dB) level associated with PAW is another undesirable factor. PAW is a more complex, costlier, difficult to operate than GTAW. Narrow width of the PAW weld can be problematic from alignment and fit-up point of view if edge preparation of the base metal is not proper.

Questions for Self-assessment

- How different parameters of pulse GTAW process are decided?
- What is advantage of hot wire GTAW process? Explain the principle of GTAW process.
- Why does hot wire GTAW offer higher deposition rate than conventional GTAW?
- Explain the role of Marangoni convection in flux-assisted GTAW.
- How does flux-assisted GTAW help in reducing angular distortion?
- How plasma arc welding is different from GTAW process?
- Explain the principle of PAW process using suitable schematic diagram.
- What are the factors determine penetration capability and energy density of PAW process?
- Distinguish the transferred and non-transferred plasma arc welding.
- How can the net heat generated by pulse GTAW be obtained?

Further Reading

Cary H (1988) Welding technology, 2nd edn. Prentice Hall

Huang H (2010) MTA, 41A, 2829

Little R (2001) Welding and welding technology, 1st edn. McGraw Hill

Metals Handbook (1993) Welding, brazing and soldering, 10th edn, vol 6. American Society for Metals, USA

Nadkarni SV (2010) Modern arc welding technology. Ador Welding Limited, New Delhi

Parmar RS Welding process and technology. Khanna Publisher, New Delhi

Welding Handbook (1987) 8th edn, vols 1 and 2. American Welding Society, USA

Chapter 17
Arc Welding Processes: Gas Metal Arc Welding: Principle, System, Parameters and Application

17.1 Fundamentals of GMA Welding

GMAW process involves consumable electrode for establishing welding, and shielding jet is used to protect the welding pool from atmospheric gases. The shielding in GMAW can be achieved either using inactive gases like carbon dioxide, mixture of carbon dioxide and argon, argon and oxygen, argon and hydrogen or inert gases like helium and argon only. In case when exclusively inert gas is used for weld pool protection then GMAW can also termed as metal inert gas welding (MIGW).

The principle of GMAW process for fusion welding is based on the melting of the faying surfaces of the base metal using heat produced by a welding arc established between base metal and a consumable electrode (Fig. 17.1). Welding arc and weld pool are well protected by a jet of shielding inert/inactive gas coming out of the nozzle. A jet of shielding gas forms a firm shroud around the arc and weld pool to protect the weld pool. However, GMA weld is not considered as clean as that of gas tungsten arc (GTA) weld. Difference in cleanliness of the weld produced by GMA and GTA welding is primarily attributed to the difference in effectiveness of shielding approach in case of these two processes. Effectiveness of shielding in these two processes is mainly determined by two characteristics of the welding arc (of these two processes), namely stability of the welding arc and length of arc, besides other welding related parameters such as type of shielding gas, flow rate of shielding gas and distance between nozzle and workpiece. The GMAW arc is relatively longer and less stable than GTAW arc. Difference in stability of two welding arcs is primarily due to the fact that in GMAW arc is established between base metal and a consumable electrode (which is consumed continuously during welding), while in GTA welding, arc is established between base metal and non-consumable tungsten electrode. Tungsten electrode offers better electron emission property (to provide free electrons for a stable welding arc) than GMAW electrode which is largely similar to the base metal. Further, consumption of the electrode during welding slightly decreases the stability of the arc. Therefore, shielding of the weld pool in GMAW is not as effective as in GTAW.

© The Author(s), under exclusive license to Springer Nature Singapore Pte Ltd. 2022
D. K. Dwivedi, *Fundamentals of Metal Joining*,
https://doi.org/10.1007/978-981-16-4819-9_17

   Main components of GMAW process are shown in Fig. 17.2. Gas metal arc welding process is similar to GTWA except that it uses the automatically fed consumable electrode; therefore it offers higher deposition rate. Consumable electrode is fed automatically, while welding arc/torch can be controlled either manual or automatically using mechanized system like robots or welding tractors (Fig. 17.2). Therefore, this process is found more suitable for welding (a) to produce good quality weld joints for industrial fabrication and (b) of comparatively thicker plates of reactive metals (Al, Mg and stainless steel). The quality of weld joints of such reactive metals otherwise is adversely affected by interaction of atmospheric gases with weld metal at high temperature.

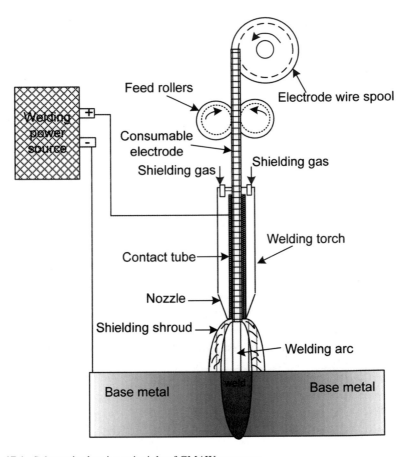

Fig. 17.1 Schematic showing principle of GMAW processes

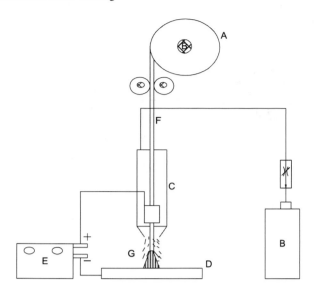

Fig. 17.2 Schematic of GMAW process showing important elements **A** welding spool, **B** shielding gas cylinder, **C** welding torch, **D** base plate, **E** welding power source and **F** consumable electrode

17.2 Power Source for GMA Welding

The welding power source for GMA welding can be either constant voltage or constant current type depending upon the electrode diameter, material and extension. For small diameter electrodes (< 2.4 mm) when electrical resistive heating predominantly controls the melting rate, constant voltage power source (DCEP) is used to take advantage of the self-regulating arc (Fig. 17.3). Whereas in case of large diameter and low electric resistivity electrode, constant current power source is used with variable speed electrode feed drive system to maintain the arc length.

17.3 Shielding Gases for GMA Welding

Like GTA welding, shielding gases such as Ar, He, CO_2 and their mixtures are used in GMAW process for protecting the welding pool from the atmospheric gases. Effect of the shielding gases on GMAW joints is similar to that of GTA welding. Inert gases are normally used welding for reactive metal like Al, Mg and stainless steel, while carbon dioxide can be used for welding of carbon and alloy steel to achieve the reasonably good quality weld joints. Application of CO_2 in welding of reactive non-ferrous metal is not preferred as decomposition of CO_2 in arc environment produces oxygen. Interaction of oxygen with reactive metals like Al and Mg (having high affinity to the oxygen) forms refractory oxides. These oxides have higher melting point than the substrate, and therefore these oxides do not melt easily and hence

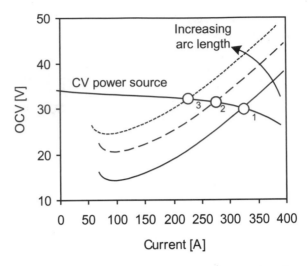

Fig. 17.3 Static characteristics of constant voltage power source showing effect of arc length on operating point

interfere with fusion of the base metal. Further, these refractory oxides increase the inclusion formation tendency in the weld metal. Moreover, shielding gases in the GMAW also affect the mode of metal transfer of the molten metal from the consumable electrode to the weld pool during the welding (Fig. 17.4). GMA welding with Ar as a shielding gas results in significant change in the mode of metal transfer from globular to spray and rotary transfer (with maximum spatter) on increasing the welding current, while application of helium mainly produces globular mode of metal transfer. GMA welding with CO_2 results in a lot of spattering during the welding. Shielding gas also affects width of weld bead and depth of penetration owing to difference in heat generation during welding. In general, helium results in deeper penetration than the argon.

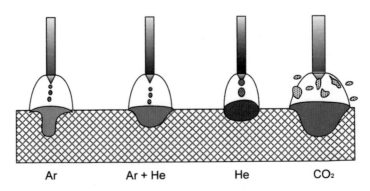

Fig. 17.4 Schematic showing influence of shielding gas on mode of metal transfer

17.4   Effect of MIG Welding Process Parameters

Among various welding parameters such as welding current, arc voltage and speed probably welding current is the most influential parameter affecting weld penetration, deposition rate, weld bead geometry and the quality of the weld. Arc voltage mainly affects the width of weld bead due to change in arc length/voltage. An increase in arc voltage in general increases the width of the weld. Welding current is primarily used to regulate the penetration, deposition rate and overall size of weld bead. Too low welding current results in pilling of weld metal on the faying surface in the form of peaked bead (due to limited fluidity of the weld metal) instead of penetrating into the workpiece. Thus, low welding current increases the reinforcement of weld bead without enough penetration. Excessive heating of the workpiece due to too high welding current can lead to undercut, weld sag, lower penetration and wider bead. Effect of welding current on melting/burn off rate of the electrode is more significant on small diameter electrode than the large diameter electrode; this difference is primarily due to high contribution of electrical resistive heating on melting of rate of small diameter electrode than large diameter electrode (Fig. 17.5). Optimum current gives optimum penetration and weld bead width. Effect of the welding speed on penetration, bead geometry and other characteristics of GMA weld is similar to that of SMAW and GTAW. Increase in welding speed reduces the penetration and weld width due to reduced heat input.

Stick out of the electrodes also called electrode extension affects the weld bead penetration and electrode melting rate (deposition rate) because it changes the electrical resistive heating trend of electrode. Increase in stick out in general increases the melting rate and reduces the penetration due to increased electrical resistive heating of the electrode itself by the welding current. Selection of the welding current is also

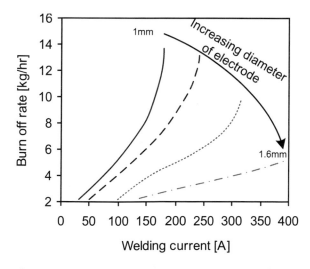

Fig. 17.5 Effect of welding current on melting of electrode of different diameters

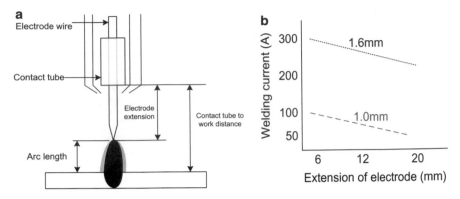

Fig. 17.6 Schematic diagram showing **a** electrode extension and **b** effect of electrode extension on welding current for different electrode diameters

influenced by electrode stick out and electrode diameter. In general, high welding current is preferred for large diameter electrodes with small electrode extension in order to obtain optimal weld bead geometry (Fig. 17.6).

17.5 Metal Transfer in GMA Welding

Metal transfer during GMA welding depending up on the welding current, electrode diameter and shielding gas can take place through different modes such as short circuit, globular, spray and rotational transfer (Fig. 17.7). Mechanisms for these metal transfer modes have already been described in Sect. 8.2.

A welding arc obtained using low welding current and low arc voltage results in low arc power and low heat input which in reduces the electrode melting rate and pinch force. Therefore, droplets of the molten metal take long time to grow of a large size enough to get detached from the electrode tip on its own (weight) under gravitation force. These welding conditions correspond to globular/short circuit transfer (Fig. 17.8a).

Increase in welding current changes mode of metal transfer from short circuiting to globular to spray transfer especially when Ar is used as a shielding gas (Fig. 17.8b). Further, increasing welding current over a narrow range leads to significant increase in drop transfer rate (number of drops per unit time) coupled with reduction volume/size of drops being transferred. Metal transfer occurring (with few tens of drops per unit second of large size) at low welding current corresponds to globular transfer; while metal transfer in form of few hundreds/thousands of drops per sec of small size at high welding current corresponds to spray transfer. A narrow welding current range over which such kind of change in drop transfer behaviour takes place is called transition current. This change in transfer behaviour is attributed to two reasons, namely (a) increase in melting rate of the electrode and (b) increase in pinch force. Thus, welding

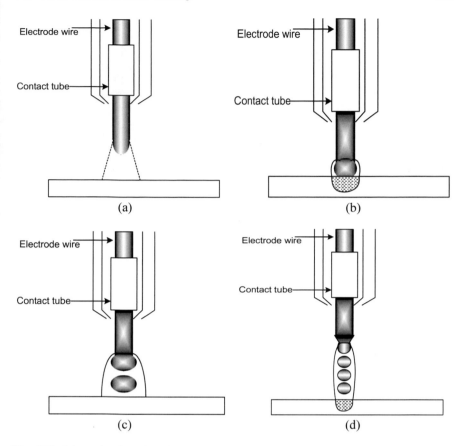

Fig. 17.7 Schematic of modes of metal transfer in MIG welding **a** typical GMAW arc, **b** short circuiting transfer, **c** globular transfer and **d** spray transfer

current at which change in mode of metal transfer from globular to spray takes place is called transition current. Typical welding current variation during different modes of metal transfer is shown in Fig. 17.8c.

17.6   Pulse GMAW Welding

Pulse GMA welding is a variant of gas metal arc welding. Pulse MIG welding is based on the principle of pulsation of welding current between a high and a low level at regular time intervals like pulse GTA welding (Fig. 17.9). However, background and peak current perform slightly different roles. The low level current also called background current is mainly expected just to maintain welding arc, while high level welding current called peak current is primarily used for (a) melting of faying surfaces

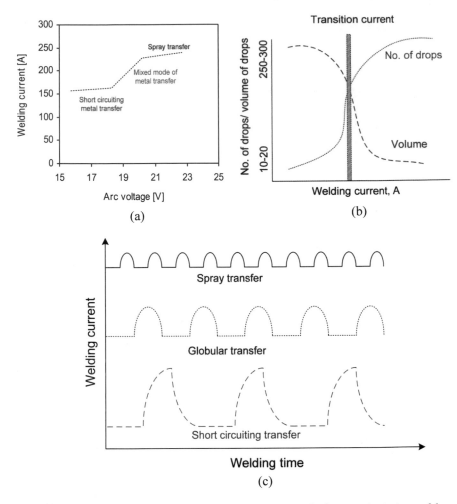

Fig. 17.8 Effect of **a** welding parameters on modes of metal transfer, **b** on number/volume of drops versus welding current during metal transfer and **c** typical variation in welding current observed for different modes of metal transfer during welding

of the base metal with desired penetration, (b) high melting rate of electrode and (c) detachment of molten droplets hanging to the tip of the electrode by pinch force to facilitate spray transfer. An optimum combination of pulse parameters results in transfer of one molten metal drop per pulse (peak current). This feature of current pulsation in pulse GMA welding reduces net heat input to the base metal during welding which in turn facilitates especially welding of thin sheets and odd position welding.

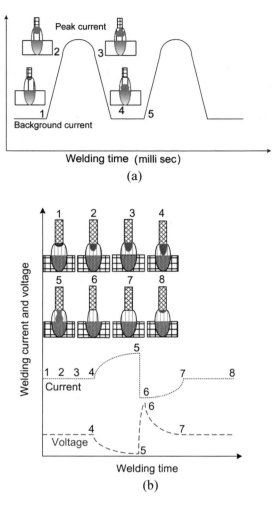

Fig. 17.9 Schematic showing **a** the relationship between the welding current and time with metal drop formation tendency in GMA, and **b** sequential steps of metal transfer in short circuiting mode of metal transfer during GMAW along with variation of current and voltage

17.7 Flux-Cored Arc Welding Process

The flux-cored arc welding (FCAW) is another variant of gas metal arc welding process. Like GMAW, this process mainly uses constant voltage power supply for generating heat by establishing a welding arc between a consumable electrode and base metal. The electrode for FCAW differs from GMAW electrode. GMAW electrode uses solid wire, while a tubular electrode is used in FCAW. Tubular electrode is filled with flux and other constituents which decompose at high temperature in arc environment to produce inactive gases for protecting the weld pool and arc zone from contamination by atmospheric gases (Fig. 17.10a). The role of flux in FCAW process is similar to that coating in shielded metal arc welding. However, the unique feature of filling the flux in continuously fed tubular electrode associated with this process for welding gives freedom from regular stoppage of welding for replacement of electrode in SMAW. Therefore, FCAW results in higher welding speed and productivity than SMAW. Since protective gases are generated in the arc environment itself, therefore ambient air flow/turbulence does not affect the protection of the weld pool appreciably.

This process is also used in two ways: (a) FCAW without shielding gas and (b) FCAW with external shielding gas like GMAW. The FCAW process with shielding gas results in somewhat more-sound weld with better mechanical properties than FCAW without shielding gas owing to the possibility of formation of few discontinuities in weld metal like porosity, slag inclusion, etc., in latter case (Fig. 17.10b). FCAW without shielding gas suffers from (a) poor slag detachability, (b) porosity formation tendency, (c) greater welding operator-skill requirement and (d) emission of harmful noxious gases and smokes which imposes the need of effective ventilation during the welding. Further, excessive smoke generation in case of FCAW without shielding gas can reduce visibility of weld pool during welding and make the welding process control difficult. FCAW with external shielding like CO_2 improves the protection of the weld pool and arc zone from atmospheric gases which in turn improves the soundness and quality of FCAW weld joints (Fig. 17.10b). FCAW is commonly used for welding of mild steel, structural steel, stainless steel and nickel alloys.

Questions for Self-assessment

- Why does GMAW process offer lesser clean weld than GTAW?
- What are important components of GMAW process and write their role?
- How do the shielding gases affect the metal transfer in GMAW?
- Describe effect of welding parameters on melting rate in GMAW.
- What are modes of metal transfer commonly observed in GMAW?
- What is transition current? Explain the effect of various factors affecting the transition current.
- Explain the principle of pulse GMAW and its advantages over conventional GMAW process.
- How does FCAW offer higher productivity than SMAW?
- Which process will you recommend for joining of thin sheets and why?
- Why does FCAW without shielding gas result in a poor weld joint than the metal inert gas welding?

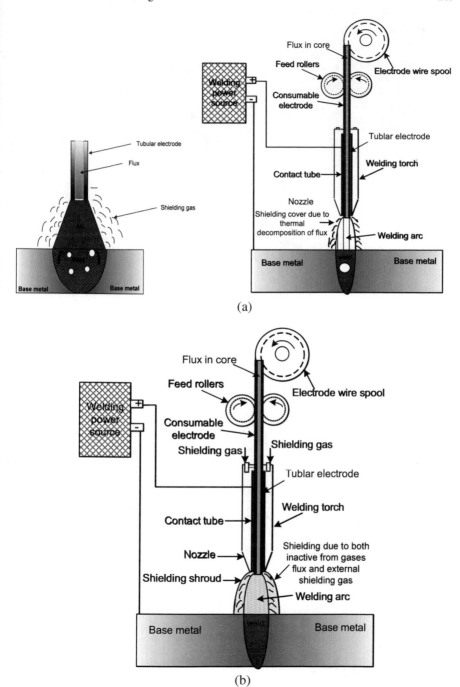

Fig. 17.10 Schematic of FCAW process **a** without shielding gas and **b** with shielding gas

Further Reading

Cary H (1988) Welding technology, 2nd edn. Prentice Hall

Little R (2001) Welding and welding technology, 1st edn. McGraw Hill

Metals Handbook (1993) Welding, brazing and soldering, 10th edn, vol 6. American Society for Metals, USA

Nadkarni SV (2010) Modern arc welding technology. Ador Welding Limited, New Delhi

Parmar RS Welding process and technology. Khanna Publisher, New Delhi

Welding Handbook (1987) 8th edn, vols 1 and 2. American Welding Society, USA

Part V
Solid Liquid Joining and Solid State Joining Processes

Chapter 18
Brazing, Soldering and Friction Stir Welding

18.1 Fundamentals of Brazing and Soldering

The joint by brazing, soldering and brazing welding is primarily developed using comparatively low melting point filler metal than base metals. The base metals are heated to a temperature high enough to facilitate melting of filler metal but significantly less than melting temperature of base metals. The molten brazing/soldering filler metal acts in two ways for developing a joint with moderate load carrying capacity by (a) mechanical interlocking by filler in irregularities present at the surface of base metals and (b) forming intermetallic compound through interactions between soldering/brazing filler and base metal (Fig. 18.1).

Brazing, and soldering are solid–liquid joining processes primarily involve three steps (a) heating of plates to be joined using suitable heat source, (b) melting of brazing/soldering metal (with suitable fluxes) placed between two parts to be joined and (c) distribution and filling of molten filler metal between the faying surfaces of the components to be joined by capillary action and followed by solidification. The solidification of brazing/soldering metal produces in a braze/solder joint accordingly. These three steps are schematically shown in Fig. 18.2a–d. An attractive feature of these two processes is that a semi-permanent joint produced without fusion of faying surfaces of workpieces. Due to this typical feature of brazing and soldering joints, these processes are preferred under following situations.

- Metallurgical incompatibility: It is difficult to develop a fusion weld joint due to metallurgical incompatibility between the components to be joined. Metals having entirely different physical, chemical and mechanical characteristics like steel and aluminium, steel and titanium, Al–Cu etc. impose many issues (cracking, embrittlement, low strength and ductility) in developing sound and strong fusion weld joints (Fig. 18.2d).
- Poor weldability: Joining of metals having poor weldability (e.g. high hardenability) for fusion welding due to cracking tendency, chemical reactivity to ambient gases etc. Like cast iron, and high carbon steels show high hardenability

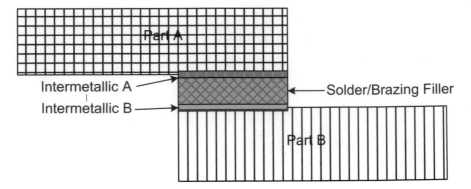

Fig. 18.1 Schematic showing principle of brazing/soldering process

resulting in very brittle and low toughness fusion weld joints. Al, Mg and stainless steels having high affinity with oxygen at high temperature form refractory oxides if not welded properly then these may cause inconsistent melting, and oxide inclusion.

- Unfavorable HAZ: Excessive metallurgical and mechanical property degradation of wide heat affected zone formed due to high heat input during fusion welding (weld thermal cycle) in case of certain metal. Deterioration in HAZ may be experienced in the form precipitate free zone formation, reversion/dissolution of precipitates, martensitic transformation, over-aging and over-tempering leading to hardening or softening making it unacceptable
- Odd position welding: Fusion welding needs melting of faying surfaces and applying molten weld metal from filler/electrode as per need. Very odd welding positions do not allow application of conventional fusion welding processes due to difficulties in melting of faying surfaces, placing molten metal in places where it is required.
- Light service conditions: Joint is not expected to take high load, high temperature, other adverse atmospheric conditions during service. The joint is primarily required to have connection between members subjected to moderate/light temperature and loading conditions during service. Solder joints are used mainly for electrical connections in electronic industry.

18.2   Joints for Brazing and Soldering

The most common joint configuration used in brazing and solder are given in Fig. 18.3a–f. Lap joint is commonly developed using both the techniques (e–f). Clearance (0.075–0.125 mm) between the members to be joined is of great importance as it affects the capillary action and so the distribution of brazing or soldering molten metal between the faying surfaces. The uniform distribution of the molten

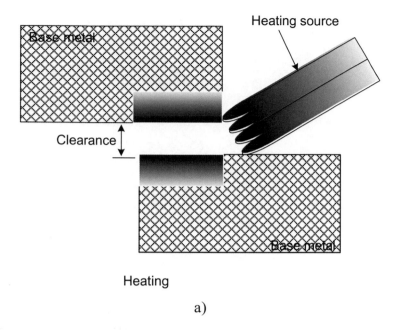

Heating

a)

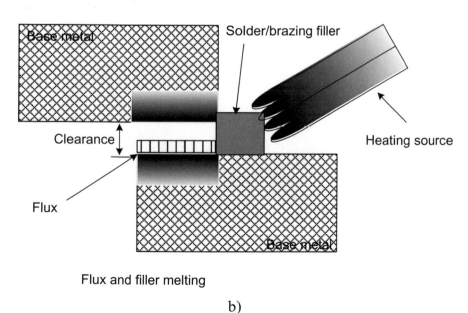

Flux and filler melting

b)

Fig. 18.2 Schematic of Step used for brazing and soldering process: **a** heating of plates, **b** placing brazing/soldering metal and heating, **c** filling of molten metal by capillary action followed by solidification and **d** formation of diffusion layer/intermetallic compound layer between brazing/soldering filler and base metal

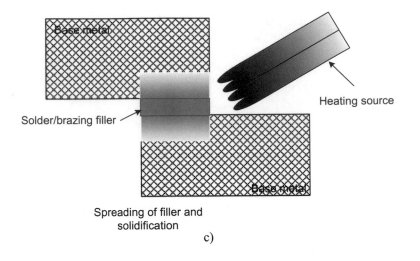

Solder/brazing filler

Heating source

Spreading of filler and
solidification

c)

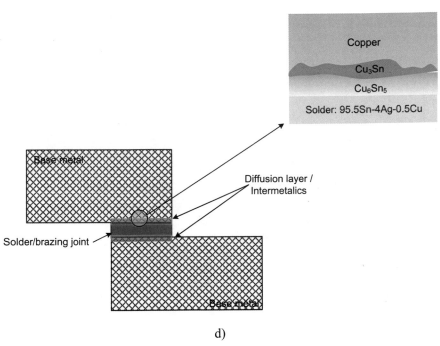

Copper

Cu$_3$Sn

Cu$_6$Sn$_5$

Solder: 95.5Sn-4Ag-0.5Cu

Diffusion layer /
Intermetalics

Solder/brazing joint

d)

Fig. 18.2 (continued)

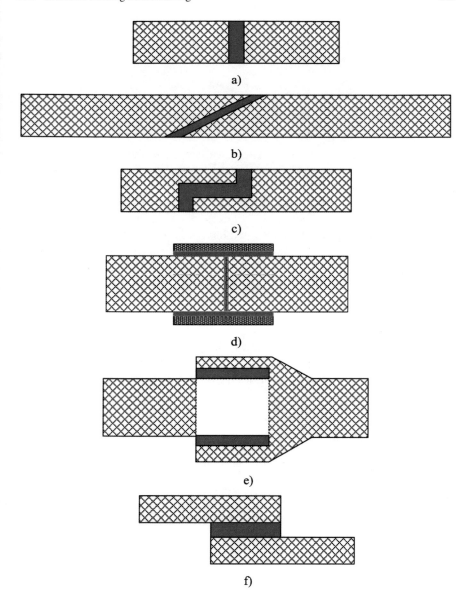

Fig. 18.3 Schematic of common brazing and soldering joints: **a, b** butt joint, **c, d** butt-lap and **e, f** lap joint configuration

filler metal at the interface results in high strength of joint. Both too narrow clearance and too wide clearances reduce sucking tendency of molten filler by capillary action. To ensure good and sound joint between the sheets, surfaces to be joined must be smooth, clean and free from impurities to ensure proper capillary action.

Butt joint can also be developed between the components with some edge preparation primarily to increase the contact area between the members to be joined so that reasonably good strength can be achieved (Fig. 18.3a–d).

18.3   Comparison of Brazing and Soldering

Brazing and soldering have common features like (a) solid/liquid joining nature, (b) no fusion of faying surface of base metal, (c) capillary action for spreading and distribution of molten filler, (d) joint is weaker than base metal and (e) both needs less heat than fusion welding. Still both brazing and soldering processes are different from each other in respect of various factors such as melting point of filler and strength of joint, ability to withstand at high temperature, heating source for developing joint and their applications.

18.3.1   Melting Point of Filler

Soldering uses the filler metals having low melting point (183–275 °C generally lower than 450 °C) called solder (alloy of lead and tin) while brazing uses comparatively higher melting point (450–1200 °C) filler metals (alloys of Al, Cu and Ni).

18.3.2   Strength of Joint

Strength of solder/braze joint is limited by the strength of filler metal. In general, brazed joints offer greater strength than solder joints. Accordingly, brazed joints are used for somewhat higher loading conditions than solder joint.

18.3.3   Ability to Withstand Under High Temperature Conditions

In general, braze joints offer higher resistance to heat than soldered joint primarily due to difference in melting temperature of solder and braze metals. Since solders have lower melting temperature than brazing metal, therefore, solder joints are preferred mainly for low temperature applications.

18.4   Application

Soldering is mostly used for joining of electronic components which are not exposed to high temperature and loading conditions during the service. Brazing is commonly used for moderate loading and temperature condition e.g. joining of tubes, pipes, wires cable, and tipped tool.

Common filler metals with brazing temperatures and applications

Filler metal	Al–Si	Cu	Cu–P	Cu–Zn	Au–Ag	Ni–Cu
Brazing temperature (°C)	600	1120	850	925	950	1120
Parent metal	Al	Ni and Cu	Cu	Steel, cast iron, Ni	Stainless steel, Ni	Stainless steel, Ni

Common soldering fillers and their applications

Solders	Applications
Tin–Lead (Sn–Pb)	General purpose
Tin–Zinc (Sn–Zn)	Aluminum
Tin–Silver (Sn–Ag)	Electronics
Tin–Bismuth (Sn–Sb)	Electronics
Lead–Silver (Pb–Ag)	Strength at higher temperatures
Cadmium–Silver (Cd–Ag)	Strength at higher temperatures
Zinc–Aluminum (Zn–Al)	Aluminum; corrosion resistance

18.5   Source of Heat for Joining

Soldering can be carried out using heat from soldering iron (20–150 W), dip soldering and wave soldering. Brazing can be performed using gas flame torch, furnace heating, induction heating, and infrared heating methods.

18.6   Limitation of Brazing and Soldering

These processes have major limitation of poor strength and inability to withstand at high temperature besides possibility of colour mismatch with the parent metals and reduced corrosion resistance.

18.7 Role of Flux in Brazing

Fluxes react with impurities present on the surface of base metal or those formed during joining to form slag apart from reducing contamination of the joints from atmospheric gases (formation of oxides and nitrides due to interaction of the atmospheric gases with the molten metal). For performing above role effectively, fluxes should have low melting point and molten filler should have low viscosity. Fluxes applied over the surface of work piece for developing joint must be cleaned from the work surface after brazing/soldering as these are corrosive in nature and can adversely affect the mechanical performance and life of joint.

18.8 Braze Welding

Braze welding combines features of both brazing and welding. Like brazing, this process is also based on principle solid–liquid joining as it involves only heating of faying surfaces of the base metals and then application of brazing filler on the heated faying surfaces to fill the gap between the members to be joined (Fig. 18.4). However, filling and distribution of the metal in braze welding is not based on capillary action. Edge preparation of faying surfaces of members to be joined is needed. Edges can be prepared in form of suitable groove geometry like in welding. Brazing joints can be issued on heavy sections where moderate strength is needed.

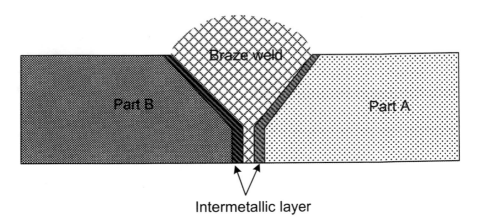

Fig. 18.4 Schematic of brazing welding process

18.9 Friction Stir Welding

The friction stir welding is comparatively a new solid-state joining process developed by The Welding Institute U.K. in 1991. This process is based on the simple principle of thermal softening of the metal of the parts to be joined followed by severe large plastic deformation to develop a weld joint. The thermal softening is facilitated by heat generation from two sources (a) friction between tool and base metal and (b) plastic deformation. The development of weld joints is facilitated by transport of metal from one (retreating) side to another (advancing) side followed by consolidation through forging action (Fig. 18.5). To ensure proper tool like and performance, tool material must be hard, tough, strong, heat and wear resistant. The typical solid-state joining feature of this process reduces undesirable effects of common fusion weld thermal cycle due to low heat input. This process is commonly applied for developing butt and lap joints. However, in the recent times, the friction stir-welding has been applied in many ways for producing other weld configuration like T joints and corner joints. Friction stir spot welding is one of the typical variants of friction stir welding used for producing lap joints. The strength of friction stir spot weld joints is found comparable or even better than resistance spot weld joints in lap weld configuration. Sequential steps involved in FSW are shown schematically in Fig. 18.6a–e. The selection of tool rotational speed, tool plunging rate, traverse speed, tool design (tool pin and shoulder diameter) are few important parameters of FSW process need to be controlled for developing sound weld joint.

Questions for Self-assessment

- Why brazing and soldering are called solid–liquid phase process?

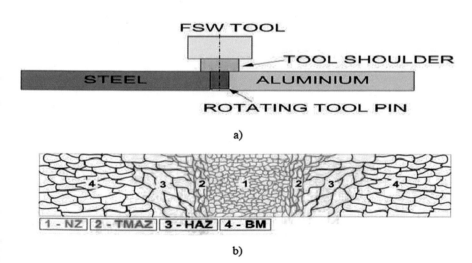

Fig. 18.5 Schematic of friction stir welding for **a** dissimilar metal joining and **b** zones of weld joints

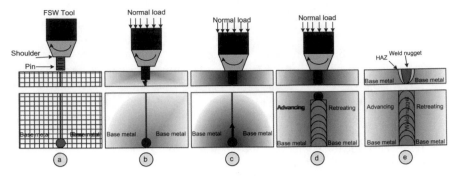

Fig. 18.6 Schematic showing step of friction stir welding showing heat generation and tool motions: **a** tool rotation at desired speed, **b** slow tool plunging at controlled speed, **c** completion of tool plunging and rotation to generate enough heat for thermal softening of base metal before tool traversing, **d** traversing speed at desired travel speed along the desired weld line and **e** developed FSW joint and its cross-section

- What are common situations where brazing and soldering processes are preferred over fusion welding processes?
- What are joint designs commonly used for brazing and soldering?
- Compare the brazing and soldering processes in respect of various technical points?
- Write application and limitations of brazing and soldering processes.
- What is role of fluxes in brazing and soldering?
- What is importance of clearance in brazing and soldering?
- What is importance of having clean faying surfaces for brazing and soldering?
- How does braze welding differ from brazing?
- Explain the mechanism of weld joint formation by friction stir welding.

Further Reading

American Society for Metals (1993) Metals handbook-welding, brazing and soldering, 10th edn., vol. 6. American Society for Metals, USA

Parmar RS, Welding process and technology. Khanna Publisher, New Delhi

Little R (2001) Welding and welding technology, 1st edn. McGraw Hill

Hensley S (ed) (2008) Friction stir welding—it is not just for aluminum. Modern Machine Shop

American Welding Society (1987) Welding handbook, 8th edn., vols. 1 & 2. American Welding Society, USA

Dwivedi DK (2018) Surface engineering. Springer, New Delhi

Dwivedi DK (2013) Production and properties of cast Al-Si alloys. New Age International, New Delhi

Part VI
Heat Flow in Welding

Chapter 19
Heat Flow and Performance of Weld Joints

Joining of metals can be realized with or without external heat application. Moreover, heat can be applied using external source (radiation, flame, arc, plasma, molten metal) or generated inherently by friction, and/or deformation during joining to facilitate fusion, plastic, deformation and diffusion as per joining process. This heat input being used for joint can be more (4–5 kJ/mm) or less (0.2–5 kJ/mm) as per the (a) power density of heat source or heat flux (10^2 to 10^{10} W/mm^2) applied, (b) fundamental approach being used for joining of metals (fusion, plastic deformation, diffusion or hybrid) and c) dimensional (thickness), mechanical (thermal softening behaviour, yield strength, ductility) and thermal (conductivity, specific heat, thermal expansion coefficient) of parent metal to be joined. Lesser the net heat input, lower will be the harmful effects of heat on for the joint performance provided a sound joint is developed.

Net heat input (kJ/mm) used for joining not only affects the weld metal/nugget microstructure, mechanical and corrosion properties but also determines the extent of changes experienced by the underlying parent metal in vicinity of weld metal/nugget commonly known as heat affected zone (HAZ). The properties of weld metal/nugget and modification in properties of parent metal are primarily dictated by the thermal cycle (peak temperature, time of high-temperature retention above certain critical value and subsequently cooling rate) experienced by these regions during the joining. The thermal cycle imposed during the joining can cause many metallurgical changes like recovery, recrystallization, grain refinement/growth, phase transformation, precipitation/dissolution of hardening constitutes as per the parent metal being joined which can in turn lead to hardening/softening of weld metal/nugget and heat affected zone. Therefore, thermal cycle of a location indicating variation in temperature as function of times gain significant importance as it predominantly dictates the mechanical and corrosion performance of the joint. Further, the localized differential heating and cooling cycle imposed during joining of metals in vicinity of the joint cause local expansion and contraction which in turn results residual stress and tendency of distortion of the joints. The typical residual stresses developed during

© The Author(s), under exclusive license to Springer Nature Singapore Pte Ltd. 2022
D. K. Dwivedi, *Fundamentals of Metal Joining*,
https://doi.org/10.1007/978-981-16-4819-9_19

joining are considered to be harmful for mechanical properties and stress corrosion cracking resistance.

19.1   Importance

A typical fusion joining process involves the melting of the faying surfaces of base metal and the filler metal, if any, followed by solidification of the molten weld metal. Melting and solidification steps of the welding are associated with the flow of heat which are affected by rate of heat transfer in and around the weld metal. Metallurgical structure of metal in weld and region close to the weld metal is mainly determined by the extent of rise in temperature followed by cooling rate experienced by the metal at particular location of HAZ and weld. Further, differential heating and cooling experienced of different zones of weld joint cause not only metallurgical hetero-geneity but also non-uniform volumetric change which in turn produces the residual stresses. These residual stresses adversely affect the mechanical performance of the weld joint besides distortion in the welded components if proper care is not taken. Since heating, soaking and cooling cycle affect the metallurgical and mechanical properties, development of residual stresses and distortion of the weld joints; there-fore, it is pertinent to study various aspects of the heat flow in welding such as weld thermal cycle, cooling rate and solidification time, peak temperature, width of heat affected zone.

19.2   Weld Thermal Cycle

Weld thermal cycle shows the variation in the temperature of a particular location (in and around the weld) as a function of welding time during the welding. As the heat source (welding arc or flame) approaches close to the location of interest first temperature increases gradually during heating regime followed by decrease of temperature in cooling regime. A typical weld thermal cycle shows (Fig. 19.1) the rate of heating (slope of a b), peak temperature and time required for attaining the peak temperature, cooling rate (slope of b c). Since distance of the point of interest away from the weld centreline directly affects all the above parameters, namely heating and cooling rate, peak temperature of the weld thermal cycle, therefore each location/point offers a different and unique weld thermal cycle (Fig. 19.2). In general, an increase in distance of point of interest away from the weld centreline:

- decreases the peak temperature
- decreases the rate of heating and cooling
- increases time to attain peak temperature
- decreases rate of cooling.

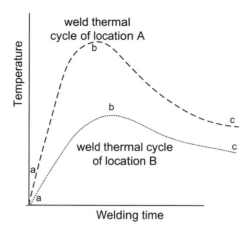

Fig. 19.1 Schematic of weld thermal cycle of two different locations away from the weld centreline

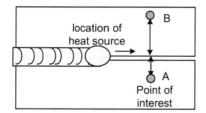

Fig. 19.2 Schematic of welding showing location of two points A & B

Additionally, the rate of cooling decreases with increase in time after passing of heat source from the point of interest.

19.2.1 Factors Affecting Welding Thermal Cycle

The weld thermal cycle varies with distance from the weld centreline, but it is also influenced by heating rate, amount of heat supplied for welding, weldment geometry, thickness of the base metal, thermal properties of base metal and initial plate temperature. The rate of heat input is primarily governed by the energy density of heat input source which to a great extent depends upon the welding process being used for development of weld joints besides the welding parameters. High energy density processes like plasma arc welding, electron beam welding and laser beam welding offer higher rate of heating, peak temperature and cooling rates than low energy density processes such as gas welding, shielded metal arc welding as shown in Fig. 19.3. Higher is the energy density of welding process, lower will be the heat input required for welding. Weld geometry parameters such as thickness of plates

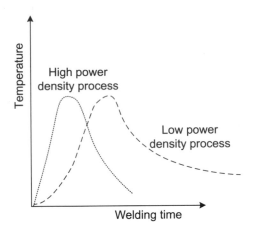

Fig. 19.3 Influence of energy density of heat source related with welding process on weld thermal cycle

being welded also affect the heating rate, soaking time and cooling rate for a given rate of heat input (welding parameters) due to change in heat transfer conditions. In general, an increase in thickness of plate increases the rate of heat transfer from the weld pool/heat affected zone to the base metal which in turn (a) decreases the high-temperature retention time of HAZ, (b) decreases the solidification time of the weld metal and (c) increases the cooling rate experienced by the HAZ and weld metal.

Thermal properties of metal like thermal conductivity and specific heat also have effect on weld thermal cycle similar to that of thickness of plates as they increase the rate of heat transfer from the weld metal and HAZ. Preheating of the plates reduces the rate of heating and cooling and increases the peak temperature and soaking period above certain temperature because preheating of the base metal to be welded reduces the rate of heat transfer away from the weld and HAZ.

Peak temperature near the weld fusion boundary decides the width of heat affected zone (HAZ). Heating and cooling rate affect the microstructure of weld metal and HAZ; therefore, weld thermal cycle of each point becomes of great interest from the mechanical properties point of view, especially in structure sensitive metals like high carbon steels.

19.3   Cooling Rate

The final microstructure of weld zone and HAZ is primarily determined by the cooling rate (CR) from the peak temperature attained due to weld thermal cycle experienced by the metal at a given location during the welding. Cooling rate above a particular temperature say 550 °C for plain carbon eutectoid steel is of great importance (and other hardenable steel) as cooling rate (CR) determines the final microstructure

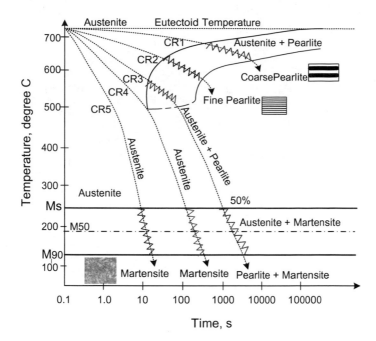

Fig. 19.4 Effect of cooling on structure of weld joints shown in form of CCT diagram

and mechanical properties of weld metal and HAZ. Since the microstructure of the hardenable steel has direct correlation with mechanical properties, therefore, structure sensitive mechanical properties are affected by the cooling rate experienced by the weld metal and heat affected zone. This is evident from the continuous cooling diagram of eutectoid steel as shown in Fig. 19.4. In the diagram, letter A, F, P, B, M indicates regions of austenite, ferrite, pearlite, bainite and martensite respectively. It can be observed that depending upon the cooling rate, austenite can transform in to pearlite, bainite, and martensite and accordingly mechanical properties are affected. Low cooling rate (as shown by line p) results in soft low strength pearlitic structure, while high cooling rates (as exhibited by line CR4 and CR5) produce hard and brittle martensitic structure.

Weld thermal cycle indicates both heating and cooling rate. Cooling rate varies as a function of time, location of point of interest and temperature (at any moment on commencement of the cooling) during cooling regime of weld thermal cycle. The cooling rate calculation for HAZ of hardenable steel weld joint is mostly made at 550 °C (corresponding to nose temperature of CCT) as cooling rate at this temperature predominantly decides the end microstructure and mechanical properties of the HAZ and weld joint. During welding, two welding parameters dictate the cooling rate (a) net heat input during the welding and (b) initial plate temperature besides the thermal and dimensional properties of the metals being welded. In general, increase in heat input and initial plate temperature decrease the cooling rate during welding of a given

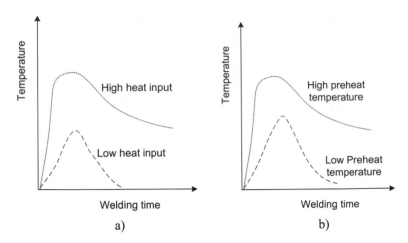

Fig. 19.5 Schematic showing effect of **a** heat input and **b** preheat temperature on weld thermal cycle

metal having specific thickness and thermal properties. An increase in both heat input and initial plate temperature raises the temperature of base metal around the weld which in turn decreases the rate of a transfer away from the weld zone primarily due to reduction in temperature difference between the weld zone and surrounding base metal (Fig. 19.5). Reduction in heat transfer rate from the weld metal to the base metal with increase of heat input and initial plate temperature in turn decreases cooling rate. In view of the above, major practical application of cooling rate equation is to determine the preheat requirement for plates to be welded so as to avoid critical cooling rate in the weld and HAZ.

Net heat input (H_{net}) during the welding is obtained using the following relationship:

$$H_{net} = f \cdot VI/S$$

where V is arc voltage (V), I welding current (A) and S welding speed mm/s and f is the fraction of heat generated and transferred to the plate.

Example Calculate the net heat input used during welding of plates if welding of steel plate is given below:

- Welding current: 150 A
- Arc voltage: 30 V
- Welding speed: 2.0 mm/s
- 80% of heat generated by the arc is used for welding.

Solution

$$\text{Net heat input: } H_{net} = f \cdot VI/S$$

$$= 0.8 \times 30 \times 150/2.0$$
$$= 1800 \text{ J/mm}$$
$$= 1.8 \text{ kJ/mm}$$

Following section describes the calculation of cooling rate, peak temperature and width of heat affected zone and solidification rate.

19.4 Calculations of Cooling Rate

Thickness of the plate to be welded directly affects the cross sectional area available for the heat flow from the weld to the base metal which in turn governs cooling rate of a specific location. Accordingly, two different empirical equations are used for calculating the cooling rate in HAZ for (a) thin plates and (b) thick plates, depending upon the thickness of plate and welding conditions. There is no clear demarcating thickness limit to define a plate thick or thin. However, two methods have been proposed to take decision whether to use thick or thin plate equation for calculating the cooling rates, and these are based on two aspects.

(1) number of passes required for completing the weld and
(2) relative plate thickness.

According to the first method, if number of passes required for the welding of two plates is less than 6, then those plates considered as thin plate else thick plate for selection of suitable equation to calculate cooling rate (Fig. 19.6). Since this method is not very clear as number of passes required for completing the weld can vary with diameter of electrode and groove geometry being used for welding, therefore a more logical second method based on relative plate thickness criterion is commonly used.

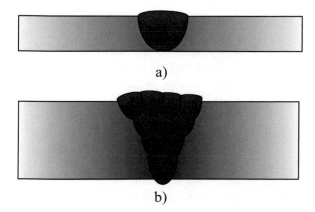

a)

b)

Fig. 19.6 Schematic showing different number of passes needed for **a** thin plate and **b** thick plates

The relative plate thickness criteria is more logical as it considers all the relevant factors which can affect the cooling rate such as thickness of the plate (h), heat input (H_{net}), initial plate temperature (T_o), temperature of interest at which cooling rate is desired (T_i) and physical properties of the base metal like specific heat (C) and density (ρ). Relative plate thickness (τ) can be calculated using following equation: $h\{\rho C(T_i - T_o)/H_{net}\}^{1/2}$.

Thin plate cooling rate equation is used when relative plate thickness (τ) is less than 0.6 and thick plate cooling rate equation is used when $\tau > 0.9$. If value of τ is in range of 0.6–0.9, then 0.75 is used as a limit value to decide the cooling rate equation to be used. Conversely, if τ is less than 0.75 than thin plate equation else thick plate equation.

Cooling rate (R) equation for thin plates:

$$\{2\pi k \rho C(h/H_{net})(T_i - T_o)^3\} \,^{\circ}C/s \tag{19.1}$$

Cooling rate (R) equation for thick plates:

$$\{2\pi k(T_i - T_o)^2\}/H_{net} \,^{\circ}C/s \tag{19.2}$$

where h is the plate thickness (mm), k is thermal conductivity, ρ is the density (g/cm^3), C is specific heat (kCal/$^{\circ}$C g), T_i is the temperature of interest ($^{\circ}$C) and T_o is the initial plate temperature ($^{\circ}$C).

Cooling rate equations can be used to (a) calculate the critical cooling rate (CCR) under a given set of welding conditions and (b) determine the preheat temperature requirement for the plates to be welded in order to avoid the CCR.

19.5 Critical Cooling Rate (CCR) Under Welding Conditions

To determine the critical cooling rate for a steel under welding conditions, bead-on-plate welds are made with varying heat inputs. On the basis of thickness of the plate (say 5 mm) to be welded, suitable electrode diameter is chosen first and then accordingly welding current and arc voltage are selected (20 V, 200 A, T_o = 30 °C) for bead-on-plate (BOP) welding. The number of BOP welds is deposited using increasing welding speed (8, 9, 10, 11, 12……mm/s). Once the BOP welds are completed at different welding speeds, transverse section of the weld is cut to measure the hardness (Fig. 19.7). Thereafter, hardness vs. welding speed plot is made to identify the welding speed above which abrupt increase in hardness of the weld and HAZ takes place. This welding speed is identified as critical welding speed (say 10 mm/min in this case) above which cooling rate of the weld and HAZ becomes greater than critical cooling rate.

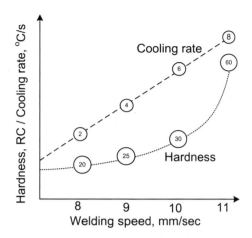

Fig. 19.7 Schematic showing effect of welding speed on cooling rate and hardness of BOP of steel

This abrupt increase in hardness of the weld and HAZ is attributed to martensitic transformation during welding as cooling rate becomes greater than critical cooling rate owing to the reduction in heat input (H_{net}) with increase of the welding speed. Using welding conditions corresponding to this critical welding speed for a given steel plate, critical cooling rate can be calculate using appropriate cooling rate equation. Assuming thermal conductivity (k) is 0.028 J/mm.s.°C and volume specific heat of steel (pC) is 0.0044 J/mm^3.°C.

Corresponding $H_{net} = f \times VI/S = 0.9 \times 20 \times 200/10 = 360$ J/mm or 0.36 kJ/mm.
where f is fraction of arc heat used for melting say 90%

Calculate relative plate thickness (RPT) parameter for these conditions: $h[(T_i - T_0)C/H_{net}]^{1/2}$: 0.31.

RPT suggests use of thin plate equation for calculating the cooling rate: $2\pi k\rho c(h/Q)(t_c - t_o)^3$.

Cooling rate (R): 5.8 °C/s, and it will be safer to consider CCR: 6 °C/s.

Similarly, these equations can also be used for calculating the cooling rate or identifying the preheat temperature to avoid CCR for a particular location under a given set of welding conditions if is known or identified.

19.6 Peak Temperature and Heat Affected Zone

The weld thermal cycle of a particular location exhibits peak temperature and cooling rate as function of time apart from other parameters.

The peak temperature study in and around the weld is important because peak temperature distribution around the weld centreline determines (a) shape of the weld pool, (b) size of heat affected zone and (c) metallurgical transformation of weld and HAZ and so mechanical properties.

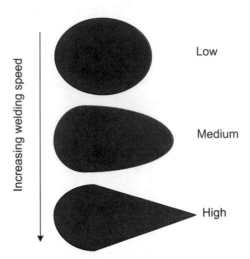

Fig. 19.8 Effect of welding parameters on weld pool profile as dictated by peak temperature

Variation in heat input and initial plate temperature affects the peak temperature distribution around the weld along the centreline during the welding. An increase in heat input by increasing the welding current (for a given welding speed) in general increases the peak temperature of a particular location and makes the temperature distribution more uniform around the welding arc (almost circular or oval shape weld pool). Increase in welding speed however makes the weld pool (and peak temperature distribution) of tear drop shape (Fig. 19.8).

Cooling rate from the peak temperature determines final microstructure of the weld and heat affected zone. Therefore, peak temperature in the region close to the fusion boundary becomes of great engineering importance as metallurgical transformations (hence mechanical properties) at a location near fusion boundary are influenced by peak temperature (Fig. 19.9). Peak temperature at any point near the fusion boundary for single pass full penetration weld can be calculated using the following equation

$$1/\left(t_p - t_o\right) = (4.13\rho chY/H_{net}) + (1/(t_m - t_o))$$ (19.3)

where t_p is peak temperature in °C, t_o is initial temperature in °C, t_m is melting temperature in °C, H_{net} is net heat input, J/mm, h is plate thickness in mm, Y is width of HAZ in mm and ρc is volumetric specific heat (J/mm^3 °C).

This Eq. (19.3) can be used for (a) calculating peak temperature at a point away from the fusion boundary, (b) estimating width of heat affected zone and (c) studying the effect on initial plate temperature/preheating and heat input on width of the HAZ. Careful observation of Eq. (19.3) reveals that an increase in initial plate temperature and net heat input will increase the peak temperature at y-distance from the fusion boundary and so width of heat affected zone considering the mechanical and metallurgical changes above a particular temperature, e.g. 727 °C, for carbon steel.

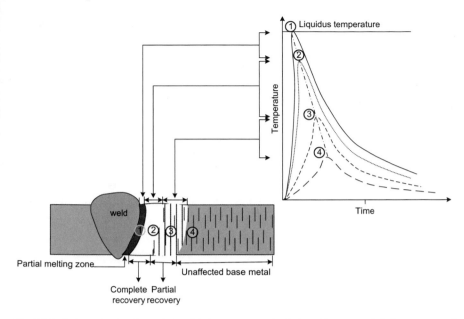

Fig. 19.9 Schematic showing relationship between a typical phase diagram and different zones of weld joints

To calculate the width of HAZ, it is necessary to mention the temperature of interest/critical temperature above which microstructure and mechanical properties of a metal will be affected by the application of welding heat (Fig. 19.10). For example, the plain carbon steels are subjected to metallurgical transformation above

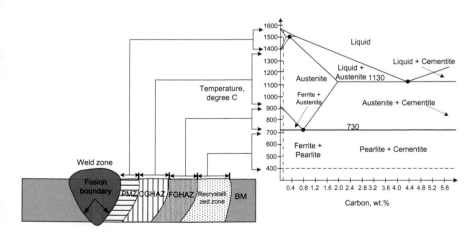

Fig. 19.10 Schematic showing relationship between a typical Fe–C phase diagram and different zones of carbon steel weld joints

727 °C, i.e. lower critical temperature; hence, temperature of interest/critical temperature for calculating of HAZ width for plain carbon steels becomes 727 °C. Similarly, a steel tempered at 300 °C after quenching treatment whenever heated to a temperature above 300 °C, it is over-tempered so the structure and properties are affected; hence, for this quenched and tempered steel, tempering temperature (300 °C) will be taken the critical temperature.

A single pass full penetration weld is made on steel plates having $\rho c = 0.0044$ J/mm^3 °C, $h = 5$ mm, $t_o = 25$ °C, $t_m = 1510$ °C, $Q = 720$ J/mm. Calculate the peak temperatures at 3.0 mm and 1.5 mm and 0 mm distance from the fusion boundary.

On replacing of values of different factors, in $1/(t_p - t_o) = (4.13\rho chY/H_{net}) + (1/(t_m - t_o))$, the peak temperature (t_p) at distance (Y) 3 mm, 1.5 mm and 0 mm is obtained.

19.7   Solidification Rate

The solidification of weld metal takes place in three stages, (a) reduction in temperature of liquid metal, (b) liquid to solid state transformation and (c) finally reduction in temperature of solid metal up to room temperature. The time required for solidification of weld metal depends up on the cooling rate. Solidification time is the time interval between start to end of solidification. Solidification time is also of great practical importance as it affects the structure, properties and response to the heat treatment of weld metal. Solidification time can be calculated using following equation:

$$\text{Solidification time of the weld } (S_t) = LQ/2\pi k\rho c(t_m - t_o)^2 \text{ in s} \qquad (19.5)$$

where L is latent heat of fusion (for steel 2 J/mm^3).

Above equation indicates that solidification time is function of net heat input, initial plate temperature and thermal properties of metal being welded such as latent heat of fusion (L), thermal conductivity (k), volumetric specific heat (ρC) and melting point (t_m).

Long solidification time allows each phase to grow to a large extent which in turn results in coarse-grained structure of weld metal. An increase in net heat input (with increase in welding current/arc voltage or reduction in welding speed) increases the solidification time. An increase in solidification time coarsens the grain structure which in turn deteriorates the mechanical properties. Non-uniformity in solidification rates in different regions of molten weld pool also brings variation in grain structure and so mechanical properties. Generally, centreline of the weld joint shows finer grain structure and better mechanical properties than those at fusion boundary primarily because of difference in solidification times. Micrographs of Al alloy weld indicate the coarser structure near the fusion boundary than the weld centre (Fig. 19.11).

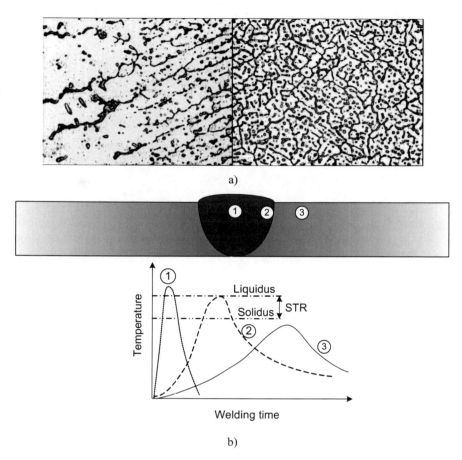

Fig. 19.11 Variation in microstructure of weld of Al-Si alloys of **a** fusion boundary and **b** weld centre owing to difference in cooling rate (200X)

Example A single pass full penetration weld pass is made using net heat input at the rate of 500 J/mm on steel having $\rho c = 0.0044$ J/mm^3 °C, $t = 5$ mm, $t_o = 25$ °C, $t_m = 1540$ °C and thermal conductivity k $= 0.025$ J/mm s °C and latent heat of fusion 2.4 J/mm^3. Determine the solidification time.

Solution

Solidification time: $LQ/2\pi k\rho c(t_m - t_o)^2$ in s.
Solidification time: $2.4 \times 500/(2\pi \times 0.025 \times 0.0044\,(1540 - 25)^2$ in s.
Solidification time: $1200/1585.54$.
Solidification time: 0.75 s.

Residual stresses in weld joints

This section describes the mechanisms of development of residual stress in fusion weld joints. Further, the influence of residual stress on performance of weld joints has been presented besides the methods of controlling the residual stresses.

19.8   Residual Stresses

Residual stresses are locked-in stresses present in the metallic components (fabricated under specific conditions) even when there is no external load. These stresses primarily develop due to non-uniform volumetric change in metallic component during production irrespective of manufacturing processes such as heat treatment, machining, mechanical deformation, casting, welding, coating, etc. However, the maximum value of residual stresses does not exceed the elastic limit of the metal because stresses higher than elastic limit leads to plastic deformation, and thus, residual stresses greater than elastic limit are accommodated/relieved in the form of distortion of components. Residual stresses can be tensile or compressive depending up on the location and type of non-uniform volumetric change taking place due to differential heating and cooling; e.g., welding and heat treatment or localized deformation are achieved like in contour rolling, machining and shot peening, etc.

19.9   Residual Stresses in Welding

Residual stresses in welded joints primarily develop due to differential weld thermal cycle (heating, peak temperature and cooling at the any moment during welding) experienced by the weld metal and region closed to fusion boundary, i.e. heat affected zone (Fig. 19.12). Type and magnitude of the residual stresses vary continuously

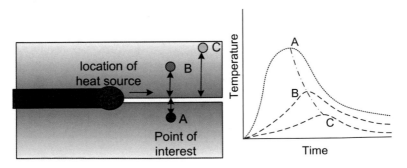

Fig. 19.12 Weld thermal cycle of **a** locations A, B, C and **b** temperature versus time relation of A, B and C

during different stages of welding, i.e. heating and cooling. During heating, primarily compressive residual stress is developed in the region of base metal which is being heated for melting due to thermal expansion, and the same (thermal expansion) is restricted by the low-temperature surrounding base metal. After attaining a peak value, compressive residual stress gradually decreases due to softening of metal being heated. Compressive residual stress near the faying surfaces eventually reduces to zero as soon as melting starts, and a reverse trend is observed during cooling stage of the welding. Tensile residual stresses develop as metal starts shrinking during cooling. However, stress develop only if shrinkage is not allowed either due to metallic continuity or constraint from job clamping. The magnitude of tensile residual stresses keeps on increasing until room temperature is attained. In general, greater is degree of constraint and elastic limit of metal higher will be the value of residual stresses.

19.10 Mechanisms of Residual Stress Development

The residual stresses in the weld joints develop mainly due to typical nature of the welding process, i.e. localized heating and cooling leading to differential volumetric expansion and contraction of metal around the weld zone. The differential volumetric change occurs at macroscopic as well as microscopic level. Macroscopic volumetric changes occurring during welding contribute to a significant proportion of residual stress development and are caused by (a) varying expansion and contraction and (b) different cooling rate experienced by top and bottom surfaces of weld and HAZ. The contribution of microscopic volumetric change on residual stresses is minor. Microscopic volumetric changes mainly occur due to metallurgical transformation (austenite to martensitic transformation) during cooling. Further, it is important to note that whenever residual stresses develop beyond the yield point limit, the plastic deformation sets in the component. The development of a particular type (tensile/compressive) of residual stress in the affected area component results in setting of balancing/neutralizing residual stress of opposite kind in neighbouring area.

Thus, if the residual stress magnitude is below the elastic limit, then a stress system having both tensile and compressive stresses for equilibrium is developed.

19.10.1 Differential Heating and Cooling

Residual stresses develop due to varying heating and cooling rate in different zones near the weld as function of time are called thermal stresses. Different temperature conditions lead to varying strength and volumetric changes in base metal during welding. The variation in temperature and residual stresses owing to movement of heat source along the centreline of weldment is shown schematically in Fig. 19.13a–c. As heat source comes close to the point of interest, its temperature increases.

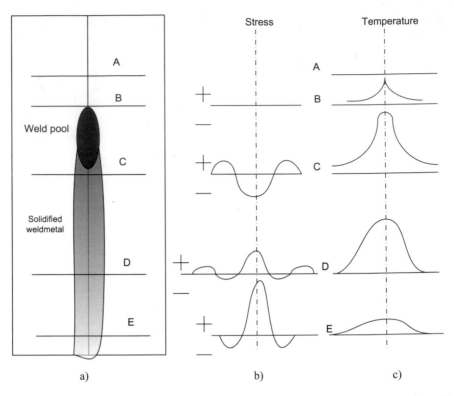

Fig. 19.13 Schematic diagram showing **a** plate being welded, **b** stress variation across the weld centreline at different locations and **c** temperature of different locations

Increase in temperature decreases the yield strength of material and simultaneously tends to cause thermal expansion of the metal being heated. However, surrounding low-temperature base metal restricts any thermal expansion which in turn develops compressive strain in the metal during heating. Compressive strain initially increases nonlinearly with increase in temperature due to variation in yield strength and expansion coefficient of metal with temperature rise. Further, increase in temperature softens the metal, and therefore, compressive strain starts reducing gradually, and eventually, it is vanished. As the heat source crosses the point of interest and starts moving away from the point of interest, temperature begins to decrease gradually. Reduction in temperature causes the shrinkage of hot metal in base metal and HAZ. Initially, at high temperature, contraction occurs without much resistance due to low yield strength of metal (at elevated temperature), but subsequently shrinkage of metal is resisted as metal gains strength owing to reduction in temperature during cooling regime of weld thermal cycle (Fig. 19.14). Therefore, further contraction in shrinking base and weld metal does not occur with reduction in temperature. This behaviour of contraction leaves the metal in strained condition which means that metal which should have contracted was not allowed to do so. This situation in turn leads to the

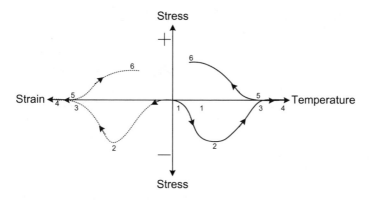

Fig. 19.14   Effect of temperature on variation in stress and strain during welding

development of the tensile residual stresses (if the contraction is prevented). The magnitude of residual stresses can be calculated from the product of locked-in strain and modulus of elasticity of metal being welded. The residual stress along the weld is generally tensile in nature while balancing compressive residual stress is developed adjacent to the weld/heat affected zone after cooling to the room temperature as evident from the Fig. 19.13b.

19.10.2   Differential Cooling Rate in Different Zone

During the welding, higher cooling rate is experienced by the top and bottom surfaces of weld joint than the core/middle portion of weld and HAZ (Fig. 19.15). The difference in cooling rates causes differential expansion and contraction through the thickness (direction) of the plate being welded. Contraction of metal near the surface starts even when metal in the core portion is still hot. This differential strain situation leads to the development of compressive residual stresses at the surface and tensile residual stress in the core.

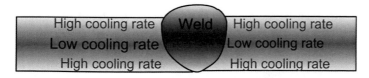

Fig. 19.15   Schematic showing different cooling rates at surface and core regions of the weld

19.10.3 Metallurgical Transformation

During the welding, heat affected zone of the hardenable steels and their weld zone invariably experience transformation of austenite into other phases phase mixture like pearlite, bainite or martensite. All these transformations occur with increase in specific volume at microscopic level. Increase in specific volume occurring due to metallurgical transformation (from austenite to pearlite and bainite) at high temperature is easily accommodated owing to low yield strength and high ductility of these phases and phase mixtures (above 550 °C). Therefore, such metallurgical transformations do not contribute much towards the development of residual stresses. Transformation of austenite into martensite takes place at very low temperature (room temperature or sub-zero temperature) with significant increase in specific volume which is not accommodated due to high yield strength and low ductility of metals at low temperature. Hence, austenite to martensite transformation contributes significantly towards development of residual stresses. Depending upon the location of the austenite to martensitic transformation, residual stresses may be tensile or compressive. For example, shallow hardening causes such transformation from austenite to martensite near the surface layers only and develops compressive residual stresses at the surface and balancing tensile stress in core, while through section hardening develops reverse trend of residual stresses, i.e. tensile residual stresses at the surface and compressive stress in the core.

19.11 Effect of Residual Stresses

The residual stresses either tensile or compressive type predominantly affect the soundness, dimensional stability and mechanical performance of the weld joints, since magnitude of residual stresses increases gradually to peak value until weld joint is cooled down to the room temperature. Therefore, effects of residual stresses are mostly observed either near the last stage of welding or after some time of welding in the form of cracks (hot cracking, lamellar tearing, cold cracking), distortion and reduction in mechanical performance of the weld joint (Fig. 19.16).

The presence of residual stresses in the weld joints can encourage or discourage failures due to external loading as their effect is additive in nature. Conversely, compressive residual stress decreases failure tendency under external tensile stresses primarily due to reduction in net tensile stresses acting on the component (net stress on the component: external stresses ± residual stresses). Residual stress of the same type as that of external one increases the failure tendency, while opposite type of stresses (residual stress and externally applied stress) decreases the failure tendency, since failure of more than 90% of mechanical component occurs under tensile stresses by crack nucleation and propagation. Therefore, the presence of tensile residual stresses in combination with externally applied tensile stress adversely affects the performance in respect of tensile load carrying capacity (tensile strength and yield

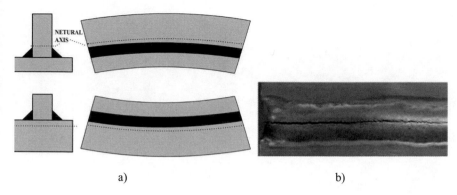

Fig. 19.16 Typical problems associated with residual stress **a** distortion and **b** solidification cracking

strength) and fatigue resistance as evident from many cases that relieving residual stresses improves the mechanical performance of weld joints (Fig. 19.17a–c), while compressive residual stress under similar loading conditions reduces the net stress and so discourages the failure tendency. Hence, compressive residual stresses are intentionally induced to enhance the tensile and fatigue performance of the mechanical components, whereas efforts are made to reduce tensile residual stresses using various approaches such as post weld heat treatment, shot peening, spot heating.

In addition to the cracking of the weld joint under normal ambient conditions, failure of weld joints exposed in corrosion environment is also accelerated in presence of tensile residual stresses by a phenomenon called stress corrosion cracking.

The presence of tensile residual stresses in weld joints causes cracking problems which in turn adversely affect their load carrying capacity. The system of residual stress (balancing various opposite type of residual stresses) is usually destabilized during machining which may lead to distortion of the weld joints. Therefore, residual stresses must be relieved from the weld joint before undertaking any machining operation or removing fixtures.

19.12 Controlling the Residual Stresses

The critical applications frequently need relieving residual stresses of weld joints by thermal or mechanical methods. Relieving of residual stresses is primarily based on releasing the locked-in strain by developing suitable conditions so as to facilitate plastic flow to relieve stresses.

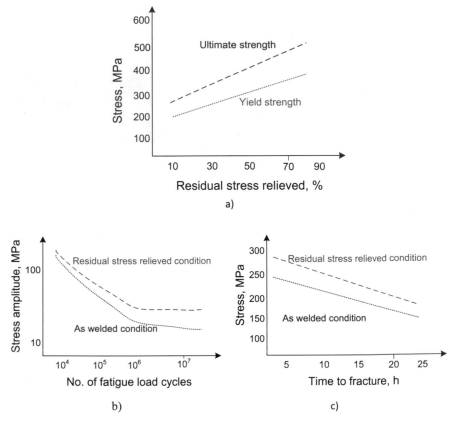

Fig. 19.17 Effect of relieving tensile residual stresses improves the mechanical performance: **a** tensile properties, **b** fatigue strength and **c** cold cracking of high carbon steel weld joints

19.12.1 Thermal Methods

These are based on the fact that the yield strength and hardness of the metals decrease with increase of temperature. The softening of metal which in turn facilitates the release of locked-in strain thus relieves residual stresses. Reduction in residual stresses depends on "how far reduction in yield strength and hardness takes place with increase of temperature". Greater is the softening, more will be the relieving of residual stresses. Therefore, in general, higher is the temperature of stress relieving treatment of the weld joint, greater will be reduction in residual stresses (Fig. 19.18). The stress relieving temperature and soaking duration are two important parameters determining the extent of residual stress relieved. In general, increase in both temperature and time increases the percentage of residual stress relieving from the weld joints.

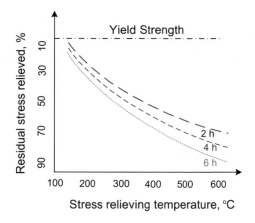

Fig. 19.18 Effect of stress relieving temperature and time on the reduction in residual stresses

19.12.2   *Mechanical Methods*

These methods are based on the principle of relieving residual stresses by reliving the locked-in strain using (a) overloading (applying external load beyond yield strength level to cause controlled plastic deformation) and (b) mechanical vibrations so as to release locked-in strain (Fig. 19.19). External load is applied in an area which is expected to have peak residual stresses. The overloading approach of residual stress relieving can be understood from three stress states of the weld joints (1) residual stress in as welded condition, (2) stress state due to overloading and (3) relieved residual stress state.

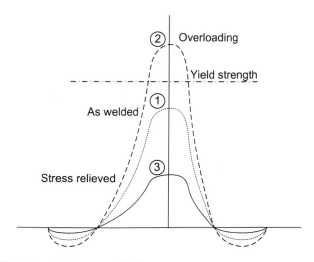

Fig. 19.19 Effect of overloading on residual stresses

Mechanical Vibration

Application of low-frequency and high amplitude vibrations decreases of peak residual stress in weld joints of metals. The vibrations of a frequency close to natural frequency of welded joint are applied on the component to be stress relieved. The vibratory stress can be applied in whole of the components or in localized manner using suitable pulsators. The development of resonance state of mechanical vibrations on the welded joints helps to release the locked-in strains so as to relieve residual stresses (Fig. 19.20a). However, the details of methods and extent relieving realized through this method is not well documented. Resonant frequency vibrations is applied for a predetermined time (10 min to 60 min or more) as per the weight of the piece and desired purpose (Fig. 19.20b).

Questions for Self-assessment

1.   What is the need to study about heat flow in welding?
2.   How does heat flow in welding affect the performance of weld joints?

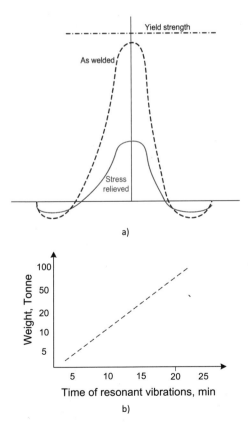

a)

b)

Fig. 19.20 Mechanical vibrations for relieving residual: **a** residual stresses state before and after stress relieving and **b** time for application of mechanical vibration for different weights of objects

3. What is weld thermal cycle and how does it affect mechanical properties of weld joints?

4. What information can be obtained from the weld thermal cycle of a location?

5. Describe effect of various factors related with welding on weld thermal cycle of a location?

6. What is importance of the cooling rate during welding of hardenable steel?

7. How can cooling rate in HAZ during welding of plates of different thicknesses be obtained?

8. Describe methodology to establish the critical cooling rate for steel under a given set of welding conditions.

9. How does welding speed affect cooling rate and temperature gradient in HAZ?

10. What are the welding parameters affecting the cooling rate during welding?

11. How can cooling rate equations be used for development of sound weld joints of steel?

12. What is importance of the peak temperature of HAZ during welding of hardenable steel?

13. How can peak temperature of HAZ during welding be obtained?

14. Explain the effect of welding parameters on heat distribution and pool shape.

15. How can peak temperature equation be used for development of sound weld joints of steel?

16. Describe factors affecting the width of HAZ?

17. What is solidification time and how does it affect soundness and performance of the weld joint?

18. How the solidification time for a weld can be obtained?

19. Describe the factors affecting the solidification time?

20. What is the relationship between solidification time and weld structure?

21. Define residual stress in weld joints?

22. Explain the mechanism of residual stress development in weld joints?

23. How do residual stresses affect the performance of weld joints?

24. Describe methods used to control residual stress in weld joints?

25. What are factors affecting the development residual stress in weld joint?

26. How do welding process and related parameters affect the residual stress development in weld joints?

27. What is the effect of weld joint design on residual stress?

28. How do preheating and post heating influence the residual stresses in weld joints?

29. How does residual stress in multi-pass welds differ from that is a single pass weld?

Further Reading

Kou S (2003) Welding metallurgy, 2nd edn. Willey, USA
Lancaster JF (2009) Metallurgy of welding, 6th edn. Abington Publishing, England
Parmar RS (2002) Welding engineering & technology, 2nd edn. Khanna Publisher, New Delhi
Little R (2001) Welding and welding technology, 1st edn. McGraw Hill
Nadkarni SV (2010) Modern arc welding technology. Ador Welding Limited, New Delhi
American Welding Society (2017) Welding handbook, 8th edn., vols. 1 & 2. American Welding
    Society, USA
Dwivedi DK (2018) Surface engineering. Springer, New Delhi
Dwivedi DK (2013) Production and properties of cast Al-Si alloys. New Age International, New
    Delhi

Part VII
Welding Metallurgy

Chapter 20
Welding Metallurgy

Metallurgical aspect of Metal Joining Processes

20.1 Introduction

The best possible way of joining metals would be the development of a joint just by putting parts to be joined together in position without doing anything further, conversely, developing a joint without applying heat, force, chemical, etc. For example, a weak bond developed between highly cleaned and finished surfaces of slip gauges. However, such slip gauge type of bonding is not found to be strong enough to sustain high real-life service load as expected from metallic joints.

Therefore, welding engineers and technologists are expected to "Design and Develop" joint as per the intended service conditions for the desired life. These joints can be developed using approaches such as adhesive joining, brazing and soldering, braze welding, solid state joining, resistance welding and fusion welding. These approaches are arranged in ascending order of heat input required for developing joints. The heat application during joining results in many undesirable metallurgical, chemical and mechanical changes in the patent metals leading to reduction in load carrying capacity, wear resistance and service life expectancy. Greater is heat input for making a joint, more will be the said issues. Therefore, efforts are always made to reduce the heat input for developing joints. The applied heat and/or force affects the metallurgical characteristics such as phases and grain structure, work hardening, thermal softening, recovery and recrystallization. The changes in metallurgical characteristics in base metals, joints and their interfaces in turn affect the mechanical properties of joints. Therefore, it becomes pertinent and important to understand the metallurgical aspects related to metal joining processes.

© The Author(s), under exclusive license to Springer Nature Singapore Pte Ltd. 2022   261
D. K. Dwivedi, *Fundamentals of Metal Joining*,
https://doi.org/10.1007/978-981-16-4819-9_20

20.2 Metallurgy and Metal Joining Processes

20.2.1 Adhesive Joining

Adhesive joining primarily involves chemical interactions with substrate surfaces, and there is no major application of heat (around 100 or 120 °C for curing) and pressure (2–4 MPa) which can alter mechanical and metallurgical characteristics of parent metal (Fig. 20.1). Therefore, parent metal properties are not compromised during adhesive joining. The strength of adhesive bonds is determined by two factors (a) strength of base metal-adhesive bond interface and (b) cohesive strength of adhesive bond itself after curing.

Thickness of adhesive layer is crucial in adhesive joining. An increase in adhesive layer first increases the adhesive bond strength, and then after reaching maxima it decreases with further increase in thickness of adhesive layer (Fig. 20.2). The thickness of adhesive at the interface of adhesive joint determines the relative importance of above two factors, i.e. adhesion at base metal-adhesive interface and cohesive strength of bond. In general, increase in thickness of adhesive layer at interface promotes the tendency of cohesive fracture of adhesive itself; thus strength of adhesive determines the joint strength. While strength of the adhesive joint with an optimum thin adhesive layer depends on weaker one among the two (adhesion at base metal-adhesive interface, cohesive strength of bond). However, adhesive joints

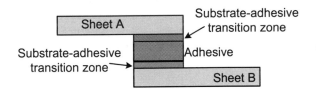

Fig. 20.1 Schematic adhesive joining process for dissimilar metals

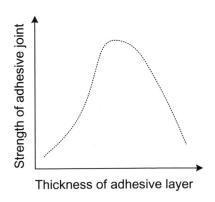

Fig. 20.2 Schematic showing effect of thickness of adhesive on strength of the joint

are not very strong, and these degrade in hostile service conditions like little high temperature (>50 °C), water, sunlight, etc. Adhesive joint still found useful for many applications under dry and non-load carrying connection.

20.2.2 Brazing and Soldering

The brazing and soldering processes involve heating of parent metal surfaces high enough to facilitate the melting of brazing and soldering metal at the interface. Soldering needs lesser heating (<420 °C) of the parent metal than brazing (420–1000 °C) depending upon the type of solder/brazing metal. Depending up on the parent metal, the heat applied during brazing and soldering may cause recrystallization without any major change in metallurgical and mechanical characteristics. Moreover, a lot will depend on the initial thermo-mechanical history of the patent metal like work hardened, quenched, tempered, annealed, etc. A parent metal in work hardened and quenched condition may experience little softening and loss of strength. The strength of soldering and brazing joint is determined by two factors (a) strength of filler brazing/soldering metal and (b) intermetallic compound formed at the soldering/brazing filler metal-base metal interface (Fig. 20.3a). In general,

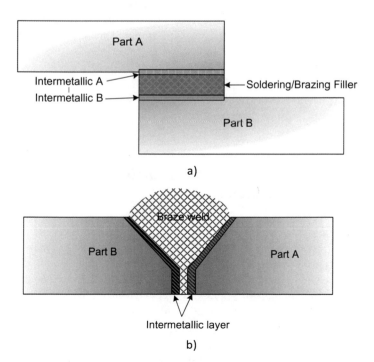

Fig. 20.3 Schematic of **a** brazing/soldering and **b** braze welding process for dissimilar metals

strength of brazing/soldering filler (with optimum clearance) determined the strength of such joints as intermetallic compound for at interfaces is usually stronger than the respective brazing/solder filler metals. Similarly, the strength of the braze weld is also determined by strength of filler brazing metal and intermetallic compound formed at the braze weld-base metal interfaces (Fig. 20.3b).

20.2.3   Fusion and Solid State Joining Processes

The remaining three approaches of joining, namely solid state joining, resistance and fusion welding, apply heat and force singly or jointly to facilitate the development of metallurgical continuity between the components being joined by (a) fusion of the faying surfaces, (b) plastic deformation and (c) diffusion at the interfaces (Fig. 20.4a–d). A lot of heat is applied to realize fusion weld joint. Therefore, many undesirable mechanical and metallurgical changes are observed in parent metals subjected to fusion welding (Fig. 20.4a). Plastic deformation-based processes like friction welding, ultrasonic welding and explosive welding apply pressure/force enough to facilitate macro-/micro-scale plastic deformation at the joint interfaces which in turn generally causes work hardening (Fig. 20.4b). However, heat generated during these processes due to friction or plastic deformation can be less or more. The high heat generation can lead to softening of nearby parent metal due to coarsening reversion, recovery, overtempering, etc. Thus, many mechanical and metallurgical changes can be experienced by the parent metals subjected to fusion welding, solid state joining and resistance welding (Fig. 20.4c). These alterations frequently degrade the performance of joint and their life expectancy during the service. Therefore, a well-thought-out suitable procedure for joining of metals for critical applications must be developed.

20.3   Metallurgy and Properties of Joints by Fusion and Solid State Joining

The metals and alloys are designed for imparting desired set of mechanical and corrosion properties considering possible engineering applications. The desired mechanical properties such as tensile strength, ductility, hardness, toughness, fatigue, fracture and creep resistance in metals and alloys are realized through various mechanism. Most of the metals for engineering applications are based on singly or combination of metal strengthening mechanisms, namely solid solution strengthening, grain refinement, precipitation and dispersion hardening, strain hardening and transformation hardening. The contribution of these metal strengthening mechanisms on mechanical properties is affected with application of heat and force (to facilitate localized fusion or plastic deformation) during joining of metals. However, the extent of effect of

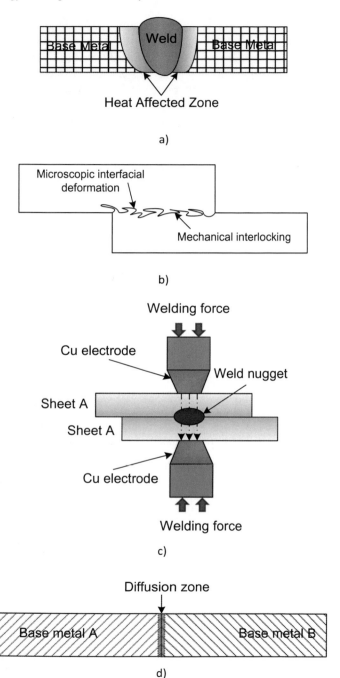

Fig. 20.4 Schematic of fusion and solid state joining processes: **a** fusion, **b** deformation, **c** resistance welding and **d** diffusion bonding

heat or force used for joining on parent metal properties depends on primary metal strengthening mechanism of the given metal.

Effect of heat and force on parent metal primarily designed using solid solution strengthening or dispersion hardening is not much, while the metals designed on the basis of precipitation hardening, grain refinement, strain hardening and transformation hardening are heavily affected by both heat and force applied for joining. The modification in characteristics of parent metals joined by using high heat processes (either for fusion or thermal softening) can be observed in the form both of HAZ softening,hardening and embrittlement. Metals strengthened by grain refinement, precipitation hardening and work hardening in general get softened with application of heat, while those strengthened by transformation hardening (like steels) usually exhibit hardening and embrittlement. Moreover, a lot will also depend on the initial condition of the parent metal (annealed, quenched and tempered, as cast, strain hardened, T6, T4, etc.).

20.4   Metal Strengthening Mechanism and Joint Properties

It is important to note that a metal will always be having effect of other than primary strengthening mechanisms also as per thermo-mechanical history imparted to the parent metal itself during its manufacturing. For example, a metal designed on the basis of solid solution strengthening will also have effect of grain refinement, strain hardening when manufacturing by rolling to produce a sheet metal. Therefore, heat or plastic deformation will not just be affecting the primary strengthening mechanism but also other secondary metal strengthening mechanisms. The effect of heat/deformation on secondary strengthening mechanism may reflect in the form of minor modification in parent metal properties.

20.4.1   Fusion Welding

The weld metal in case of fusion welding can be stronger or weaker than the parent metal as per the filler/electrode metal used, and therefore as per requirement of the application suitable filler/electrode metal with desired set of properties can be chosen for joining purpose. Failure from a sound weld metal hardly occurs, while failure is generally observed from a region in vicinity of the weld metal, i.e. heat affected zone. The autogenous fusion welding, except in case of transformation hardening metals, mostly results in a weaker weld than the parent metal due to development of cast weld structure and possibility of the weld discontinuities, stress raisers and loss of effect of all other strengthening mechanism like grain refinement, work hardening, etc.

The fusion welding involves heating of faying surfaces of the parent metal to the molten state followed by solidification. The mechanism of solidification of the weld metal in fusion welding largely governed by weld metal composition which in turn depends on filler/electrode composition and dilution (%). Fusion weld joints can be developed (a) without any filler using autogenous and (b) with filler of matching type or completely different type.

In case of autogenous welding or welding with fillers of matching composition, the solidification of the weld metal takes place directly by growth mechanism on partially melted grains without any nucleation stage. Since the composition of the solidifying molten weld metal is similar to that of partially melted grains of the parent metal, therefore, on cooling during the solidification the molten weld metal directly starts depositing on the partially melted grain at the fusion boundary. This type of solidification is known as epitaxial solidification.

Fusion welding using filler of completely different composition is also performed to deal with issue related to dissimilar metal joining or to realize specific set of properties in weld metal. In such cases, the solidification of the weld metal takes place in two stages, i.e. nucleation and growth. Initially, the solidifying weld metal nucleates at fusion boundary and followed by growth stage of the solidification.

The weld metal solidification occurs in four modes, namely planar, cellular, dendritic and equiaxed, from the fusion boundary to weld centre depending upon weld composition and cooling conditions. The modes of solidification in turn govern shape/morphology of grains in weld. Size of grain is primarily determined by the cooling rate experienced by the weld metal during the solidification as per heat input and other prevailing thermal conditions.

At macroscopic level, the weld metal generally shows three types of grain morphologies (a) curved grain, (b) columnar grain and (c) axial grain. While at the microscopic level, weld metal can exhibit planar, cellular, dendritic and equiaxed grains.

20.4.2 Solid State Joining

The joint produced using plastic deformation-based joining technique (like ultrasonic welding, friction stir welding) is generally expected to offer a stronger weld nugget than parent metal due to work hardening. However, if heat generated during plastic deformation-based joining technique is high enough to cause the reversion, recovery and recrystallization and grain growth, then the weld nugget can be weaker also than parent metal leading to reduced joint efficiency.

The joining processes such as ultrasonic welding (USW), explosive welding (EW), forged welding, friction welding and its variants are primarily based on principle of bringing the components (to be joined) close to each other at atomic level to develop metallurgical continuity through diffusion and mechanical interlocking and form a joint (Fig. 20.5). The localized macro-/micro-scale plastic deformation at joining interface is realized through the controlled application of suitable compressive/shear force.

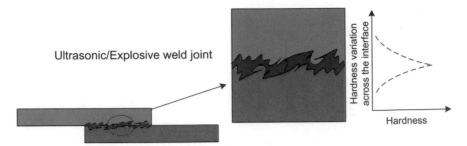

Fig. 20.5 Schematic showing interfacial deformation and hardness distribution in ultrasonic/explosive weld joint

During the joining heat generated may be more or less depending up on the process. The heat generation in joining processes like USW, EW is very limited as observed from the minor increase in temperature of the base metal, while huge heat is generated during friction welding which affects the structure and properties of the joint interface and nearby heat affected zone. Heat generated due to friction and deformation during joining can increase the temperature of the base metal up to 0.5 to 0.7 times of Tm (the melting temperature in K). Such high temperatures can cause many metallurgical transformations like recrystallization, recovery, reversion grain growth, phase transformations, etc., in base metals.

The joints developed by solid state joining processes involving mainly plastic deformation with limited or no heat generation offer higher joint efficiency due to strain hardening effect due to development dislocations. Ratio of strength of joint to that of the base metal is termed as joint efficiency. However, depending of strain hardening exponent of metal being joint efficiency of the sound joint may differ significantly. A base metal like austenitic stainless steels with high strain hardening exponent offers much stronger joints than base metals like aluminium, magnesium (having low strain hardening exponent) under identical conditions (Fig. 20.6). The base metal properties are not much affected on joining by such low heat generation processes.

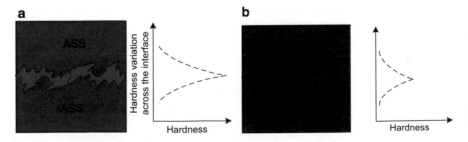

Fig. 20.6 Schematic showing interfacial deformation and hardness distribution in ultrasonic/explosive weld joint of **a** austenitic stainless steel and **b** magnesium

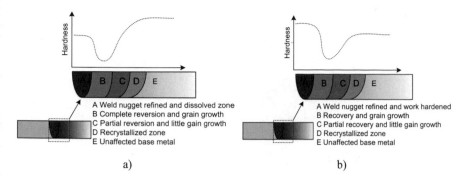

Fig. 20.7 Schematic showing formation of different zones in high heat input solid state joining process like FSW on formation of different zones and hardness distribution of weld joints of **a** precipitation hardening metals and **b** work hardening metals

High heat input

The solid state joining processes like friction welding which inherently generate a lot of heat during joining affect not just weld nugget properties but also nearby parent metal characteristics leading to formation of either singly or both thermo-mechanically affected zone (TMAZ) and heat affected zone (HAZ). High heat input in general results in coarser grain structure in weld nugget and heat affected zones due to lower cooling rates which in turn usually degrades the joint properties in metal strengthened by precipitation hardening, work hardening and grain refinement due to reversion, recovery and grain growth, respectively (Fig. 20.7).

Moreover, low cooling rate imposed in weld and HAZ regions due to high heat input can be favourable in joining of transformation hardening metals like steel due to reduced tendency of cracking and embrittlement. Low cooling rate in steel joining promotes the soft phases like ferrite, pearlite as compared to hard phases like martensite (Fig. 20.8). Therefore, as per metal system, solid state joining process (high or low heat input) can affect the joint properties in different ways. Post-weld heat treatment of the joint developed using SSJ processes in general leads to softening and weakening of the joints due to loss of work hardening effect imparted during the joining except in case of PH hardening metal where restoration of hard precipitates during PWHT can improve the joint efficiency.

20.4.3 Brazing and Soldering

The performance of brazed and solder joint is determined by brazing or soldering filler selected for development of joint. Filler metal affects the two important aspects of brazing/soldering, namely (a) wetting and spreading and (b) interaction between base metal and filler metal leading to the intermetallic formation. Filler metals (for brazing/soldering) are mostly alloys. An alloy of eutectic composition is considered

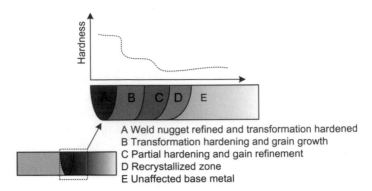

A Weld nugget refined and transformation hardened
B Transformation hardening and grain growth
C Partial hardening and gain refinement
D Recrystallized zone
E Unaffected base metal

Fig. 20.8 Schematic showing formation of different zones in high heat input solid state joining process like FSW on formation of different zones and hardness distribution of weld joints of transformation hardening metals

the best as a filler because that melts and solidifies at single and minimum temperature, while alloys of other than eutectic composition have a solidification temperature range leading to the formation of a mushy zone during the brazing/soldering which in turn adversely affects the wetting and fluidity of filler at the interface. Sometimes, even molten metal separates out from the unmelted filler. These characteristics of filler lead to the development of joints with discontinuities.

Further, the filler should have at least one such element which interacts/dissolves with the surface of base metal. Such kind of interaction between the filler and base metal may lead to (a) formation of an alloy with very low melting point on the surface of the base metal which improves the fluidity and spreading of the molten filler during joining and (b) a more effective metallurgical interaction between the filler and base metal to develop a strong joint by forming reaction or intermetallic layer. However, formation of thick, hard and brittle intermetallic layer at filler–base metal interface results in joints of poor joint strength and ductility. The formation of thin (2–10 μm) intermetallic layer is considered to be good for joint strength, but a thick (10–50 μm), hard and brittle intermetallic layer degrades the mechanical performance of joints (Fig. 20.9).

For example, phosphorus-containing filler may lead to formation of brittle phosphides of iron and nickel. So phosphorus-containing alloys are not preferred for brazing of nickel and ferrous alloys. Similarly, in case of boron-containing filler, the boron may diffuse into the base metals to hard and brittle borides along the grain boundaries. The formation of such undesirable intermetallic layers can be avoided by (a) selection of suitable filler which is compatible with base metals and (b) proper control of process parameters like clearance between the base metals, heat input and thermal cycle associated with brazing/soldering process.

The solidification condition during brazing and soldering mostly favours the formation of either planar or cellular structure at braze/solder-base metal interface; while at other locations of the joint away from the interface, the morphology of the micro-structure may be dendritic/equiaxed as per cooling condition and thermal

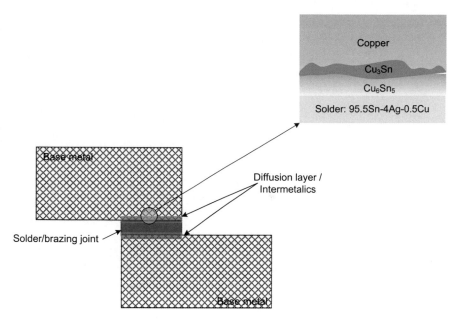

Fig. 20.9 Schematic of brazing/solder processes showing formation of intermetallic compounds at the interfaces

cycle imposed during the brazing/soldering (Fig. 20.10). Molten filler can interact with atmospheric gases and impurities to form inclusion and pores. These discontinuities reduce strength of the joint by reducing load resisting cross-sectional area and acting as stress raisers.

Questions for Self-assessment

(a) Explain the factors determining the strength of adhesive joint.
(b) How does the heat applied for brazing and soldering affect the parent base metal properties?
(c) What is the mechanism of joint formation in ultrasonic welding?
(d) How does the high heat generation in solid state joining of metals affect the parent metal properties?

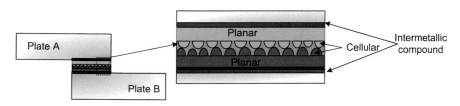

Fig. 20.10 Schematic showing micro-structure of brazed/solder joint

(e) Using suitable schematic explain the hardness distribution across the joint interface developed by explosive welding of metals.
(f) How does the filler metal affect the brazing/soldering process?
(g) Describe the typical micro-structure of soldered/brazed joint using suitable schematic.

Further Reading

Kou S (2003) Welding metallurgy, 2nd edn. Willey, USA
Lancaster JF (2009) Metallurgy of welding, 6th edn. Abington Publishing, England
American Society for Metals (1993) Metals handbook-welding, brazing and soldering, 10th edn., vol. 6. American Society for Metals, USA
Parmar RS (2002) Welding engineering & technology, 2nd edn. Khanna Publisher, New Delhi
Little R (2001) Welding and welding technology, 1st edn. McGraw Hill
Nadkarni SV (2010) Modern arc welding technology. Ador Welding Limited, New Delhi
American Welding Society (2017) Welding handbook, 8th edn., vols. 1 & 2. American Welding Society, USA
Mandal NR (2005) Aluminium welding, 2nd edn. Narosa Publications
Avner SH (2009) Introduction to physical metallurgy, 2nd edn. McGraw Hill, New Delhi.
Dwivedi DK (2018) Surface engineering. Springer, New Delhi
Dwivedi DK (2013) Production and properties of cast Al-Si alloys. New Age International, New Delhi

Chapter 21
Welding Metallurgy

Weld Metal: Solidification, Grain Refinement, Chemical Reactions and Gases in Weld

21.1 Solidification of Weld Metal

The weld metal solidification begins at the fusion boundary of partially melted base metal. The solidification starts from the fusion boundary and progresses towards the weld centre in the direction opposite that of heat flow (Fig. 21.1a) The kinetics of liquid–solid interface during the solidification of the weld metal is described by the ratio of actual temperature gradient at solid–liquid interface (G) and travel speed of liquid–solid interface (R) as shown in Fig. (Fig. 21.1b).

Value of G and R is generally found in range of 100–1000 K/m and 10^{-3} to 10^3 m/s, respectively. Solidification mode is determined by G & R for the given solidification condition. There can be four modes of the solidification (as per G and R value), namely (a) planar (high G and low R), (b) cellular, (c) columnar dendritic and (d) equiaxed dendritic (low G and high R). These four modes are shown in Fig. 21.2. Ratio of G and R determines the mode of solidification, and product of two G and R indicates the cooling rate. Thus, these two parameters (G & R) affect the fineness and grain structure (Fig. 21.3).

21.2 Types of Solidification of Weld Metal

Solidification of the weld metal can take place in two ways, namely (a) epitaxial and (b) non-epitaxial, depending up on the composition of the weld metal. In a weld pool, temperature gradient is observed right from the centre of the weld pool to fusion boundary of the base metal, and isothermal temperature lines exist around the weld. These isotherms determine the boundaries of heat affected zone, mushy zone and liquid weld metal zone (Fig. 21.4). The peak temperature is found at the centre of the weld and then decreases gradually on approaching towards weld fusion boundary. Grains grow from the fusion boundary towards the weld centre. The growth

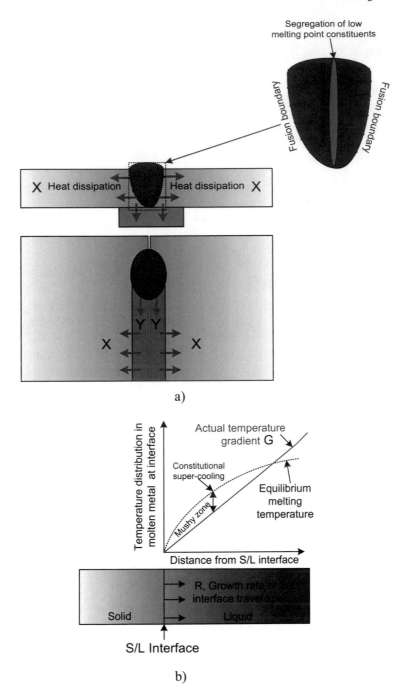

Fig. 21.1 Schematic showing **a** heat transfer from weld pool and grain growth in weld zone and **b** temperature gradient G at S/L interface and growth rate R of S/L interface affecting modes of solidification

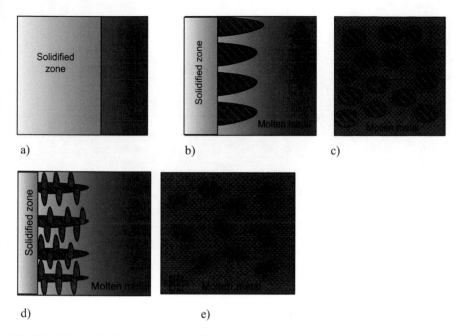

Fig. 21.2 Schematic of modes of the solidification **a** planar, **b** and **c** cellular, **d** columnar dendritic and **e** equiaxed dendritic

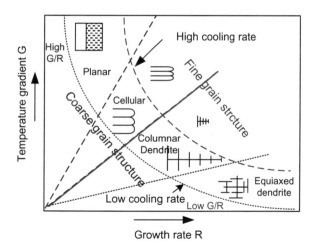

Fig. 21.3 Influence of G & R on mode of solidification and grain structure

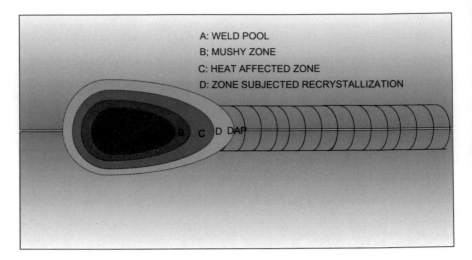

Fig. 21.4 Schematic showing different zones

generally occurs at a faster rate in the direction perpendicular to the fusion boundary and opposite to that of the heat flow than other directions.

21.2.1 Epitaxial Solidification

The transformation of the molten weld metal from liquid to solid state is called solidification of weld metal, and it occurs due to loss of heat from weld pool. Generally, solidification in casting takes place by nucleation and growth mechanism. However, solidification of weld metal can occur either by nucleation and growth mechanism or directly through growth mechanism depending upon the composition of the filler/electrode metal with respect to base metal composition. In case, composition of the filler/electrode is completely different from the base metal, solidification occurs by nucleation and growth mechanism, e.g. use of nickel electrode for joining steel. And when filler/electrode composition is similar to the base metal, solidification is accompanied by growth mechanism only by deposition of metal on partially melted grain of the base metal directly. This type of solidification is known as epitaxial solidification (Fig. 21.5). The growth of grain on either newly developed nuclei or partially melted grain of the base metal occurs by consuming liquid metal, i.e. transforming the liquid into solid to complete the solidification sequence.

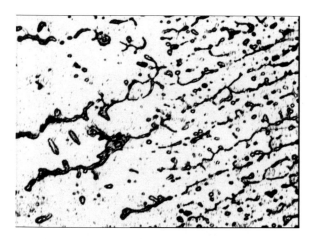

Fig. 21.5 Typical micro-graph showing epitaxial solidification in Al weld joint

21.2.2 Modes of Solidification

The structure of grain in growth stage is governed by mode of solidification. There are four types of grain commonly observed in solidified metal, namely planar, cellular, dendritic and equiaxed corresponding to the respective modes of the solidification (Fig. 21.6). Moreover, the mode of solidification in weld depends on composition and cooling conditions experienced by weld metal at a particular location during the solidification. Thermal conditions of the weld metals during solidification are determined by heat transfer in weld pool which in turn affect the temperature gradient (G) at solid–liquid metal interface (°C/mm) and growth rate of solidification front (R) which is indicated by the growth rate (mm/s) of solid–liquid metal interface.

The shape of solid–liquid metal interface determines morphology of microstructural features of the weld metal. A stable plane solid–liquid metal interface results in planar solidification. The condition for stability of plane solid liquid metal interface is given by $(G/R) >$ or $= (\Delta T/D)$. Where G is the temperature gradient in liquid near solid–liquid metal interface, R is growth rate of solidification front, ΔT is the solidification temperature range for a given composition, and D is the diffusion coefficient of solute in liquid metal (Fig. 21.7a).

Moreover, the stability of the solid–liquid metal interface is governed by thermal and constitutional supercooling condition prevailing in the liquid metal near the solid–liquid metal interface. Destabilization of solid–liquid metal interface results in the growth of interface in cellular or dendritic form. The constitutional supercooling for instability of plane solid–liquid metal interface is expressed by the following relationship: $(G/R) \leq (\Delta T/D)$.

A combination of high actual temperature gradient (G) and low growth rate (R) results in planar solidification, i.e. where liquid–solid interface is plane. A combination of low actual temperature gradient (G) and high growth rate (R) results in

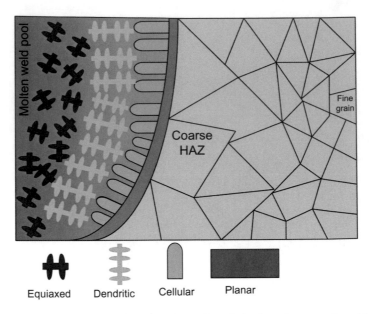

Fig. 21.6 Different modes of solidification change from fusion boundary towards weld centre in general in order of planar at fusion boundary followed by cellular, columnar dendritic and equiaxed dendritic structure

equiaxed solidification as shown in Fig. 21.9. A combination of intermediate G and R values results in cellular and dendritic mode of solidification. Product of G and R indicates the cooling rate. A high value of $G.R$ produces finer grain structure than low $G.R$ value. During welding, weld pool near the fusion boundary experiences high value of G and low value of R which in turn results in planar solidification, while at the weld centre reverse conditions of G and R exist which lead to the development of equiaxed grains. In fact, G and R vary continuously from the weld fusion boundary to the weld centre; therefore all common modes of the solidification can be seen in weld metal structure in sequence of planar at the fusion boundary, cellular, dendritic and equiaxed at the weld centre (Fig. 21.7b). In general, equiaxed grain structure is the most favourable weld structure as it results in the best mechanical performance of weld. Therefore, attempts are made to achieve the fine equiaxed grain structure in the weld by different approaches, namely inoculation, controlled welding parameters and application of external force such as electromagnetic oscillation, arc pulsation and mechanical vibrations. In following sections, these approaches will be described in detail.

In addition to micro-structural variations in the weld, macroscopic changes also occur in weld, which are largely governed by welding parameters such as heat input (as determined by welding current and arc voltage) and welding speed. Macroscopic observation of the weld reveals two types of grains based on their orientation, namely (a) columnar grain and (b) axial grain (Fig. 21.8). As reflecting from their names, columnar grains generally grow perpendicular to the fusion boundary in direction

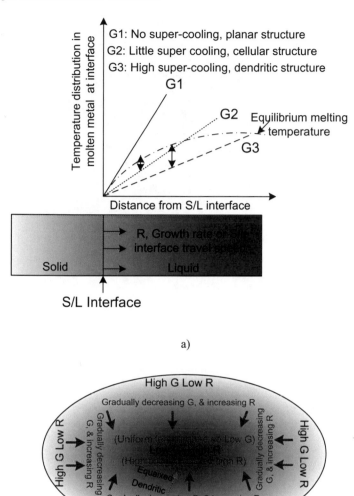

Fig. 21.7 **a** Schematic of solid–liquid metal interface conditions with actual temperature gradient *G* and growth rate *R* temperatures distribution during solidification near solid–liquid metal interface and **b** gradually changing condition of *G* & *R* within the weld pool

opposite to the heat flow, while axial grains grow axially in the direction of welding (Fig. 21.8). The axial grains weaken the weld and increase the solidification cracking tendency; therefore effort should be made to modify the orientation of axial grains.

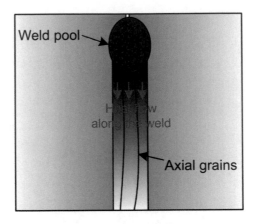

Fig. 21.8 Schematic of axial grain in weld joints

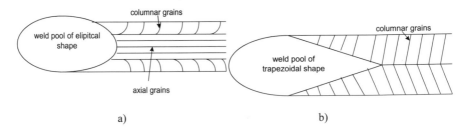

Fig. 21.9 Effect of welding speed on shape of weld pool and grain structure at **a** low speed and **b** high speed (S Kou, Welding metallurgy)

21.3 Effect of Welding Speed on Grain Structure of the Weld

Welding speed appreciably affects the orientation of columnar grains due to difference in the shape of weld puddle under varying welding conditions. Low welding speed results in elliptical shape weld pool and produces curved columnar grain with better distribution of low melting point phases and alloying elements in the weld metal which in turn lowers solidification cracking tendency of the weld as compared to those welds produced using high welding speed (Fig. 21.9). At high welding speed, the shape of the trailing end of weld pool becomes like a tear drop shape, and so the grains mostly grow perpendicular to the fusion boundary of the weld. In this case, low melting point phases and alloying elements are mostly segregated along the weld centreline, thereby making the weld more sensitive for solidification cracking.

21.4 Common Methods of Grain Refinement

It is generally preferred to have fine grain structure in the weld metal to improve the mechanical properties (except creep) and reduce solidification cracking tendency and segregation. There are many mechanisms playing a crucial role in grain refinement, namely heterogeneous nucleation, grain detachment, grain fragmentation and surface nucleation (Fig. 21.10). Many methods based on these mechanisms have been devised.

21.4.1 Inoculation

This method is based on increasing the heterogeneous nucleation at nucleation stage of the solidification by adding alloying elements in weld pool. These elements either themself or their compounds act as nucleants. Increased number of nucleants in the weld metal eventually on solidification results in fine grains in the weld (Fig. 21.11). It is understood that elements/compounds having (a) melting point higher than the liquidus temperature of the weld metal and (b) lattice parameter similar that of base metal can perform as nucleants. For aluminium, titanium Zr, and boron based compound as such as TiB_2, TiC, Al–Ti–B, Al–Zr are commonly used as grin refiners. Addition of grain refiners in molten metal can lower the surface energy to facilitate the

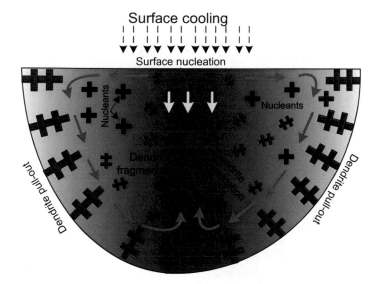

Fig. 21.10 Schematic showing fundamental mechanisms of grain refinement, namely heterogeneous nucleation, surface nucleation, dendrite fragmentation and dendrite pull-out due to weld pool convention current

Fig. 21.11 Schematic of grain refinement by inoculation

nucleation even with limited under-cooling. Increase in the nucleation rate facilitates the grain refinement. For steel, Ti, V and Al are commonly used grain refiners.

21.4.2   Arc Pulsation

This method of grain refinement of the weld metal is based on the principle of creating many nucleants in the weld puddle through rapid cooling as high cooling rate is known to lower the effective phase transformation temperature leading to high nucleation rate and low growth rate. Further, high cooling rate reduces the solidification time, so very limited time is available for grain to grow during solidification. The gas metal arc and gas tungsten arc welding process generally use DC constant voltage and constant current power source, respectively. Pulsation of welding current (between base current and peak current) is effectively used to reduce the heat input for welding which in turn offers high cooling rate. Base current is the minimum current primarily used to just a have stable arc, and it supplies least amount of the heat to the weld; solidification of the weld is expected to take place during the base current period (Fig. 21.12). The peak current is the maximum current supplied by the power source to the weld arc to generate the heat required for melting of the faying surfaces. The cycle of alternate heating and cooling results in smaller weld puddle, and so rapid cooling of the weld metal which in turn results in finer grain structure than the conventional welding, i.e. without arc pulsation (Fig. 21.13). It is believed that abrupt cooling of the weld pool surface during base current period can also lead to development of few nucleants at the surface which will tend to settle down gradually. Gradual settlement makes distribution of nucleants more uniform in the molten weld pool. Thus, increased availability of nucleants due to surface nucleation can also be assisting in refinement of grain structure of the weld.

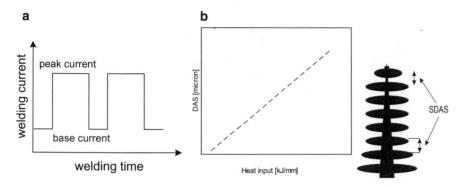

Fig. 21.12 Schematics of **a** pulse current vs time welding and **b** effect of heat input on dendrite arm spacing

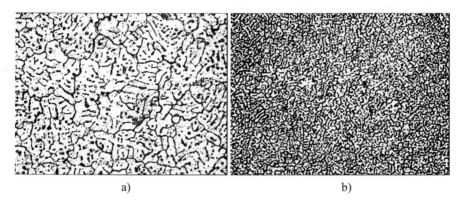

Fig. 21.13 Micro-structure of aluminium weld developed **a** without arc pulsation using 160 A current and **b** arc pulsation between 120 and 160 A (200X)

21.4.3 Mechanical Vibrations and Electromagnetic Force

These methods of grain refinement of the weld metal are based on the principle of creating many nucleants in the weld puddle through dendrite fragmentation and detachment from fusion boundary by creating disturbance in molten weld pool. In these methods, external excitation force is used to disturb solidifying weld metal so as to create more number of the nucleants in weld metal. The external disturbance causes forced flow and turbulence in the viscous semi-solid weld metal carrying dendrites and nucleants. Relative motion of the molten metal with respect solid dendrite can result in (a) fracture of partially melted grains at the fusion boundary of the base metal, (b) fragmentation of solidifying dendrites and (c) improved distribution of chemical composition and the nucleants (Fig. 21.14). The fractured dendrites are pulled out of partially melted grains and act as nucleants for solidifying weld metal as they are of the same composition as in solid state.

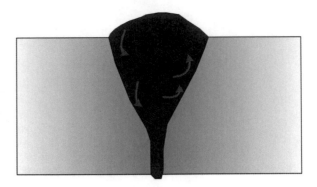

Fig. 21.14 Refinement using external excitation force

21.4.4 Magnetic Arc Oscillation

A welding arc is composed of charged particles that can be deflected using magnetic field. Arc oscillation affects the weld pool in two ways, namely (a) reduction in the size of weld pool due to intermittent supply of heat at specific location and (b) alternate heating and cooling of weld (similar to that of arc pulsation) decreases the weld pool size and increases the cooling rate as shown in Fig. 21.15. A combination of above two factors leads to rapid cooling of weld metal which in turn reduces grain size owing to increased nucleation rate and reduced growth rate. As cooling rate of the solidifying weld metal increases, the effective liquid to solid state transformation temperature decreases which is known to increase the nucleation rate and lower the growth rate.

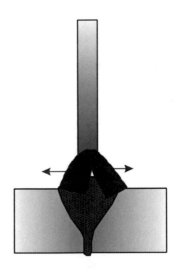

Fig. 21.15 Arc oscillation due to electromagnetic filed around welding arc

21.4.5 Welding Parameter

Heat generated (kJ) by the welding arc is determined from the product of welding current and arc voltage $(V \cdot I)$ for given welding conditions such as type and size of electrode, arc gap, base metal and shielding gas (if any). While the exact amount of heat supplied to base metal for melting of the faying surfaces is significantly determined by the welding speed. Increase in welding speed for a given welding current and voltage results in reduced heat input per unit length of welding (kJ/mm) which is also termed as net heat input for sake of clarity. Cooling rate experienced by the weld metal and heat affected zone is found inversely proportional to net heat input (Fig. 21.16). Increase in net heat input lowers the cooling rate. Low cooling rate results in (a) longer solidification time (needed to extract complete sensible and latent heat from the molten weld pool) and (b) high effective solid to liquid state transformation temperature. Longer solidification time permits each grain to grow to a greater extent, and this leads to a coarse grain structure (Fig. 21.17). Further, high heat input causing high effective liquid–solid transformation temperature leading to

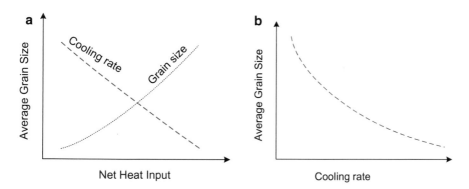

Fig. 21.16 Schematic showing effect of **a** heat input and **b** cooling rate on grain size

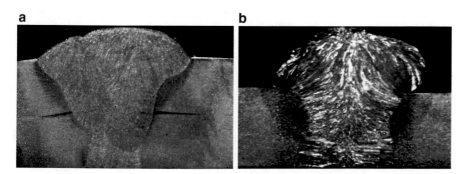

Fig. 21.17 Macro-photographs of weld joints produced using **a** 3.0 kJ/mm and **b** 6.0 kJ/mm heat input with help of submerged arc welding

low nucleation rate and high growth rate which in turn results in coarse grain structure. Increase in welding current or reduction in welding speed generally increases the grain size of weld metal as these increase the net heat input and lower the cooling rate experienced by the weld metal during solidification.

21.5 Typical Metallurgical Discontinuity of the Weld

Due to typical nature of welding process, common metallurgical discontinuities observed in the weld are banding and micro-segregation of the elements. In the following section, these have been described in detail.

21.5.1 Micro-Segregation

Micro-segregation refers to non-uniform distribution of elements in the weld which primarily occurs due to inherent nature of solidification mechanism, i.e. transformation of high temperature alpha phase from liquid to solid by rejection of alloying elements into the liquid metal, thereby lowering its solidification temperature. Except planar mode, other modes of solidification, namely cellular, dendrite and equiaxed, involve segregation. Therefore, intercellular, interdendritic and interequiaxed region are generally enriched of alloying elements compared to cells (Fig. 21.18). Segregation of the alloying elements increases the solidification cracking tendency, reduces corrosion resistance and increases heterogeneity in mechanical properties including embrittlement.

21.5.2 Banding

Welding arc is never in steady state as very transient conditions exit during arc welding which in turn leads to severe thermal fluctuations in the weld pool. Therefore, cooling conditions vary continuously during the solidification of the weld metal. Variation in cooling rate of weld pool causes changing growth rate of the grain in weld and fluctuating velocity of solid–liquid metal interface. Abrupt increase in growth rate decreases the rate of rejection of alloying elements in liquid metal near the solid–liquid metal interface due to limited time for diffusion of alloying elements, while low cooling rate increases the rejection of elements near the solid–liquid metal interface as long time available for diffusion to occur. This alternate enrichment and depletion of alloying elements produces band like structure as shown in Fig. 21.19. This structure becomes very notch sensitive and therefore known to adversely affect fatigue and notch toughness properties of weld joints.

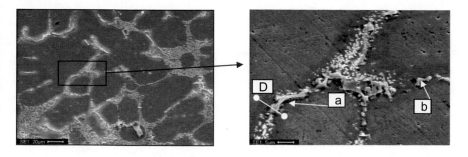

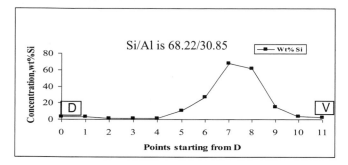

Fig. 21.18 Segregation of alloying elements at grain boundary

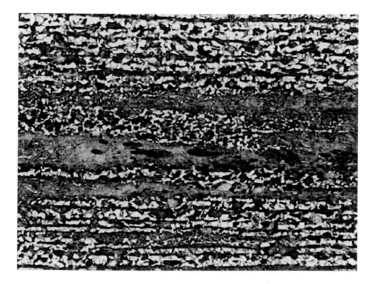

Fig. 21.19 Banded structure showing bands for ferrite in steel

21.6   Chemical Reaction in Welds

This section presents the need of protecting the weld and rationale behind variations in cleanliness of the weld developed by different welding processes. The gas metal reactions and slag metal reactions have also been described besides their effect on elemental transfer efficiency.

21.6.1   Welding Process and Cleanliness of the Weld

In fusion welding, the application of heat of the arc or flame results in the melting of the faying surfaces of the plates to be welded. At high temperature, metals become very reactive to atmospheric gases such as nitrogen, hydrogen and oxygen present in and around the arc environment. These gases either get dissolved in weld pool or form their compound. In both the cases, gases adversely affect the soundness of the weld joint and mechanical performance. Therefore, various approaches are used to protect the weld pool from the atmospheric gases such as developing envelop of inactive (GMAW, SMAW) or inert gases (TIGW, MIGW) around arc and weld pool, welding in vacuum (EBW) and covering the pool with molten flux and slag (SAW, ESW). The effectiveness of each method for weld pool protection is different. That is why adverse effect of atmospheric gases in weld produced by different arc welding processes is different (Fig. 21.20).

Among the most commonly used arc welding processes, the cleanest weld (having minimum nitrogen and oxygen) is produced by gas tungsten arc welding (GTAW) process due to two important factors associated with GTAW: (a) short arc length and (b) very stable arc produced between non-consumable tungsten electrode and base metal. A combination of short and stable arc with none-consumable tungsten

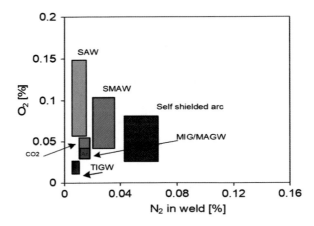

Fig. 21.20   Schematic diagram showing nitrogen and oxygen content in different welding processes

electrode results in a firm shielding of arc and protection of the weld pool by inert gases; these conditions restrict the entry of atmospheric gases in the arc zone. Gas metal arc welding (GMAW) also offers clean weld but not as clean as produced by GTAW because in case of GMAW arc length is somewhat greater and arc stability is poorer than GTAW. Submerged arc weld (SAW) joints are usually high in oxygen and less in nitrogen because SAW uses flux containing mostly metallic oxides. These oxides decompose and release oxygen in arc zone. The self-shielded fluxed cored metal arc welding processes use electrodes having fluxes in core act as de-oxidizer and slag former to protect the weld pool. However, weld produced by the self-shielded fluxed arc welding processes is not as clean as those produced by GMAW.

21.6.2 Effect of Atmospheric Gases on Weld Joint

The gases present in weld zone (atmospheric gases or those dissolved in liquid metal) affect the soundness of weld joint. Gases such as oxygen, hydrogen and nitrogen are commonly present in and around the liquid metal. Both oxygen and hydrogen are very important (from soundness and mechanical properties point of view) in welding of ferrous and non-ferrous metals. Gases are mostly produced by decomposition of water vapours (H_2O) and carbon dioxide (shielding gas) due to high temperature of arc. Oxygen reacts with carbon in case of steel to form CO or CO_2. These dissolved gases should get enough time to escape out from the molten weld metal during the solidification; however, owing to high solidification rate encountered during fusion welding processes these gases may not be able to come up to the surface of molten metal and may get trapped to produce gaseous defects in the weldment like porosity, blowhole etc. Chances for these defects further increase if the difference in solubility of the gases in liquid and solid state is high. Oxygen reacts with aluminium to form refractory alumina. Alumina interferes with melting and results in inclusions which in turn reduce the weldability. The formation can be reduced by proper shielding of arc zone either by inert or inactive gases. Only source of nitrogen is atmosphere, and it may form nitrides. Moreover, nitrogen creates fewer problems in aluminium and steel welding. Hydrogen is a main problem creator in welding of steel and aluminium alloy due to high difference in liquid and solid state solubility. In case of steel, besides the porosity and blow holes hydrogen cause the problem of cold cracking even if it is present in very small amount (~2 ppm), whereas in case of aluminium hydrogen causes pin hole or micro-porosity.

Oxides and nitrides formed by these gases if not removed from the weld act as site of weak zone in form of inclusions which in turn lower the mechanical performance of the weld joint, e.g. iron reacts with nitrogen to form hard and brittle needle shape iron nitride (Fe_4N) as shown in Fig. 21.21a, b. These needle-shaped micro-constituents offer high stress concentration at the tip of particle–matrix interface which under external tensile stresses facilitate the easy nucleation and propagation of crack, and therefore fracture is accompanied with limited elongation (ductility). Similar logic can be given for reduction in mechanical performance of weld joints

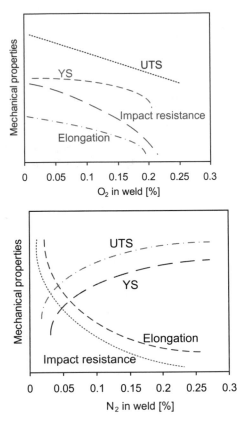

Fig. 21.21 Influence of oxygen and nitrogen as impurities on mechanical properties of steel weld joints

having high oxygen/oxide content. However, the presence of N_2 in weld metal is known to increase the tensile strength due to the formation of hardness and brittle iron nitride needles.

Additionally, these inclusions formed by oxygen, nitrogen and hydrogen break the discontinuity of metal matrix which in turn decreases the effective load resisting cross-sectional area. Reduction in load resisting cross-sectional area lowers the load carrying capacity of the welds. Nitrogen is also an austenite stabilizer which in case of austenitic stainless steel (ASS) welding can play crucial role. Chemical composition of ASS is designed to have about 5–8% ferrite in austenite matrix to control solidification cracking of weld. Presence of nitrogen in weld metal either from atmosphere or with shielding gas (Ar) stabilizes the austenite as the cost of ferrite (so increases the austenite content) and reduces ferrite content in weld which in turn increases the solidification cracking tendency because ferrite in these steels acts as sink for impurities like P and S which otherwise increase cracking tendency

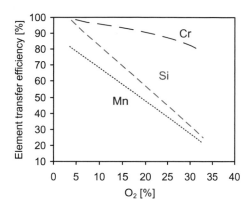

Fig. 21.22 Influence of oxygen concentration on element transfer efficiency of common elements

of weld by lowering the melting point and increasing the solidification temperature range.

21.6.3 Effect on Weld Compositions

Presence of oxygen in arc environment not only increases chances of oxide inclusion formation tendency but also affects the element transfer efficiency from filler/electrode to weld pool due to oxidation of alloying elements (Fig. 21.22). Sometimes composition of the weld is adjusted to get desired combination of mechanical, metallurgical and chemical properties by selecting electrode of suitable composition. Melting of electrode and flux involves transfer of the elements from the electrode across the arc zone. The retention of molten metal drop at the electrode tip and its flight across the arc gap causes the oxidation of some of the highly reactive elements which may be removed in form of slag. Thus transfer of especially reactive elements to weld pool is reduced which in turn affects the weld metal composition (usually by lowering the concentration of alloying elements) and so deterioration in mechanical and other performance characteristics of the weld.

21.6.4 Hydrogen in Weld Metal and Fluxes

This section presents influence of hydrogen in weld joints on the soundness and performance of weld joints. Further, different types of fluxes, their stability and effect on weld metal properties have been described. The concept of basicity index of the fluxes and its effect on weld has also been elaborated.

21.6.4.1   Effect of Hydrogen on Steel and Aluminium Weld Joints

Hydrogen

Hydrogen in weld joints of steel and aluminium is considered to be very harmful as it increases the cold cracking tendency in hardenable steel and porosity in aluminium welds. Hydrogen-induced porosity in aluminium welds is formed mainly due to high difference in solubility of hydrogen in liquid and solid state. The hydrogen rejected by weld metal during the solidification of the weld metal if doesn't get enough time for escaping then it is entrapped in the weld and produces hydrogen induced fine porosity. Welds made using different processes contain varying hydrogen concentration owing to difference in solidification time, moisture associated with consumables and protection of the weld pool from atmospheric gases, use of different consumables (Fig. 21.23). Hydrogen in steel and aluminium weld joint is found mainly due to high difference in solubility of hydrogen in liquid and solid state (Figs. 21.22 and 21.24).

Cold cracking is caused by hydrogen especially when hard and brittle martensitic structure is formed in the weld and HAZ of hardenable steel. Many theories have been advanced to explain the cold cracking due to hydrogen. Accordingly to one hypothesis, hydrogen diffuses towards the vacancies, grain boundary area and other crystallographic imperfections and at these locations, segregation of the hydrogen results in first transformation of atomic hydrogen into gaseous molecules which gradually builds up the pressure until it is high enough to cause growth of void by propagation of cracks in one of directions having high stress concentration as shown in Figs. 21.23 and 21.25. Thereafter, process of building up of the pressure and growth of crack is repeated until complete fracture of the weld without any external load occurs. Existence of external stress or residual tensile stresses further accelerates the

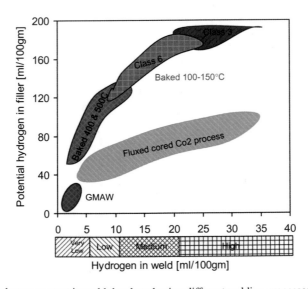

Fig. 21.23 Hydrogen content in weld developed using different welding processes

crack growth rate which in turn lowers the time required for failure by cold cracking. Presence of both of above discontinuities (cracks and porosity) in the weld decreases mechanical performance of weld joint. Hydrogen in arc zone can be introduced from variety of sources, namely

- moisture (H_2O) in coating of electrode or on the surface of base metal
- hydrocarbons present on the faying surface of base metal in the form of lubricants, paints, etc.
- inert gas (Ar) mixed with hydrogen to increase the heat input
- hydrogen in dissolved state in metal (beyond limits) being welded such as aluminium and steel.

The proper baking of electrodes directly reduces the cold cracking tendency and increases the time for failure by delayed cracking. Therefore, attempt should be made to avoid the hydrogen from above sources by taking suitable corrective action (Figs. 21.24 and 21.25).

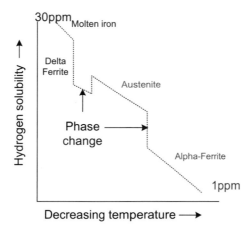

Fig. 21.24 Schematic of hydrogen solubility as a function of temperature of iron

Fig. 21.25 Hydrogen-induced crack

21.7 Flux in Welding

Fluxes are commonly used to take care of problems related to oxygen and nitrogen. Variety of flux is used to improve the quality of the weld by reducing impurities and gases. These fluxes are grouped in three categories, namely halide fluxes (mainly composed of chlorides and fluorides of Na, K, Ba, Mg), oxide fluxes (oxides of Ca, Mn, Fe, Ti, Si) and mixture of halide and oxide fluxes. Halide fluxes are free from oxides, and therefore these are mainly preferred for welding of highly reactive metals having good affinity with oxygen such as Ti, Mg and Al alloys, while oxide fluxes are used for welding of low strength and non-critical welds joints of steel. Halide-oxide type fluxes are used for semi-critical application in welding of high strength steels. In general, calcium fluoride in flux reduces hydrogen concentration in weld (Fig. 21.26).

21.7.1 Basicity of the Flux

The composition of the flux is adjusted so as to get proper basicity index as it affects the ability of flux to remove impurities like sulphur and oxygen from the melt. The basicity index of the flux refers to ratio of sum of all basic oxides (wt. %) and that of non-basic oxides (wt. %). Basic oxides (CaO is most common) are donors of the oxygen ions (O^{2-}), while acidic oxides (such as SiO_2) are acceptor of oxygen ions (O^{2-}). Common acidic and basic oxides are shown in table below. A flux having BI < 1 is categorized as acidic flux, neutral fluxes have 1 < BI < 1.2,

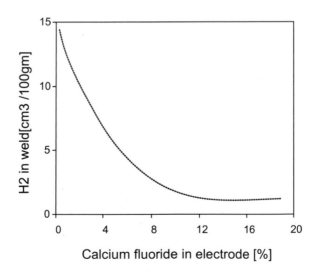

Fig. 21.26 Influence of calcium fluoride on hydrogen concentration in weld joints

while basic fluxes have BI > 1.2. Increase in BI of the flux (Cao/SiO$_2$) from 0 to 5 results in significant decrease in sulphur content of the weld. The basic oxides act as desulphurizer as sulphur is removed from the weld in the form of SO$_2$ by reaction between oxygen released by basic oxides and S. Thus, the weld is desuphurized. These oxides get decomposed at high temperature in arc environment. Stability of each oxide is different. Oxides with decreasing stability are as follows: (i) CaO, (ii) K$_2$O, (iii) Na$_2$O and TiO$_2$, (iv) Al$_2$O$_3$, (v) MgO, (vi) SiO$_2$ and (vii) MnO and FeO. On decomposition, these oxides invariably produce oxygen and result in oxidation of reactive elements in weld metal (Table 21.1).

In general, an increase in basicity of the flux up to 1.5 decreases the S and oxygen concentration (from about 900 to 250 ppm) in weld joints as shown in Fig. 21.27a, b. Thereafter, further increase in basicity does not affect the oxygen content appreciably. Oxygen content remains constant at about 200–250 PPM level despite of using fluxes of high basicity index. Further, there is no consensus among the researchers on the mechanism by which an increase in basicity index decreases the oxygen content.

Questions for Self-assessment

1. With help of schematic diagram explain the mechanism of epitaxial solidification.
2. Describe the modes of solidification in fusion welds?
3. Write the fundamental mechanisms of grain refinement of weld metal.
4. How do welding speed and heat input affect the grain structure of the weld metal?

Table 21.1 Type of oxide fluxes used in submerged arc welding

Type of oxide	Decreasing strength						
	1	2	3	4	5	6	7
Acidic	SiO$_2$	TiO$_2$	P$_2$O$_5$	V$_2$O$_5$			
Basic	K$_2$O	Na$_2$O	CaO	MgO	BaO	MnO	FeO
Neutral	Al$_2$O$_3$	Fe$_2$O$_3$	Cr$_2$O$_3$	V$_2$O$_3$	ZnO		

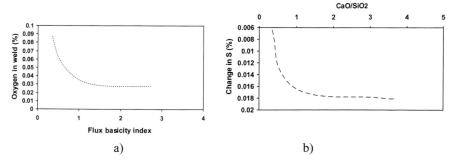

a)   b)

Fig. 21.27 Influence of basicity of flux on **a** oxygen and **b** sulphur concentration in weld

5. Describe common methods of grain refinement of weld metal.
6. What is heterogeneous nucleation? Explain the principle of inoculation for refinement of weld metal.
7. How does arc pulsation help in grain refinement of weld metal?
8. How does application of mechanical vibrations and electromagnetic forces in weld metal during solidification refine the grain structure?
9. What is the effect of welding parameters on the grain structure of the weld metal?
10. Explain the fundamental mechanism of grain refinement using magnetic arc oscillation method.
11. Explain the metallurgical discontinuities in weld, namely segregation and banding and factors, lead to their development.
12. What is the need to protect the weld pool from atmospheric gases during welding?
13. Describe common approaches used to protect the weld pool from atmospheric gases?
14. What are factors affecting protection of the weld pool in general?
15. How does type of shielding gas affect the cleanliness of the weld?
16. Explain the effect of welding parameters on protection of the weld pool.
17. Explain the factors affecting protection of the weld pool associated with the following processes:

    a. SMAW
    b. SAW
    c. GMAW
    d. ESW
    e. GTAW

18. Cleanliness of the weld metal produced using different welding processes is found different. why?
19. Describe the effect of atmospheric gases, namely oxygen, hydrogen and nitrogen, on composition and mechanical performance of welds?
20. How does hydrogen affect the weld joints of steel and aluminium alloys?
21. What are different types of fluxes and write about their stability and application.
22. What is basicity index (BI) of the flux?
23. How does BI affect the quality of the weld?

Further Reading

Kou S (2003) Welding metallurgy, 2nd edn. Willey, USA
Lancaster JF (2009) Metallurgy of welding, 6th edn. Abington Publishing, England
American Society for Metals (1993) Metals handbook-welding, brazing and soldering, 10th edn., vol. 6. American Society for Metals, USA
Parmar RS (2002) Welding engineering & technology, 2nd edn. Khanna Publisher, New Delhi

Little R (2001) Welding and welding technology, 1st edn. McGraw Hill

Nadkarni SV (2010) Modern arc welding technology. Ador Welding Limited, New Delhi

American Welding Society (2017) Welding handbook, 8th edn., vols. 1 & 2. American Welding Society, USA

Mandal NR (2005) Aluminium welding, 2nd edn. Narosa Publications

Avner SH (2009) Introduction to physical metallurgy, 2nd edn. McGraw Hill, New Delhi.

Dwivedi DK (2018) Surface engineering. Springer, New Delhi

Dwivedi DK (2013) Production and properties of cast Al-Si alloys. New Age International, New Delhi

Chapter 22
Welding Metallurgy: Physical Metallurgy of Welding: Steel, PH Hardenable and Work Hardening Metals

22.1 Relevance of Physical Metallurgy of Steel Welding

To understand the behaviour of steel in connection with welding, it is important to look into the different phases, phase mixtures and intermetallic generally found in steel besides the changes in phase that can occur during welding due to heating and cooling cycles. All these aspects can be understood by going through following sections presenting significance of Fe–C diagram, time–temperature transformation diagram and continuous cooling transformation diagram.

The relevance of Fe–C diagram in fusion welding/solid state joining is limited to:

- Understand the temperature at which phases transformation is taking place primarily lower critical and upper critical temperatures affecting the HAZ structure and properties at different locations adjacent to the fusion boundary and
- Solidification temperature range determining the size of partial melting zone for a steel of given composition.

22.2 Fe–C Equilibrium Phase Diagram

Fe–C diagram also called iron–iron carbide diagram because these are the two main constituents observed at the room temperature in the plain steel while the presence of other phases depends on the type and amount of alloying elements. This diagram shows the presence of various phases and phase transformations in plain carbon steel and unalloyed cast irons as a function of temperature during heating/cooling under equilibrium conditions (Fig. 22.1).

Metallurgical aspects of phases observed in Fe–C phase diagram along with their mechanical properties relevant to the welding are presented in following section.

D. K. Dwivedi, *Fundamentals of Metal Joining*,
https://doi.org/10.1007/978-981-16-4819-9_22

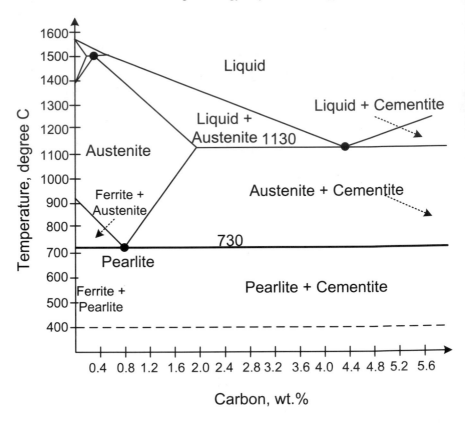

Fig. 22.1 Fe–C phase diagram

Ferrite

Ferrite is an interstitial solid solution of carbon in iron having BCC structure. Solubility of carbon in iron having BCC structure at room temperature is about 0.005% while that at eutectoid temperature (727 °C) is 0.025%. Ferrite is a soft, low strength, tough and ductile (50% elongation) phase.

Austenite

Austenite is an interstitial solid solution of carbon in iron having FCC structure. This phase is not stable below eutectoid temperature (727 °C) in plain Fe–C system as it transforms into pearlite below this temperature. Solubility of carbon in iron having FCC structure at temperature 1130 °C is 2.0% while that at eutectoid temperature (727 °C) is 0.8%. Austenite is a comparatively harder, stronger, tougher but of lower ductility (%elongation) than the ferrite.

Cementite

Cementite is an intermetallic compound of iron and carbon i.e. iron carbide (Fe_3C). Cementite contains 6.67% of carbon and has orthorhombic structure. It is the hardest constituents amongst all the phases appearing in Fe–C diagram. The hardness of cementite is extremely high while tensile strength is very poor.

Pearlite

Pearlite is a phase mixture of ferrite and cementite and is a result of eutectoid transformation. The pearlite has alternate layers (lamellas) of cementite and ferrite. Strength of pearlite is more than any of the individual phases of which it is made. Mechanical properties, i.e. strength, ductility, toughness and hardness of pearlite, depend on the interlamellar spacing. Thinner plates (layers) of alpha ferrite and Fe_3C result in better mechanical properties.

Ledeburite

Ledeburite is also a phase mixture of austenite and cementite and is formed as a result of eutectic transformation. However, it is observed above the eutectoid temperature in Fe–C diagram only as below this temperature austenite present in ledeburite transforms into pearlite.

22.3   Effect of Phases on Mechanical Properties

It is important to note that every phase or phase mixture has its own mechanical properties. Some of the phases are very soft (ferrite), and some are extremely hard (cementite). Therefore, variation in proportions/relative amounts of these phases will affect the mechanical properties of steel as a whole. Increase in carbon content in steel linearly increases the proportion of pearlite but at the cost of ferrite. Since ferrite is of low strength, soft and ductile while pearlite is hard, strong and of poor ductility and toughness, hence increase in percentage of pearlite increases strength and hardness and reduces the ductility and toughness of steel as a whole. Cementite appears as an individual phase only above eutectoid composition (steel having carbon > 0.8%). It tends to form a network along the grain boundary of pearlite depending upon carbon content. Complete isolation of pearlite colonies with the cementite (because of continuous network of cementite) decreases tensile strength and ductility because mechanical properties of the alloy/steel largely depend upon the properties of phase, which is continuous in alloy. Increase in the carbon content above the eutectoid compositions (0.8% C) therefore reduces the strength and ductility because in hypereutectoid steel, network of cementite is formed along the grain boundaries of pearlite and cementite has very low tensile strength (3.0 MPa).

22.4   Phase Transformation

Steel having 0.8% carbon is known as eutectoid steel. Steels with carbon lesser than 0.8% are known as hypoeutectoid and those having more than 0.8% and less than 2% carbon are called hypereutectoid steels. Fe–C systems having carbon more than 2% are called cast irons. Cast iron having 4.3% carbon is known as eutectic cast iron. Cast irons with carbon less than 4.3% are known as hypoeutectic and those having more than 4.3% are called hypereutectic cast iron.

22.4.1   Time Temperature Transformation (TTT) Diagram

This diagram shows the transformation of metastable austenite phase at constant temperature into various phases as a function of time (Fig. 22.2a). Therefore, it is also known as isothermal transformation diagram. Transformation of austenite into various phases such as pearlite, bainite and martensite depends on the transformation temperature. Time needed to start the transformation of austenite into pearlite or bainite is called incubation period which is initially low at lower transformation temperature and then increases as transformation temperature increases. TTT diagram for eutectoid C-steel has C shape due to variation in time needed to start and end the transformation of austenite at different transformation temperature. This curve has a nose at about 550 °C. Transformation of austenite into pearlite takes place on exposure at any constant temperature above the nose. It is observed that higher the transformation temperature (T1) more is the time required for starting and completing the transformation. Transformation of austenite into pearlite or bainite occurs by nucleation and growth process. Hence, at high-temperature high growth rate, low nucleation rate coupled with longer transformation time results in coarse pearlitic structure, whereas at low transformation temperatures fine pearlitic structure is produced because of low growth rate, high nucleation rate and short transformation time (Fig. 22.2b). High transformation temperature lowers the strength and hardness of steel owing to the coarse pearlitic structure.

Transformation of austenite (T3–T5) at a temperature below the nose of the curve results in bainitic structure. Bainite is a very fine intimate mixture of ferrite and cementite like pearlite. However, pearlite is a mixture of lamellar ferrite and cementite. That is why bainite offers much better strength, hardness and toughness than the pearlite. Degree of fineness of bainite also increases with the reduction in transformation temperature like pearlitic transformation. Bainite formed at high temperature (T3) is called feathery bainite while that formed at low temperature (T5) near the M_s temperature is called acicular bainite.

Transformation of austenite at a temperature (T6, T7) below the M_s temperature results in hard and brittle phase called martensite structure. Martensite is a supersaturated solid solution of carbon in iron having body-centred tetragonal (BCT) structure. This austenite to martensite transformation is a thermal transformation

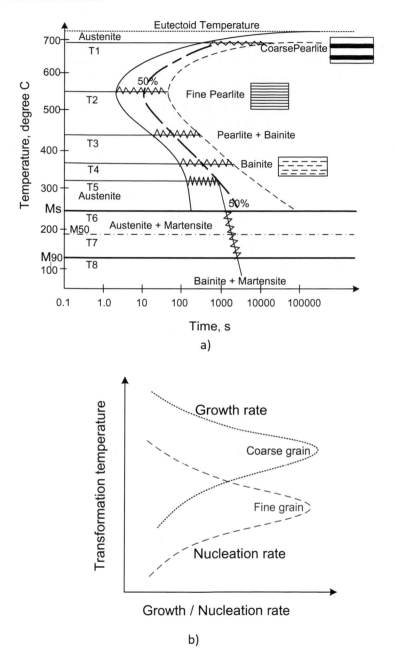

Fig. 22.2 **a** Time–temperature transformation diagram showing isothermal transformation, and **b** effect transformation temperature on growth and nucleation rate

as it takes place by diffusionless process. Rapid quenching/cooling from austenitic temperature to a temperature below the "M_s" prevents any kind of atomic diffusion. Therefore, carbon atoms which are easily accommodated within FCC unit cell (in austenite) at high temperature should be rejected at low temperature because of reduction in solid solubility. But at such a low temperature (below "M_s"), diffusion is prevented and that leads to the formation of supersaturated solid solution of carbon in iron having BCC structure. This supersaturation of carbon in iron (BCC) causes the distortion of the BCC lattice structure and makes it BCT by increasing the c/a ratio more than 1. Degree of distortion is measured in terms of c/a ratio. This ratio depends upon the carbon content. Increase in carbon content up to 0.8% increases the c/a ratio. The c/a ratio can be directly related to the increase in hardness, as there is linear relation between the two up to 0.8% carbon content.

Position of transformation lines and nose depends on steel composition, homogeneity of austenite and grain size. For each composition of steel, there will be just one TTT diagram. Steel other than eutectoid composition will have one more line initiating from the nose in TTT diagram corresponding to transformation of austenite into proeutectoid phase. In case of hypoeutectoid steel, first austenite forms ferrite as a proeutectoid phase, subsequently it transforms into pearlite, whereas the proeutectoid phase for hypereutectoid steel is cementite.

Addition or reduction in carbon % in steel with respect to eutectoid composition shifts the nose of TTT diagram. For hypoeutectoid steels, nose of the curve is shifted towards the left (reduced incubation time), whereas for hypereutectoid steels nose is shifted to right as compared to that for eutectoid steel. Temperature corresponding to start (M_s) and end (M_f) of martensite transformation is found as a function of alloy compositions. Addition of alloying elements lowers these temperatures.

22.4.2 Continuous Cooling Transformation (C.C.T.) Diagram

The TTT diagram superimposed with cooling curves gives idea of different temperatures at which various phases can be formed transformation (Fig. 22.3a). However, actual phase transformation occurs under continuous cooling conditions observed in CCT diagram at slightly different time and temperature and apart from missing below the nose region of TTT diagram.

CCT curve shows the transformation of austenite into various phases as a function of time at different cooling rates but not at constant temperature like in TTT diagram (Fig. 22.3b). This diagram is similar to that of TTT diagram except that under continuous cooling conditions (when temperature changes continuously) nose of the curve is shifted to right in downward direction and bainite transformation part as obtained in TTT diagram (below the nose) is absent in CCT diagram. Hence, continuous cooling diagram for eutectoid steel has only two lines above M_s, corresponding to the start and end of pearlite transformation (Fig. 22.3b). Various lines CR1, CR2, CR3, CR4 and CR5 show the reduction in temperature with time representing the different cooling

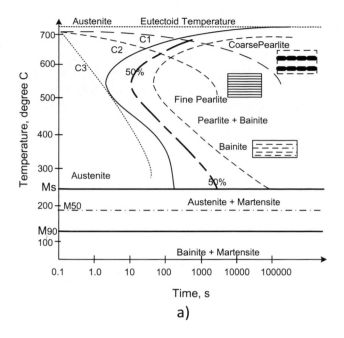

a)

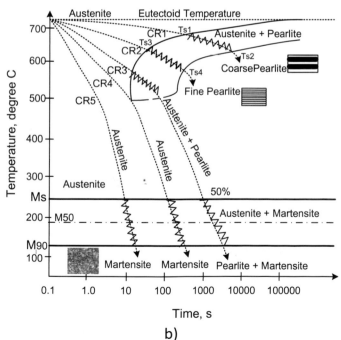

b)

Fig. 22.3 **a** TTT diagram superimposed with different cooling curves and **b** continuous cooling transformation (CCT) diagram

rates called cooling curves. A very low cooling rate "CR1" results in the transformation of austenite into coarse pearlite starts at T_{s1} temperature and ends at T_{s2} temperature. Therefore, transformation takes place over a range of temperature from T_{s1} to T_{s2}. Such low cooling rates are used for annealing of steels which increases the softness. At somewhat higher cooling rate (KN), the transformation of austenite into pearlite starts at T_{s3} temperature and ends at T_{s4} temperature. Therefore, effective transformation takes place over a range of temperature from T_{s3} to T_{s4}. Grain size depends on the transformation temperature. High transformation temperature produces coarse grain. Since the continuous cooling conditions result in transformation over a range of temperature say T_{s1} to T_{s2}, the grain size also varies accordingly. Therefore, at the start of transformation, coarser pearlite grains are formed than that at the end of transformation. High cooling rate reduces the effective transformation temperature, hence fine grain structure is produced. Such cooling rates are used for normalizing of steel. Normalizing increases the strength, hardness and toughness due to finer grain structure.

Cooling curve "CR3" is tangential to 50% transformation line, where 50% austenite has transformed into pearlite and 50% austenite is yet to transform. Thereafter, transformation of austenite stops, and no further transformation takes place until its temperature goes down to M_s. As austenite crosses the M_s temperature, remaining 50% austenite begins to transform into the martensite. This transformation line is expected to result in about 50% of martensite and approx. 50% pearlite.

Further, higher cooling rate line "CR4" which is tangential to the nose of the CCT diagram does not cause any transformation of austenite into pearlite. Austenite remains stable until M_s temperature is reached. Further reduction in temperature transforms the austenite into the martensite. Moreover, complete transformation of austenite into martensite depends on the quenching temperature. If quenching temperature is below the "M_f", then only whole austenite is expected to transform into the martensite; otherwise, some untransformed austenite is left in steel called retained austenite. Retained austenite is comparatively soft, therefore, its presence reduces the hardness of steel. Amount of retained austenite depends on the quench temperature between M_s and M_f. Lower the transformation temperature (between M_s and M_f), smaller is the amount of retained austenite in steel. There is a nonlinear relationship between the amount of austenite transforming into martensite and quenching temperature in range of M_s and M_f. Minimum cooling rate that ensures complete transformation of austenite into martensite and avoids the formation of soft phases/phase mixtures (like pearlite) is called critical cooling rate (CCR).

Critical cooling rate depends on the position of the nose, which is governed by the alloy composition, grain size and homogeneity of austenite. To take into account the effect of all alloying elements on the critical cooling rate, carbon equivalent is used.

High carbon equivalent lowers the critical cooling rate, hence less drastic cooling is required for hardening. In general, presence of all alloying elements (except Co) shifts the nose of CCT diagram towards right (conversely increases the incubation period to begin transformation) which in turn reduces the critical cooling rate and increases hardenability. Reduction in carbon content increases the critical cooling rate and makes hardening of steel more difficult. Fine-grained austenite starts the transformation earlier, so nose of curves is shifted to left (reduces the incubation period for transformation). This increases the critical cooling rate. On the other hand, coarse grain structure reduces the critical cooling rate which in turn increases the hardenability. Similarly, inhomogeneous austenite (due incomplete transformation/austenitizing during heating) also reduces the transformation time, shifting nose of CCT curve to left and so increases the critical cooling rate.

22.5 Metallurgical Transformation in Fe–C System During Fusion Welding

A typical fusion welding of Fe–C system like wrought iron, steel, alloys and cast irons forms three zone fusion weld zone, heat affected zone and unaffected base metal. The fusion zone and heat affected zone undergo many metallurgical transformations as per composition of parent metal and weld thermal cycle experienced during welding.

22.5.1 Fusion Weld Zone (Autogenous Welding/Matching Filler and Electrode)

Fusion weld zone at fusion boundary is heated to a maximum liquids temperature while the maximum temperature of weld centre can vary as per power density of heat sources being used for welding purpose from 3000 °C for gas welding to more than 20,000 °C for laser welding. Similar maximum cooling rate occurs at the weld centre and minimum at fusion boundary. The cooling rate directly affects the grain size and phase transformation. High cooling rate at the centre results in finer grain than those at the fusion boundary imminently after solidification. In addition to grain size, cooling rate of the weld also dictates the solid state phase transformation, wherein phases can be estimated using, respectively, CCT diagram.

The power density of welding process determines the net heat input which in turn governs the cooling rate imposed. Increasing power density for the welding reduces the net heat input required which in turn increases the cooling rate. High-power density welding process like laser welding (10^8 W/mm^2) imposes a cooling rate in the weld which is found to greater than critical cooling rate of even low carbon steel leading to partial martensitic transformation apart from fine pearlite, while power density welding processes (like gas welding, submerged arc welding) can result in fine ferrite, pearlite, etc. The most of the steels and cast irons undergo solid state transformation from austenite to the other phases as per composition and cooling rate, therefore typical planar, cellular, dendritic and equiaxed grain structure are not observed in weld metal. Solid state phase transformation in weld metal after solidification is considered to be a more problematic especially in case of steel having carbon equivalent CE > 0.4 due to high cracking tendency and embrittlement caused by martensitic transformation in the weld metal developed using either autogenous welding or matching filler/electrode. As cooling rate experienced by the weld metal during fusion welding is found to be much higher than the critical cooling rate for most of steel having carbon equivalent CE > 0.4. Moreover, the welding conditions like process, parameters, net heat input and preheat leading to a cooling rate lower than critical cooling rate of the weld metal will be resulting in somewhat coarser and softer phases like pearlite, ferrite and bainite.

Hardness profile of weld zone of hardenable steel developed using arc fusion welding processes invariably experiences martensitic transformation irrespectively of the base metal condition due to the fact that cooling rate experienced in weld pool is higher than the critical cooling rate. However, the response of HAZ may be varying significantly depending upon the heat treatment of conditions hardenable steel (Fig. 22.4). Hardenable steel in cold/hot rolled, annealed, normalized and hardened conditions usually exhibits the hardening of HAZ (due to martensitic transformation) in arc welding without preheat while the same hardenable steel in Q & T condition when subjected to arc welding shows the HAZ softening primarily due to overtempering caused by weld thermal cycle during fusion welding.

22.5.2 Fusion Welding Dissimilar Filler/Electrode for Steel and Cast Iron

The fusion welding of high CE steel and cast iron is something performed using filler/electrode like nickel alloys, austenitic stainless steel, etc., which are completely different from parent metal to realize desired properties like toughness, abrasive wear and corrosion resistance and avoid cracking and residual stress-related issues. In such cases, the solidification proceeds through nucleation and growth stages, and weld metal exhibits typical planar, cellular, dendritic and equiaxed structure per the cooling condition. Grain size increases from the fusion boundary to weld centre as in such cases hardly solid state phase transformation during the cooling of the weld metal is observed.

Filler/electrode for fusion welding of hardenable steel and cast iron can be different from the base metal itself to deal with issues related to cracking, distortion and residual stresses. Therefore, filler/electrode for fusion welding can be stronger/weaker and according to hardness profile, the weld zone obtained can be harder/softer than base metal and HAZ. However, the response of the HAZ will depend on conditions of hardenable steel. Q & T steel on fusion welding invariable shows the HAZ softening irrespective of the weld metal characteristics (Fig. 22.5a), while the hardenable steel in annealed, normalized and hardened conditions usually shows hardening of the HAZ (Fig. 22.5b). Further, the factors governing the cooling

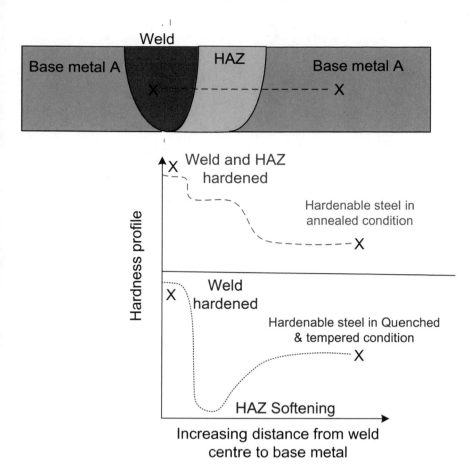

Fig. 22.4 Schematic showing different zones in autogenous fusion weld joint of hardenable steel and hardness profile from weld centre to hardenable steel base metal in annealed and quenched and tempered (Q & T) conditions

rate like initial base metal temperature and heat input can bring in variation in hardness distribution as per structural transformations occurring in weld and HAZ of hardenable steels.

22.5.3 *Weld Zone Developed Using Solid State Joining Processes*

Solid state joining processes like ultrasonic welding (USW), eemperature and zero sxplosive welding (EW) and friction stir welding (FSW) result in a combination of

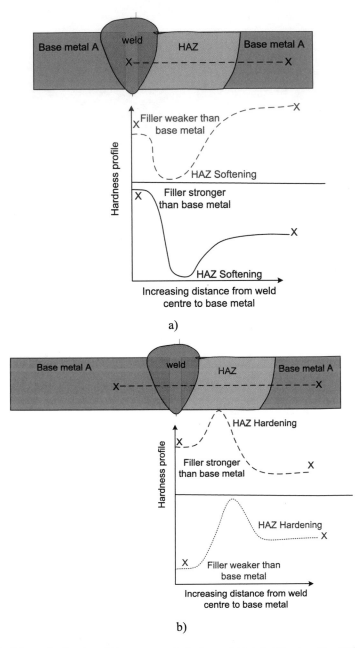

Fig. 22.5 Schematic showing different zones in fusion weld joint of hardenable steel developed using both filler/electrode stronger and weaker than base metal and hardness profile from weld centre to hardenable steel base metal experiencing **a** HAZ softening with base metal in Q & T condition and **b** HAZ hardening with base metal in normalized condition

heat generation under in situ condition due to friction and deformation. If the heat generated during joining causes minor increase in interfacial temperature of base metal (USW, EW), then chances for metallurgical transformation due to heating and subsequent cooling are very less. However, in case of all other solid state joining processes (like FSW, resistance welding) where heat generated during welding of steel can increase the temperature of weld zone well above the upper critical temperature, and therefore solid state metallurgical transformation occurs as per cooling condition and carbon equivalent of metal. Therefore, FSW weld nugget of steels commonly shows both martensite and pearlite phases with significant increase in hardness and strength of weld leading joint efficiency greater than 100%. Further, many ferrous systems like high manganese steel, austenitic steel and Hadfield steel show deformation-assisted austenite to martensite transformation; such steels after solid state joining may exhibit combined pearlite, martensite and austenite structure in the weld zone leading to higher hardness and strength of joining (Fig. 22.6a). However, the solid state joining of precipitation hardenable and work hardenable steel shows entirely different hardness profile of weld and HAZ due to dissolution/reversion of hardening precipitates, grain refinement in welds and coarsening in HAZ, work hardening and deformation-assisted transformation. In case of precipitation, hardening metals mostly weakening and softening of weld and HAZ is caused by reversion/dissolution of precipitates and grain coarsening in HAZ while work hardenable metals coupled with deformation-assisted transformation usually result in hardening of weld and may cause softening of HAZ as per weld thermal cycle experienced by HAZ (Fig. 22.6b).

22.6 Heat Affected Zone in Weld Joint of Fe–C System

Heat affected zone of weld joint developed by fusion welding and solid state joining processes may differ in many ways.

The partially melted zone is formed adjacent to the fusion boundary in fusion welding, while thermomechanical affected zone is developed in case of deformation-based solid state joining processes like friction stir welding.

22.6.1 The Partial Meting Zone

Pure metals and eutectic alloys have single melting temperature and zero solidification temperature range, therefore PMZ is not formed while alloys having a range of solidification temperature result in the PMZ (Fig. 22.7a, b). The partial melting zone depends on solidification temperature range (difference of liquidus and solidus temperature) of base metal. A high solidification temperature range causes wider partial melting zone (PMZ) forming a mushy zone due to heat supplied for fusion welding (Fig. 22.7c). The PMZ under the influence of residual tensile stress causes liquation cracking in the region adjacent to fusion boundary. Stainless steel,

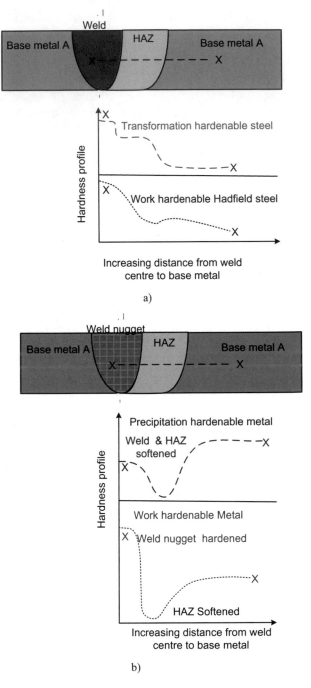

Fig. 22.6 Schematic showing different zones of weld joint developed by solid state joining process (like FSW) of **a** transformation hardenable metal like alloy steel and **b** work hardening metals like Hadfield steel

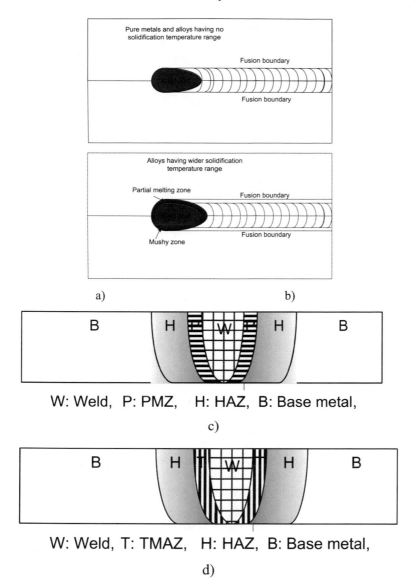

Fig. 22.7 Schematic showing different zone formed in a weld joint developed by **a, b** formation of zones around the weld pool in case of pure metal and alloys, **c** transverse section of fusion weld joint of an alloy showing PMZ and HAZ and **d** solid state joining like FSW

aluminium alloys and all metal exhibiting high solidification temperature range form PMZ and show cracking tendency.

22.6.2   Thermo-Mechanically Affected Zone

Solid state joining processes like FSW involving significant plastic deformation of faying surfaces of base metals form a zone adjacent to the weld nugget called thermo-mechanically affected zone (TMAZ). The TMAZ experiences both effect of "stress induced" and "heat generated" during joining, which in turn leads to formation of zone having highly strained and elongated grain (Fig. 22.7d). Although width of the TMAZ is very narrow, e.g. 2–5 μm, this zone is expected to have effect of work hardening while HAZ primarily shows effect of metallurgical changes.

22.6.3   Heat Affected Zone

Further, the severity of change observed in HAZ of a weld joint primarily depends on the net heat input. Higher the net heat input, greater will be undesirable alteration in HAZ properties in the form grain growth, mechanical properties and corrosion resistance. In general, wider heat affected zone is observed in case of fusion welding than that in case of the solid state joining. Moreover, the extent of grain coarsening in HAZ of weld joint developed by solid state joining processes is limited as compared to fusion welding. A typical fusion weld joint of an alloy steel shows weld zone, partial melting zone, coarsening grain heat affected zone, fine grain heat affected zone, recrystallized zone and base metal (Fig. 22.8).

The metallurgical transformation taking place in the HAZ depends on the thermal cycle experienced by metal. The weld thermal cycle of a location based on the peak temperature and duration of high-temperature retention above certain critical temperature determines the recrystallization, recovery, homogeneity of solid solution, dissolution, reversion of precipitates and gain growth. The peak temperature and high-temperature retention period both increase on approaching from base metal

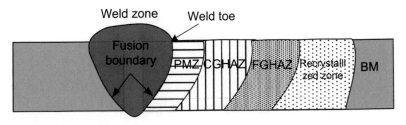

Fig. 22.8 Schematic showing formation of different zones a fusion weld joint of steels

to weld fusion boundary or weld nugget. Therefore, increasing severity of mostly undesirable metallurgical changes takes place on moving from the base metal to fusion boundary/weld nugget. The kind of metallurgical change occurring in HAZ and corresponding effect on mechanical properties depends on the type of metal system (WH, PH, TH, DH, SSH) being joined; however, grain coarsening of HAZ is common in all the cases. The relative effect of grain coarsening on HAZ properties is not same for all the types of the metal systems. For example, if relative effect of main metal strengthening mechanism of a given base metal (SSH, DH, TH) is more dominating over effect of grain refinement, then despite the grain, coarsening hardening/strengthening of HAZ can occur like in case of welding of many steel and cast irons. On the other hand, coarsening of grain structure in HAZ can cause significant weakening and deterioration in mechanical properties.

22.7   HAZ of Base Metal Strengthened by Work Hardening

Heat affected zone of a joint made of work hardened base metals (Like AA 5xxx alloys) either by a fusion welding or a solid state joining process (if raised temperature of the base metal is high enough) shows the significant softening/weakening. The HAZ softening mainly occurs due to (a) loss of work hardening effect caused by recovery and (b) grain coarsening. Further, the degree of softening and width of softened heat affected zone mainly depend on net heat input imparted during welding (Fig. 22.9). In general, the higher is the heat input, more severe will be softening and width of weak zone of HAZ.

The recovery and grain coarsening in HAZ decrease on moving away from fusion boundary/weld nugget zone accordingly, and hardness increases gradually until the unaffected base metal. These variations in structure and hardness of HAZ occur due to weld thermal cycle experienced during welding/joining. The zone close to fusion boundary/weld nugget experiences higher peak temperature for longer time, which in turn causes more severe recovery, recrystallization and grain growth. While on moving away from fusion boundary/weld nugget in the HAZ, both peak temperature and high-temperature retention period decrease causing lesser and lesser recovery, recrystallization and grain growth leading to gradually increasing hardness of HAZ (Fig. 22.10).

The softening of HAZ makes it sensitive for fracture as well as preferred fracture location, because in case of fusion welding, choice of suitable filler/electrode can provide a weld stronger than HAZ and base metal itself. The weld nugget developed by solid state joining will have work hardening effect due to localized plastic deformation so weld nugget would be stronger than HAZ and even base metal also.

The typical 5xxx series aluminium alloys are strengthened by work hardening. Almost all metals manufactured by deformation-based processes possess some strain hardening effect, which is eliminated/reduced after welding. Therefore, all metals after the welding may experience little weakening unless other metal strengthening mechanisms (such as TH, PH) are not offsetting such deterioration.

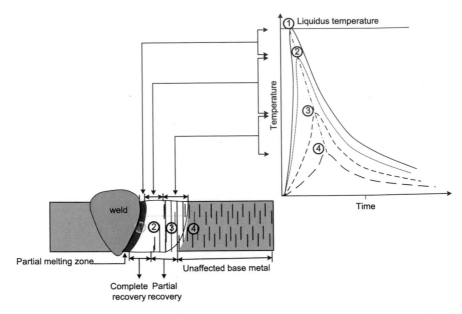

Fig. 22.9 Schematic showing relationship between weld thermal cycle and structural changes observed in HAZ of work hardenable metals

22.8   HAZ of Base Metals Strengthened by Precipitation Hardening

To understand the HAZ characteristics of PH metals, we need to be aware of stability of precipitates associated with metal as a function of temperature. Strengthening precipitates get destabilized or dissolved in the matrix above certain critical temperature depending upon the composition as evident figure below. Formation and dissolution of precipitates affect hardness and strength of metal during the welding.

Heat affected zone of weld joint of precipitation strengthened base metals (like AA2xxx, AA6xxx, AA7xxx, PH SS, Mg and Cu alloy) developed either by fusion welding or solid state joining process (if raised temperature of the base metal is high enough) also shows the significant softening/weakening. The HAZ primarily occurs due to (a) loss of strengthening precipitates caused by reversion/dissolution, (b) coarsening of grain and precipitates. Further, the degree of softening and width of softened heat affected zone in case of PH metals also determined by net heat input are imparted during welding. Increasing net heat input causes more softening and wider HAZ.

The reversion (dissolution of precipitates) and grain coarsening in HAZ decrease on moving away from fusion boundary/weld nugget zone accordingly, and hardness increases gradually until the unaffected base metal. These variations in structure and hardness of HAZ occur due to weld thermal cycle experienced during welding/joining. The zone close to fusion boundary/weld nugget experiences higher

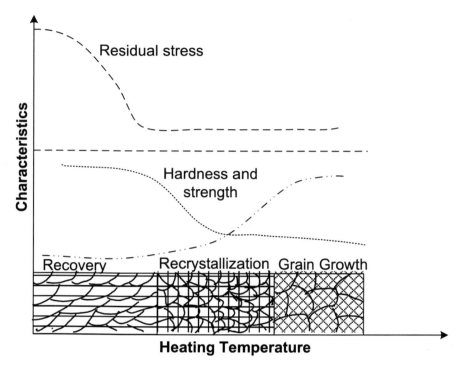

Fig. 22.10 Schematic showing effect of heating temperature during welding on structure and mechanical properties in HAZ

peak temperature for longer time, which in turn causes more severe reversion and grain growth. While on moving away from fusion boundary/weld nugget in the HAZ, both peak temperature and high-temperature retention period decrease causing lesser and lesser reversion and grain growth leading to gradually increasing hardness of HAZ. A combination of decrease in reversion (increasing the availability of hardening precipitates) and finer grains and precipitates in HAZ increases the hardness at locations away from fusion boundary and approaching to the unaffected base metal.

Softening of HAZ makes it prone to fracture. In fusion welding, choice of suitable filler/electrode can provide a weld stronger than HAZ and base metal itself. However, the weld metal/nugget can be further weaker than HAZ of PH metal especially in case of autogenous fusion welding or high heat input solid state joining process like friction stir welding where in all strengthening precipitates are dissolved from weld metal/weld nugget. The fracture under tensile loading in such weld joints frequently occurs from weld zone itself.

PH metals are of two types, namely naturally aged (AA 2xxx and AA 6xxx) and artificially aged (AA 7xxx). Metals (like AA 7xxx) showing natural ageing tendency regain the lost hardness and strength within 7–10 days after welding. While other PH metals need post-weld heat treatment (PWHT) like artificial ageing T6 treatment to restore the properties (Fig. 22.11).

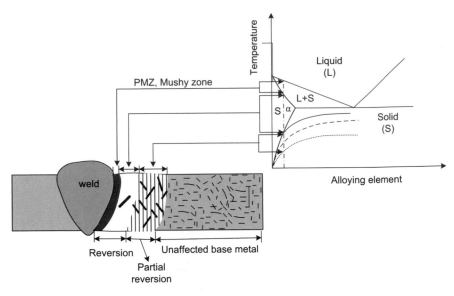

Fig. 22.11 Schematic showing relationship between weld thermal cycle and structural changes observed in HAZ of precipitation hardenable metals

22.9 HAZ of Transformation Hardening Metals

HAZ of transformation hardening metals (like steel, cast iron) experiences more complex changes than work hardening and precipitation hardening metal during the welding. Modification in microstructure and mechanical properties in heat affected zone of transformation hardening metals can be understood from careful correlation of weld thermal cycle, Fe–C diagram and continuous cooling transformation diagram. The weld thermal imposed during welding is found to be completely different from conventional heat treatment of steel in respect of (a) peak temperature and its distribution in HAZ, (b) high-temperature retention/soaking period and (c) cooling rate.

22.9.1 Peak Temperature

The peak temperature in vicinity of fusion boundary is very high and then keeps on decreasing on moving away from fusion boundary towards the unaffected base metal. In case of Fe–C system, based on peak temperature experienced by HAZ, five different zones can exist (Fig. 22.12):

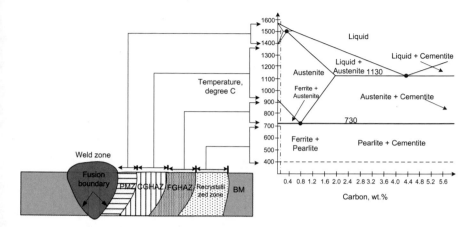

Fig. 22.12 Schematic showing relationship between various phase transformation temperatures of Fe–C phase diagram and formation of different zones in weld joints of carbon steel

- A zone next to fusion boundary having temperature between liquidus and solidus forming a mushy zone of delta ferrite/austenite and molten metal, thus this zone is called partial melting zone and is found sensitive to liquation cracking.
- High-temperature gamma loom zone forms only coarse austenite, and it promotes the coarse martensite/pearlite as per cooling conditions, thus this zone forms coarse grain HAZ (CGHAZ)
- Intercritical temperature zone forms both austenite and ferrite, and it leads to the formation of fine grain ferrite, pearlite/martensite/bainite as per composition and cooling rate, thus this zone forms fine grain HAZ (FGHAZ)
- Zone experiences temperature between sub-critical temperature and recrystallization temperatures, wherein phases remain unaffected, but recrystallization takes place.
- Unaffected base metal zone wherein temperature is lower than recrystallization temperature.

22.9.2 High-Temperature Retention Period/Soaking Time

The high-temperature retention period during welding in HAZ is very short to develop homogeneous solid solution of austenite through complete dissolution and phase transformation especially in alloy steel having stable phase, compound. Therefore, very inhomogeneous solid solution of austenite along with the presence of other stable phases formed in HAZ which in turn affects the transformation behaviour of austenite completely different from that is expected from CCT diagram. In general, increase in soaking temperature decreases time required for the formation of homogeneous austenite from the different phases of carbon and alloy steels. The HAZ region is rich in alloying elements leading to high CE and austenite to martensite

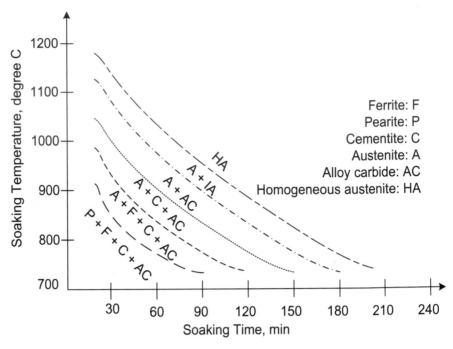

Fig. 22.13 Schematic showing effect soaking temperature on time required for transformation of various phases commonly found in carbon and alloy steel in homogeneous austenite

transformation, while zones deficient of alloying elements result in softer phases like ferrite, pearlite and bainite. The region next to the fusion boundary experiences very high temperature for a short period leading to coarsening of grain for formation of usually inhomogeneous austenite for transformation in different phases as per cooling rate encountered during the welding (Fig. 22.13). A coarse austenite enriched with alloying elements promotes the martensitic transformation.

22.9.3 Cooling Rate

Like peak temperature, the cooling rate experienced by HAZ is maximum at fusion boundary, and then it keeps on decreasing gradually on moving away from fusion boundary towards the unaffected base metal. Therefore, coarse austenite in CGHAZ (corresponding to CR1) tends form martensite, while austenite in FGHAZ (corresponding to CR2 and CR3) may form fine pearlite/martensite along with pre-existing proeutectoid ferrite (Fig. 22.14). Thus in a very narrow zone (HAZ) next to the fusion, a large variation in microstructure and mechanical properties can be observed. Cooling rate of experienced by steel having peak temperature (in sub-critical range) does not affect the final microstructure and so mechanical properties appreciably.

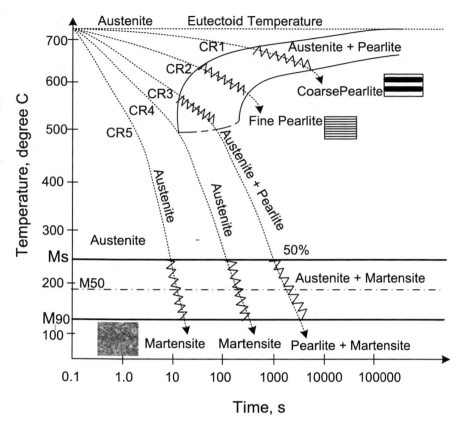

Fig. 22.14 CCT diagram showing effect of cooling on transformation of austenite into different phases as the cooling rate encountered in weld metal and HAZ of hardenable steel

Further, it is important to note that base metal composition (CE) and condition (heat treatment condition, e.g. annealed, normalized, quenched and tempered "Q & T", work hardened significantly dictate the changes that can be experienced in heat affected zone. For example, HAZ of a medium carbon steel welded in annealed condition shows significant hardening while the same steel in Q & T condition exhibits HAZ softening due to overtempering.

22.10 HAZ of Solid Solution and Dispersion Strengthened Metals

The HAZ of solid solution strengthened metals (like many alloys) and dispersion hardened metal (like the most of metal matrix composites developed using reinforcement of hard constituents in soft and tough metal matrix) show very marginal variation in microstructure and properties. The weld thermal cycle leading to very high temperature for little longer time causes coarsening and recovery (if metal is in work hardened condition) as shown in Fig. 22.15. These changes may result in minor softening of HAZ.

Questions for Self-assessment

(a) How does the transformation temperature affect the grain structure of fusion weld joint?

(b) Explain the importance of physical metallurgy of steel in welding.

(c) How can Fe–C diagram and CCT diagram can be used to explain microstructural transformation in carbon steel weld joint?

(d) How do the micro-constituents found in steel (as evident from Fe–C diagram) affect the mechanical properties?

(e) How does the selection of filler metal affect the hardness distribution in fusion weld joint of hardenable steel?

(f) How do HAZ characteristics of a hardenable steel fusion weld joint be related to Fe–C diagram?

(g) Explain the weld and HAZ properties of joints developed by fusion and solid state joining techniques of following types of metals.

    I.   Work hardenable

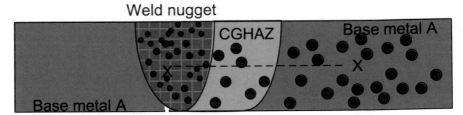

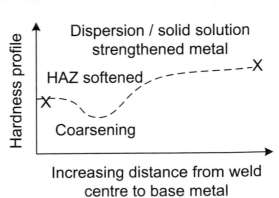

Fig. 22.15 Schematic showing different zones of weld joint developed by solid state joining (like FSW)/autogenous fusion process (like GTAW) of solid solution strengthened/dispersion hardened metals

II.   Transformation hardenable
III. Precipitation hardenable
IV.   Dispersion hardenable.

Further Reading

Avner SH (2009), Introduction to physical metallurgy, 2nd edn. McGraw Hill, New Delhi
Dwivedi DK (2013) Production and properties of cast Al-Si alloys. New Age International, New Delhi
Dwivedi DK (2018) Surface engineering. Springer, New Delhi
Kou S (2003) Welding metallurgy, 2nd edn. Willey, USA
Lancaster JF (1999) Metallurgy of welding, 6th edn. Abington Publishing, England
Little R (2001) Welding and welding technology, 1st edn. McGraw Hill
Mandal NR (2005) Aluminium welding, 2nd edn. Narosa Publications
Metals Handbook (1993) Welding, brazing and soldering, 10th edn, vol 6. American Society for Metals, USA
Nadkarni SV (2010) Modern arc welding technology. Ador Welding Limited, New Delhi
Parmar RS (2002) Welding engineering and technology, 2nd edn. Khanna Publisher, New Delhi
Welding Handbook (1987) 8th edn, vols 1 and 2. American Welding Society, USA

Part VIII
Design of Weld Joints

Chapter 23
Design of Welded Joints: Weld Failure Modes, Welding Symbols: Type of Welds, Joints, Welding Position

23.1 Introduction

Weld joints may be subjected to variety of loads ranging from a simple tensile load to a complex combination of torsion, bending and shearing loads depending upon the service conditions. The capability of weld joints to take up a given load comes from metallic continuity across the members being joined. Mechanical properties of the weld metal and load resisting cross-sectional area of the weld are the two most important parameters which need to be established for designing a weld joint. Moreover, in some cases HAZ properties gain more significance in the design of weld joints if softening of the HAZ due to weld thermal cycle is encountered (Fig. 23.1). The weld zone and heat affected zone can weaker or stronger than base metal depending upon the strengthening mechanism of the base metal, choice of filler/electrode, if any, welding procedure, etc. Further, the joints invariable comprises stress raisers in the form of discontinuities, changing cross-sectional area which in turn encourage the failure of the joints. These aspects must be kept in mind while designing weld joints.

23.2 Modes of Failure of the Weld Joints

A poorly designed weld joint can lead to the failure (Fig. 23.2a–c) in three ways: (a) elastic deformation of weld joint beyond acceptable limits (like bending or torsion of shaft and other sophisticated engineering systems like precision measuring instruments and machine tools), (b) plastic deformation of engineering component across the weld joint leading to change in dimensions and geometry beyond acceptable limits as-decided by application of the service load and (c) fracture of weld joint into two or more pieces under external tensile, shear, compression, impact creep and fatigue loads. Thus, depending upon the expectation from the weld joints for an

© The Author(s), under exclusive license to Springer Nature Singapore Pte Ltd. 2022   327
D. K. Dwivedi, *Fundamentals of Metal Joining*,
https://doi.org/10.1007/978-981-16-4819-9_23

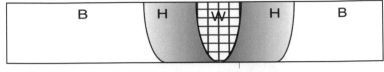

W: Weld, P: PMZ,   H: HAZ, B: Base metal,

Fig. 23.1 Schematic of a typical weld joint of metal having weld, HAZ and base metal

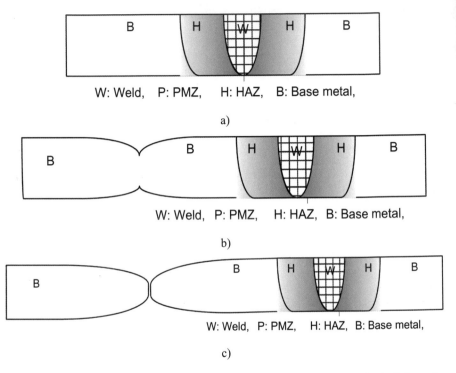

W: Weld, P: PMZ,   H: HAZ, B: Base metal,

a)

W: Weld, P: PMZ,   H: HAZ, B: Base metal,

b)

W: Weld, P: PMZ,   H: HAZ, B: Base metal,

c)

Fig. 23.2 Schematic showing different modes of failure **a** elastic deformation, **b** plastic deformation and **c** fracture of a weld joint

application failure of may occur in different ways. Hence, different approaches are needed for designing the weld joints as per application and service requirements.

23.3   Design of Weld Joints and Mechanical Properties

Stiffness and rigidity are two important parameters for designing weld joints when failure can occur due to elastic deformation. Under such conditions, weld metal of high modulus of elasticity (E) and rigidity (G) is deposited for producing weld joints besides selecting suitable load-resisting cross-sectional area. When the failure criterion for a weld joint is plastic deformation, then weld joints are designed on the basis of yield strength of the weld metal. When the failure criterion for weld joint is to avoid fracture, then it is required to consider the type of loading such as static load, fatigue and impact loading. For static loading, ultimate strength of the weld metal is used as a basis for design while under fatigue and creep conditions design of weld joints is based on specialized approaches which will be discussed in the later stages of this chapter. Under simplified conditions, design for fatigue loads is based on endurance limit.

Weld joints invariably possess different types of weld discontinuities of varying sizes. These discontinuities can be very crucial in case of critical applications, e.g. weld joints used in nuclear reactors, aerospace and spacecraft components. Therefore, weld joints for critical applications are designed using fracture mechanics approach. The fracture mechanics approach takes into account the size of discontinuity (in form of crack, porosity or inclusions), applied stress and weld material properties (yield strength and fracture toughness) in the design of weld joints.

23.4 Factors Affecting the Performance of the Weld Joints

It is important to note that the mechanical performance of the weld joints is governed by not only mechanical properties of the weld metal and its load-resisting cross-sectional area (as mentioned above) but also on HAZ properties. The properties of the weld and HAZ are influenced by the welding procedure used for developing a weld joint which includes the edge preparation, weld joint design, and type of weld, number of passes, preheat and post-weld heat treatment, if any, welding process and welding parameters (welding current, arc length, welding speed) and method used for protecting the weld contamination from atmospheric gases. As most of the above-mentioned steps of welding procedure influence weld bead geometry affecting stress concentration, metallurgical properties and residual stresses in weld joint which in turn determine the mechanical (tensile and fatigue) performance of the weld joint.

23.5 Design of Weld Joints and Loading Conditions

Design of weld joints for static and dynamic loads needs different approaches because in case of static loads the direction and magnitude become either constant or changes very slowly while in case of dynamic loads such as impact and fatigue load conditions, the rate of loading is usually high and direction may change (Fig. 23.3). In case of fatigue loading both magnitude and direction of load may fluctuate in a regular or irregular pattern. Under the static load condition, low rate of loading allows a lot of time available for localized yielding to occur in the area of high stress concentration in the weld joints (Fig. 23.3b). The localized yielding across the section in turn results in

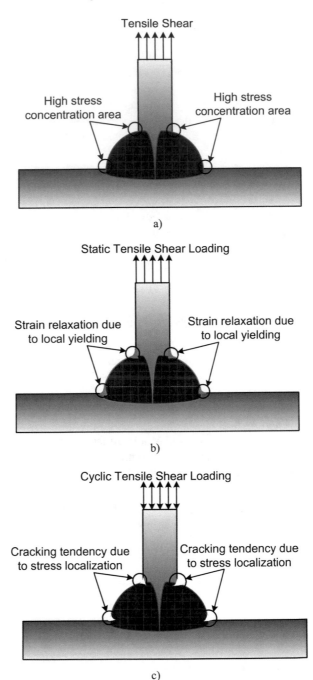

Fig. 23.3 Schematic showing effect of loading on stress localization and cracking tendency **a** weld joint with stress raiser at toe of weld, **b** stress raiser at toe of weld undergoing stress relaxation due to local yielding under static loading and **c** stress raiser at toe of weld getting cracked under cyclic loading

stress relaxation by redistribution of stresses throughout the cross section while under dynamic loading conditions, due to the lack of availability of time, yielding across the entire section of weld does not take place and only localized excessive deformation occurs near the site of a high concentration stress. The excessive localized plastic deformation provides an easy site for nucleation and growth of cracks like defect in case of fatigue loading (Fig. 23.3c).

23.6   Need of Welding Symbols

The welding procedure must be communicated without ambiguity to all those who are involved in various stages of weld fabrication ranging from edge preparation to final inspection and testing of welds. To assist in this regard, standard symbols and methodology for representing the welding procedure and other conditions have been developed. Symbols used for showing the type of weld to be made are called **weld symbols**. Some common weld symbols are shown below in Figs. 23.4 and 23.5. The common weld symbols are given below.

Welding Symbols

These are used to show not only the type of weld but all aspects related to welding like size and location of weld, welding process and parameters, cleaning, edge preparation, welding sequence, bead geometry and weld inspection process and location of the weld to be fabricated and method of weld testing, etc. The following section presents standard terminologies and joints used in the field of welding (Fig. 23.6). Different components usually exhibited by welding symbols include

- A reference line
- An arrow
- Tail
- Basic weld symbols
- Dimensions and other data
- Supplementary symbols
- Finish symbols
- Specifications, process and other references.

23.7   Types of Weld Joints

The classification of weld joints is based on the orientation of plates/members to be welded. Common types of weld joints and their schematics are shown in Fig. 23.7a–e.

- Butt joint: plates are in the same plane and are aligned with maximum deviation of 5°.

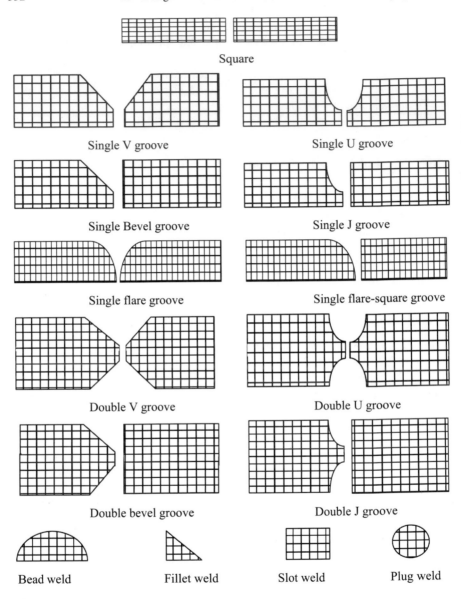

Fig. 23.4 Common groove geometries and weld symbols

- Lap joint: plates overlapping each other and the overlap can be just one side or both the sides of plates to be welded.
- Corner joint: joint is made by melting corners of two plates to be welded, and therefore, plates are approximately perpendicular (with angle range from 75°–90°) to each other at one side of the plates being welded.

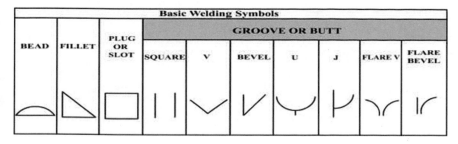

Fig. 23.5 Basic weld symbols

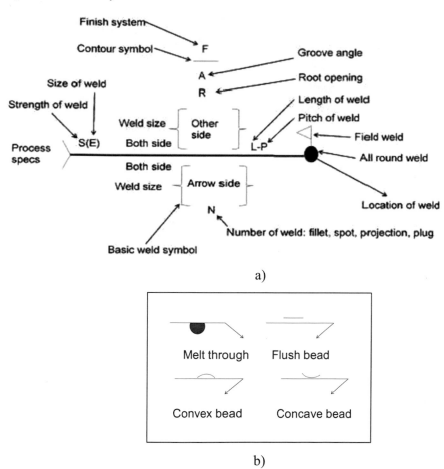

Fig. 23.6 Standard method of representing **a** welding symbol and **b** symbols for type of bead geometries

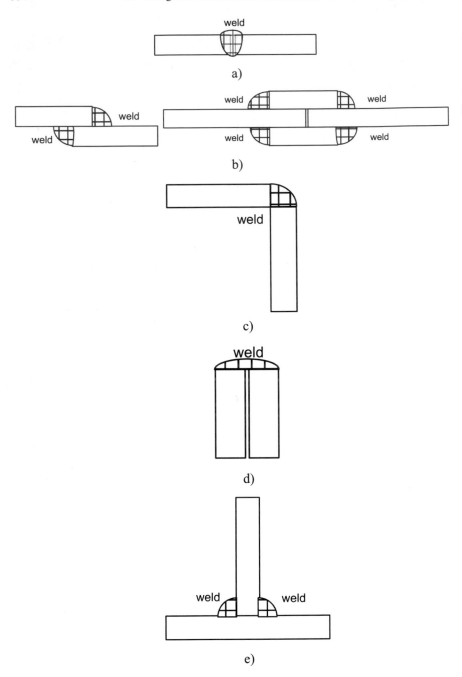

Fig. 23.7 Schematic of different types of weld joints **a** butt, **b** lap, **c** corner, **d** edge and **e** T joint

- Edge joint: joint is made by melting the edges of two plates to be welded, and therefore, the plates are almost parallel (0°–5°)
- T joint: one plate is approximately perpendicular to another plate (85°–90°) but not at the edges or corners.

23.8 Types of Weld

This classification in based on the combined factors like "how weld is made" and "orientation of plates" to be welded. Common types of weld joints are groove weld, fillet weld, plug weld and bead weld. Schematics of these welds are shown in Fig. 23.8a–d.

Welding Position and Selection of Weld Joints

The following section presents common types of welding position and various difficulties associated with them. Further, need of the edge preparation and the rationale for the selection of suitable groove design have also been presented.

23.9 Welding Position

The welding positions are classified on the basis of the plane in which weld metal is deposited.

Flat Welding

In flat welding, plates to be welded are placed on the horizontal plane and weld bead is also deposited horizontally (Fig. 23.9). This is one of most commonly used and convenient welding positions. Selection of welding parameters for flat welding is not very crucial for placing the weld metal at desired location in flat welding.

Horizontal Welding

In horizontal welding, plates to be welded are placed in vertical plane while weld bead is deposited horizontally (Fig. 23.10). This technique is comparatively more difficult than flat welding. Welding parameters determining the heat input for horizontal welding should be selected carefully for easy manipulation/placement of weld metal at the desired location.

Vertical Welding

In vertical welding, plates to be welded are placed on the vertical plane and weld bead is also deposited vertically (Fig. 23.11). It imposes great difficulty in placing the molten weld metal from electrode in proper place along the weld line due to the tendency of the melt to fall down under the influence of gravitational force. Viscosity and surface tension of the molten weld metal which are determined by the composition of weld metal and its temperature predominantly control the tendency

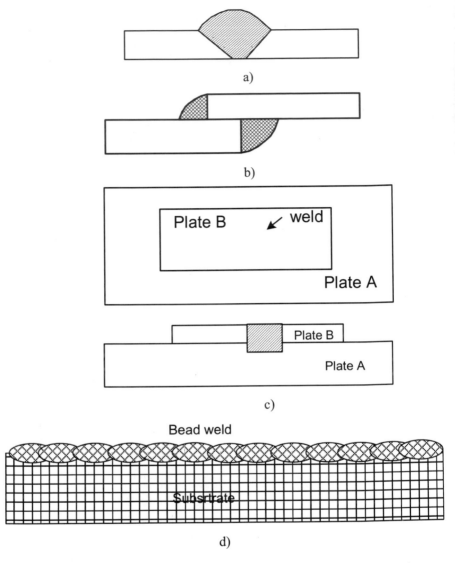

Fig. 23.8 Schematic of different types of weld **a** groove, **b** fillet, **c** plug and **d** bead-on-plate

of molten weld metal to fall down due to gravity. Increase in concentration of certain alloying elements/impurities and rise in temperature of molten weld metal in general decrease the viscosity and surface tension of the weld metal and thus making the liquid weld metal more thin with greater fluidity. High fluidity in turn increases tendency of weld metal to fall down conversely it increases difficulty in placing weld metal at the desired location. Therefore, selection of welding parameters (welding current, arc manipulation during welding and welding speed influencing the heat generation),

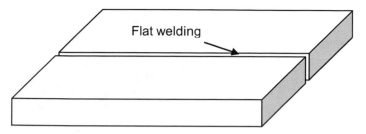

Fig. 23.9 Scheme of placement of components to be welded for flat welding

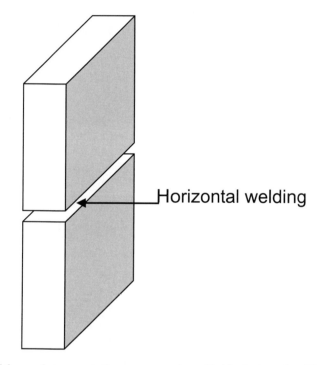

Fig. 23.10 Scheme of placement of components to be welded for horizontal welding

electrode coating (affecting composition of weld metal) and dilution become very crucial for placing the weld metal at desired location in vertical welding.

Overhead Welding

In overhead welding, weld metal is deposited in such a way that face of the weld is largely downward so there is high tendency of falling down of molten weld metal during the welding (Fig. 23.12). Molten weld metal is transferred from the electrode (lower side) to base metal (upper side) with great care and difficulty. Hence, it imposes problems similar to that of vertical welding but with greater intensity. Accordingly, the selection of welding parameters, arc manipulation and welding

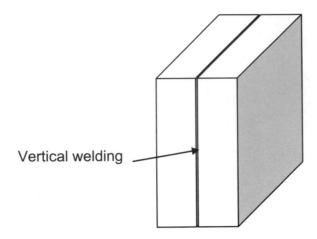

Vertical welding

Fig. 23.11 Scheme of placement of components to be welded for vertical welding

consumable should be done after considering all factors which can decrease the fluidity of molten weld metal so as to reduce the weld metal falling tendency. Overhead welding is the most difficult welding position, and therefore, it needs great skill and care to place the weld metal at desired location with close control.

23.10   Rationale Behind Selection of Weld and Edge Preparation

Groove Weld

Groove weld is called so because first a groove is made between plates to be welded. This type of weld is used for developing butt joint, edge and corner joint. The groove preparation especially in case of thick plates ensures proper melting of the faying surfaces (via through thickness penetration) by providing proper access of heat source up to the root of the plates and so as develop a sound weld joint (Fig. 23.13). It is common to develop grooves of different geometries for producing butt, corner and edge joint. The most common groove geometries include square, U (single and double), V (single and double), J (single and double) and bevel (single and double). The following sections describe various technical aspects of different types of groove welds suggesting the conditions for the selection of appropriate groove geometry.

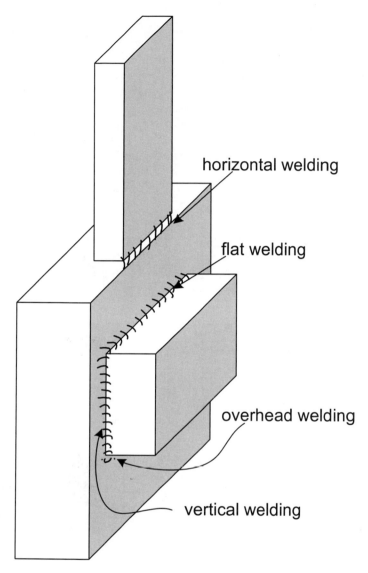

Fig. 23.12 Scheme of placement of components to be welded for different types of welding positions including overhead welding

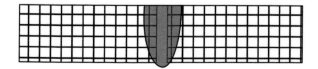

Fig. 23.13 Schematic of butt weld joint

23.10.1   Single Groove Weld

Single groove means edge preparation of the plates to produce desired groove from one side only resulting in just one face and one root of the weld (Fig. 23.14). While in case of double groove, edge preparation is needed on both sides (top and bottom) of the plates to be welded; it results in two faces of the weld, and welding is needed from both sides of the plates to be welded. Single groove weld is mainly used in case of plates of thickness more than 5 mm but less than 15 mm. Moreover, this range is not very hard and fast as it depends on the penetration capability of welding process used for welding besides welding parameters. Welding parameters, namely current and speed, affect the depth up to which melting of plates can be achieved from the top.

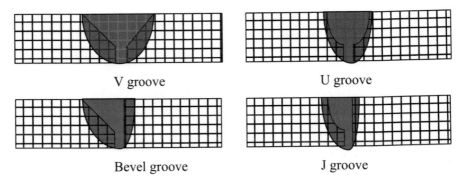

V groove   U groove

Bevel groove   J groove

Fig. 23.14 Schematic of butt weld joint developed using different single groove geometries

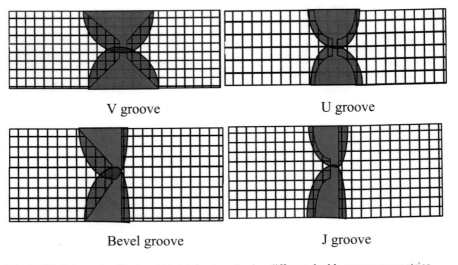

V groove   U groove

Bevel groove   J groove

Fig. 23.15 Schematic of butt weld joint developed using different double groove geometries

23.10.2   Double Groove Weld

Double groove edge preparation is used especially under two conditions of welding a) when thickness of the plate to be welded is more than 25 mm, (so the desired through thickness penetration is not achievable) and b) distortion of the weld joints is to be controlled (Fig. 23.15). Further, double groove edge preparation lowers the volume of weld metal to be deposited by more than 50% as compared to that for the single groove weld especially in case of thick plates. Therefore, the selection of double groove welds helps to develop weld joints more economically, at much faster welding speed than the single groove weld for thick plates.

23.11   Comparative Features of U and J Groove Geometry

U and J groove geometries are better than V and bevel grooves in terms of

- (low) volume of weld metal to be deposited
- (less) distortion and residual stress-related problems
- (high) welding speed.

However, these suffer from

- difficulty in machining
- poor arc accessibility for achieving desired penetration
- poor fusion of the faying surfaces.

Questions for Self-assessment

a. What is the need to design the weld joint systematically?
b. Describe different modes of failure of weld joints?
c. How are the weld joints designed for realizing different mechanical performances?
d. Explain the factors affecting the mechanical performance of the weld joints?
e. Why does designing of the weld joints for static loading become easier than fatigue loading?
f. What is need to study the welding symbols?
g. A T joint is to be made at the site using SMAW process having intermittent double fillet welds of 90 mm length. The centre distance between the welds is about 160 mm. The leg length of the weld is about 10 mm. Draw the welding symbol for the joint.
h. Schematically show the different types of weld joints.
i. Explain different types of weld with help of suitable schematic diagram?
j. Explain the following term with suitable sketch:

    i.   Backing

ii.   Base metal
iii. Bead or weld bead
iv.   Crater
v.   Deposition rate
vi.   Puddle
vii. Root
viii. Toe of weld
ix.   Weld face
x.   Weld pass

k.   With the help of schematic diagram describe different types of welding position?
l.   Does welding position affect the selection of welding parameters?
m.   Why vertical and overhead welding positions are found more difficult than horizontal and flat positions?
n.   What precautions should be taken in odd position welding?
o.   What is rationale behind the selection of single and double groove weld joints?
p.   Why is double groove weld joint design preferred for welding of thick plates?

Further Reading

Hicks J (1999) Weld joint design, 3rd edn. Abington Publishing, England
Kou S (2003) Welding metallurgy, 2nd edn. Wiley, USA
Maddox SJ (1991) Fatigue strength of welded structures. Woodhead Publishing
Parmar RS (2002) Welding engineering and technology, 2nd edn. Khanna Publisher, New Delhi
Radaj R (1990) Design and analysis of fatigue resistant welded structures. Woodhead Publishing
Welding Handbook (1987) 8th edn, vols 1 and 2. American Welding Society, USA

Chapter 24
Design of Welded Joints: Weld Bead Geometry: Selection, Welding Parameters

24.1 Groove Weld

Selection of a particular type of groove geometry is influenced by the compromise between two factors (a) machining cost to obtain desired groove geometry and (b) cost of weld metal (on the basis of volume) need to be deposited, besides other factors such as welding speed, accessibility of groove for depositing the weld metal, residual stress and distortion control requirement.

U and J groove geometries are more economical (than V and bevel grooves) in terms of volume of weld metal to be deposited, and offer less distortion and residual stress-related problems. These geometries allow higher welding speed due to low weld metal requirement to complete the joint. However, these groove geometries suffer from (a) difficulty in machining, (b) poor accessibility of heat sources up to root of the groove for achieving desired penetration and fusion of the faying surfaces (Fig. 24.1).

On contrary, V and bevel groove geometries can be easily and economically produced by machining or flame cutting and also provide good accessibility for applying heat up to root of groove. However, these groove geometries need comparatively more volume of weld metal which in turn causes more residual stress and distortion-related problems than U and J groove geometries.

Square groove geometry does not need any edge preparation except making edges clear and square, but this geometry can be used only maximum up to 10 mm plate thickness. However, this limit can vary significantly depending upon the penetration capability of welding process and welding parameters. Square groove is usually not used for higher thicknesses (above 10 mm) mainly due to the difficulties associated with poor penetration, poor accessibility of root and lack of fusion tendency at the root side of the weld. Therefore, it is primarily used for welding of thin sheets by GTAW/GMAW or thin plates by SAW.

Groove butt welds are mainly used for general purpose and critical applications where tensile and fatigue loading is expected during the service. Since butt groove geometry does not cause any stress localization (except those which are caused by

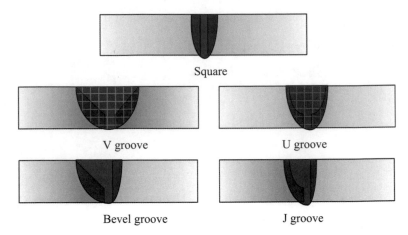

Square

V groove   U groove

Bevel groove   J groove

Fig. 24.1 Schematic showing different types of groove weld geometries

poor weld geometry at the weld toe and weld defects), therefore stresses developing in weld joints due to external loading largely become uniform across the section of the weld hence fatigue crack nucleation and subsequent propagation tendency are significantly reduced in groove butt weld as compared with fillet and other type of welds.

24.2   Fillet Weld

Fillet welds are used for producing lap joint, edge joint and T joint primarily for non-critical applications. Generally, these do not require any edge preparation; hence, these are more economically produced especially in case of thin plates as compared to groove weld. However, to satisfy the penetration requirement and better loading-carrying capacity sometimes groove plus fillet weld combination is also used. An increase in size of weld (throat thickness/leg length of the weld) increases the volume of weld metal in case of fillet welds significantly during of welding thick plates; hence, fillet welds become uneconomical for large size weld compared to groove weld (Fig. 24.2).

Due to inherent nature of fillet weld geometry, stresses are localized and concentrated near the toe of the weld. Therefore, weld toe frequently becomes an easy site for nucleation and growth of tensile/fatigue cracks. The stress concentration in the fillet weld near the toe of the weld occurs mainly due to abrupt change in load-resisting cross-sectional area from the base metal to weld. To reduce the adverse effect of stress localization on mechanical performance, efforts are made to have gradual transition/change in load-resisting cross-sectional area from the base metal to weld using (a) controlled deposition of the weld metal using suitable weld parameters (so as to have as low weld bead angle as possible), and manipulation of molten

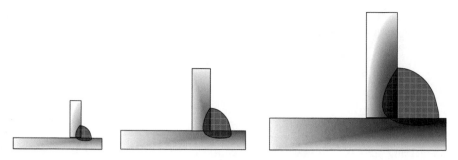

Fig. 24.2 Schematic showing effect of plate thickness on volume of fillet weld, i.e. increasing volume with thickness of plate

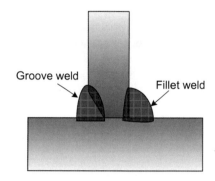

Fig. 24.3 Schematic of combined fillet and groove weld joint

weld metal while depositing the same and (b) controlled removal of the weld metal by machining/grinding. Sometimes to take the advantage of both groove and fillet welds are used to develop joints (Fig. 24.3).

24.3 Bead Weld

The bead weld is mainly used to apply a layer of a good quality metal over the comparatively poor quality base metal so as to have functional surfaces of better characteristics such as improved hardness, wear and corrosion resistance (Fig. 24.4). To reduce degradation in characteristics of weld bead of good quality materials during welding, it is important that intermixing of molten metal of weld bead with fused base metal is as less as possible while ensuring good metallurgical bond between the bead weld and base metal. The intermixing of bead weld metal with base metal during welds is called dilution. Higher dilution leads to greater degradation in quality of weld joint. Control over the dilution can be achieved by reducing extent of melting of base metal using suitable welding procedure such preheating, welding parameter,

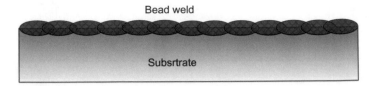

Fig. 24.4 Schematic of combined fillet and groove weld joint

and welding process. For example, plasma transferred arc welding (PTAW) causes less dilution than SAW primarily due to the difference in net heat input which is applied during welding in two cases. PTAW supplies lesser heat compared to other processes, namely MIGW, SMAW and SAW. Bead welds are also used just to deposit the weld metal same of the same composition as that of base metal so as to regain the lost dimensions. This process is called reclamation. The loss of dimensions of the functional surfaces can be due to variety of reasons such as wear and corrosion. These bead welds are subsequently machined out to get the desired dimensional accuracy and surface finish.

24.4   Plug Welds

The plug welds are used for comparatively less critical applications and static loading but not for severe impact and fatigue conditions. For developing plug weld first a through thickness slot (of circular/rectangular shape) is cut in one of plates to be joined (Fig. 24.5). The plate with cut hole is placed over another plate to be welded

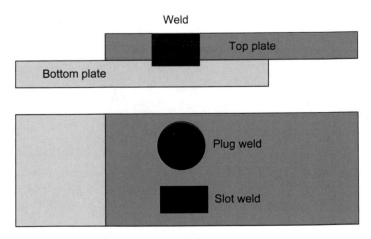

Fig. 24.5 Schematic of plus and slot weld in front and top view

and then weld metal is deposited in the slot so that joint is formed by fusion of both bottom plate and lower edges of slot in upper plate.

24.5 Welding and Weld Bead Geometry

For developing a fusion weld joint, it is necessary that molten metal from both sides of base metals and that from electrode/filler, if any, mixes together properly. Heat of arc/flame must penetrate the base metal up to sufficient depth for proper melting of base metal and then mixing with fused filler/electrode metal to develop metallurgical weld joint (Fig. 24.6a). Heat input predominantly affects the parameters of weld bead geometry, namely penetration depth, width of weld, reinforcement of weld bead, bead angle in groove weld (Fig. 24.6b) and leg length, theoretical/effective throat thickness in fillet weld (Fig. 24.6c). Capital (A, B, C) small case alphabets (a, b, c) in Fig. 24.6b show the depth of penetration and width of weld bead, respectively. These parameters change with heat input. Heat generation in arc welding and its application for fusion welding is determined by welding current, and voltage and welding speed. An optimum value of all three parameters is needed for through thickness penetration sound weld joint.

24.5.1 Welding Current

In general, welding current increases the weld cross-sectional area. Moreover, increase in welding current initially increases both depth of penetration and width of weld bead but not in the same order. It has been reported that initially increase in current results deeper penetration weld while at range of current, the increase in current produced wider weld (Fig. 24.7).

Low welding current results in less heat generation. Low heat of the arc increases lack of fusion and poor penetration tendency besides the formation of too high reinforcement of weld bead owing to poor fluidity caused by low-temperature molten weld metal. On the other hand, too high welding current may lead to undercut in the weld joint near the toe of the weld due to excessive melting of base metal and flattened weld bead apart from increased tendency of weld metal to fall down during vertical, horizontal and overhead welding owing to high fluidity of weld meal caused by low viscosity and surface tension. Increase in welding current in general increases the depth of penetration/fusion. Therefore, an optimum value of welding current is important for producing sound weld joint.

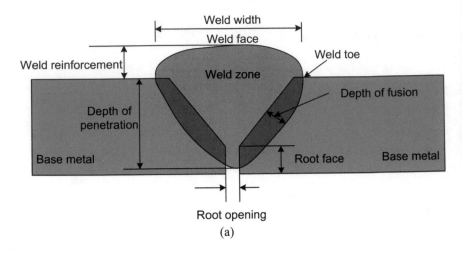

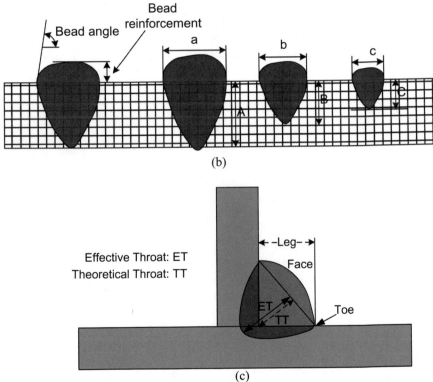

Fig. 24.6 Schematic showing **a** common use terms of groove weld, **b** groove weld geometry parameters and **c** fillet weld nomenclature

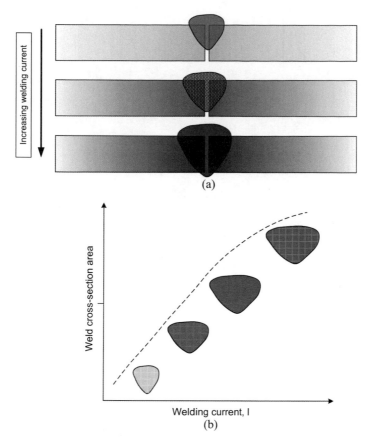

Fig. 24.7 Schematic showing effect of welding current on **a** weld bead geometry and **b** weld cross section

24.5.2 Arc Voltage

Similar to the welding current, an optimum arc voltage also plays a crucial role in the development of sound a weld as low arc voltage results in unstable arc which in turn results in poor weld bead geometry while to high voltage causes increased arc gap and wide weld bead and shallow penetration (Fig. 24.8).

24.5.3 Welding Speed

Welding speed influences both fusion of base metal and weld bead geometry. Low welding speed causes flatter and wider weld bead while excessively high welding speed reduces heat input which in turn lowers penetration and weld bead width

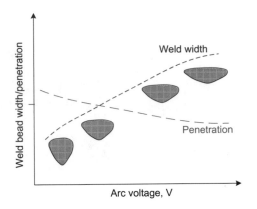

Fig. 24.8 Schematic showing effect of arc voltage on weld bead geometry, i.e. weld width and depth of penetration

and increases weld reinforcement and bead angle. Therefore, an optimum value of welding speed is needed for producing sound weld with proper penetration and weld bead geometry. A more general trend of variation in weld cross-sectional area, reinforcement and penetration with welding speed due to changing net heat input is shown in Fig. 24.9a, b.

Questions for Self-assessment

a.   What are the factors that need to be considered for the selection of groove geometry?
b.   When fillet weld is preferred for developing weld joints?
c.   Why do fillet weld joints offer lower fatigue performance than butt weld joints?
d.   What for bead weld is used and write the precaution to be taken while developing it?
e.   How are welding parameters and weld joint design interrelated for developing sound weld joint?
f.   What is the fundamental approach of weld joint design for given external loading?

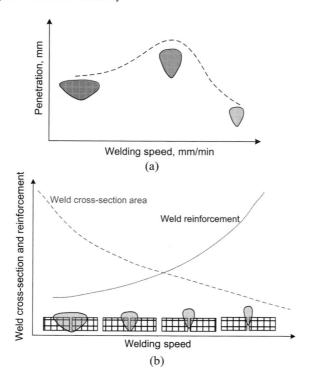

Fig. 24.9 Schematic showing effect of welding speed on parameters of weld bead geometry **a** depth of penetration and **b** weld cross section and weld reinforcement

Further Reading

Hicks J (1999) Weld joint design, 3rd edn. Abington Publishing, England
Kou S (2003) Welding metallurgy, 2nd edn. Wiley, USA
Maddox SJ (1991) Fatigue strength of welded structures. Woodhead Publishing
Parmar RS (2002) Welding engineering and technology, 2nd edn. Khanna Publisher, New Delhi
Radaj R (1990) Design and analysis of fatigue resistant welded structures. Woodhead Publishing
Welding Handbook (1987) 8th edn. vols 1 and 2. American Welding Society, USA

Chapter 25
Design of Welded Joints: Weld Joint Design for Static and Fatigue Loading

25.1 Design Aspects of Weld Joint

Strength of weld joints is determined by not only the properties of weld metal but also characteristics of heat affected zone (HAZ) and weld bead geometry (due to stress concentration effect). Generally, properties of HAZ are degraded to such an extent that they become even lower than weld metal due to increase in (a) softening of the heat affected zone and (b) corrosion tendency of HAZ. Assuming that the effect of weld thermal cycle on properties of HAZ is negligible, suitable weld dimensions are obtained for given loading conditions through weld joint design. Design of a weld joint mainly involves establishing the proper load-resisting cross-sectional area of the weld which includes throat thickness of the weld and length of the weld. In case of groove butt weld joints, throat thickness is the shortest length of the line passing through the root across the weld (top to bottom) as shown in Fig. 25.1.

Conversely, throat thickness becomes the minimum thickness of weld or thickness of thinner plate when joint is made between plates of different thicknesses. While in case of fillet welds, throat thickness is the shortest length of line passing root of the weld and weld face. Any extra material (due to convexity of weld face) in weld does not contribute much towards load-carrying capacity of the weld joint.

In practice, however, load-carrying capacity of the weld is dictated not just by weld cross-sectional area but also by properties of weld metal and HAZ apart from the stress concentration effect induced by weld bead geometry and weld discontinuities under the static as well as fatigue loading conditions.

25.2 Design of Weld Joint for Static Loading

The designing of a weld joint needs determination of the throat thickness and length of the weld. Measurement of throat thickness is easier for groove butt weld joint than fillet weld joint because root is not accessible in case of fillet weld. Throat thickness of

D. K. Dwivedi, *Fundamentals of Metal Joining*,
https://doi.org/10.1007/978-981-16-4819-9_25

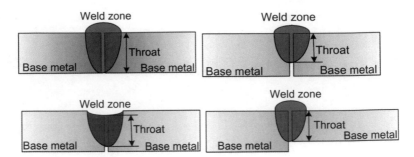

Fig. 25.1 Schematics showing throat thickness in case of groove butt weld joints of different plates thickness and penetration conditions

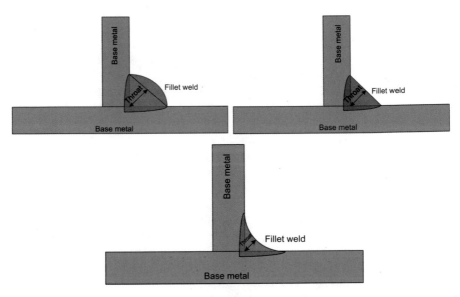

Fig. 25.2 Schematics showing throat thickness in case of fillet weld joints of different weld bead geometries, e.g. convex, straight, concave

fillet welds is obtained indirectly (mathematically) from leg length: $2^{1/2}$X leg length. Leg length of fillet weld can be measured directly using metrological instruments. Further, minimum throat thickness for different plate thicknesses has been identified (by the American welding society) keeping in mind cracking tendency of fillet weld due to tensile residual stresses in the weld joints. Small fillet weld developed on thick plates exhibits cracking tendency primarily due to inability of small fillet weld to sustain heavy residual tensile stresses developed during welding of thick plates.

Depending upon the expected service load, a weld joint can be designed by considering tensile, compressive and shear stresses. A weldment joint design program starts with recognition of a need to design like new design or failure of existing design of weld joints followed by main steps of weldment design procedure including:

1. Determination or estimation of expected service load on the weld joint
2. Collecting information about working condition and type of stresses
3. Based on the requirement identify design criteria (ultimate strength, yield strength, modulus of elasticity, creep rate, endurance limit, fracture toughness)
4. Using suitable design formula calculate length of weld or throat thickness as per need or data given
5. Determine length and throat thickness required to take up given load (tensile, shear, bending load, etc.) during service.

Methodology

In general, weld joint design methodology involves the following procedure (a) select the type of weld joint and edge preparation for design as per the service requirements, (b) establish the maximum load for which a weld joint is to be designed, (c) throat thickness is identified which is generally fixed for a given thickness of the plate; e.g. for full penetration fillet weld, throat thickness is about 0.707 time of leg length of the weld while in case of the groove weld throat thickness is generally equal to thickness of thinner plate (in case of dissimilar thickness weld) or thickness of any plate for equal thickness of plates (Fig. 25.3) and (d) using suitable factor of safety and suitable design criteria determine the allowable stress for the weld joint, (e) calculate length of the weld using external maximum load, allowable stress, throat thickness and allowable stress.

25.2.1 Design of Fillet Welds

(a) Stress on fillet weld joint can be obtained by using the following relationship:

Load/weld throat cross-sectional area
Load/(throat thickness × length of weld joint × number of welds)
Load/(0.707 × leg length of the weld × length of the weld × number of welds)

25.2.2 Design of Butt Weld Joint

Stress on butt weld joint between equal thickness plates (Fig. 25.4) is obtained using the following relationship: Stress: Load/weld throat cross-sectional area = Load/(throat thickness × length of weld joint × number of welds) = Load/(thickness of any plate × length of the weld × number of welds).

Stress (σ) on the butt weld joint between plates of different thicknesses ($T1$ and $T2$) subjected to external load (P) experiences eccentricity (e) owing to difference

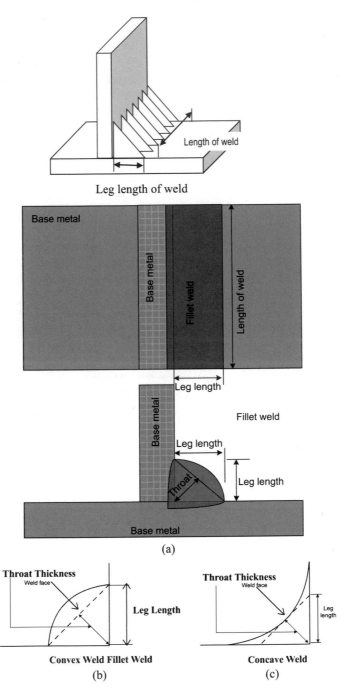

Fig. 25.3 Schematic diagram showing **a** leg length and length of weld, **b** throat thickness for convex and **c** throat thickness for concave fillet welds

Fig. 25.4 Schematic
diagram of butt weld joint
between plates of equal
thickness showing throat
thickness and length of weld

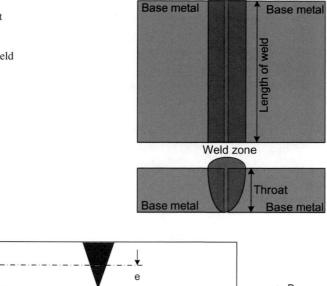

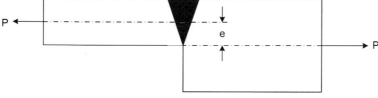

Fig. 25.5 Schematic diagram of butt weld when both the plates are of different thickness

in thickness of plates and T1 thickness of thinner plate of the joint (Fig. 25.5). Even axial loading due to eccentricity causes the bending stress in addition to axial stress. Therefore, stress on the weld joint becomes sum of axial as well as bending stress and can be calculated as under.

$$\text{Stress in weld} = \text{Axial Stress} + \text{Bending Stress}$$

$$\sigma_{\text{total}} = \frac{P}{T_1} + \frac{P \cdot e \cdot \frac{1}{2} \cdot T_1}{\frac{T_1^3}{12}}$$

25.3 Design of Weld Joints for Fatigue Loading

The approach for designing weld joints for fatigue load conditions is different from that of static loading primarily due to high tendency of the fracture by crack nucleation and growth phenomenon. A weld joint can be categorized in a specific class depending upon the severity of stress concentration, weld penetration (full or partial penetration weld), location of weld, type of weld and weld constraint. The class of a weld joint to be designed for fatigue loading is used to identify allowable stress range for a given life of weld joint (number of fatigue load cycles) from stress range vs. number of load cycle curves developed for different loading conditions and metal systems (Fig. 25.6). Thus, allowable stress range obtained on the basis of the class of the weld and fatigue life of weld (for which a joint is to be designed) is used to determine the weld-throat-load-resisting cross-sectional area (throat thickness, length of weld and number of weld).

A weld joint to be designed needs to be identified for the suitable class as per weld joint configuration, weld finish condition, severity of fatigue loading (Fig. 25.7). The class of weld joint suggests allowable range for the desired fatigue life cycles.

Procedure of weld joint design for fatigue loading

Weld joints for fatigue loading condition are designed using the following steps:

- Identify the class of the weld joint based on severity of loading, type of weld, penetration and criticality of the joint for the success of the assembly.
- As per class of the weld joint, obtain a value of the allowable stress range using fatigue life (number of cycles) for which a joint is to be designed.
- The allowable stress range and service loading condition (maximum and minimum load) are used to determine load-resisting cross-sectional area of the weld joint (Fig. 25.4).

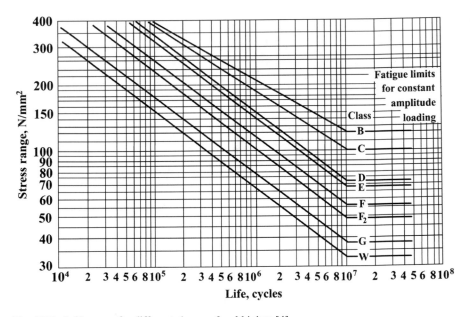

Fig. 25.6 S–N curves for different classes of weld joints [4]

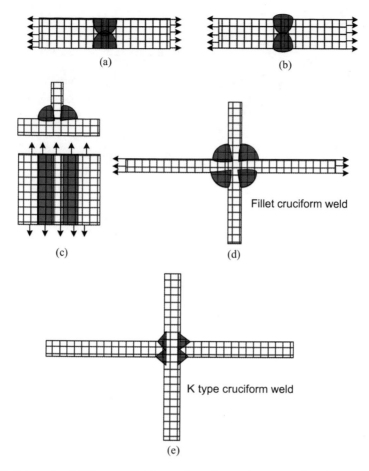

Fig. 25.7 Schematic of different weld joint configurations and corresponding classes as under: A flush ground butt joint (**a**) is suggested to be in Class B. As-welded butt joint (**b**) is in Class C. A transverse gusset joint (**c**) can be in Class F. A longitudinal gusset joint (**d**) in Class F2. Joints (**e**) with a rather primitive design geometry in class G

- For given set of loading condition and identified class of the weld joint various details like throat thickness, length of weld joint and number of welds can be obtained from calculated load-resisting cross-sectional area desired.
- Generally, the maximum length of the weld becomes the same as the length of the plate to be welded and maximum number of welds for butt welding is one and that for fillet weld can be two in case of uninterrupted fillet welds. This suggests that primarily throat thickness of the weld is identified if length and number of weld are fixed else any combination of the weld parameters such as throat thickness, length of weld and number of welds is obtained in such a way that their product is equal to the required load-resisting cross-sectional area.

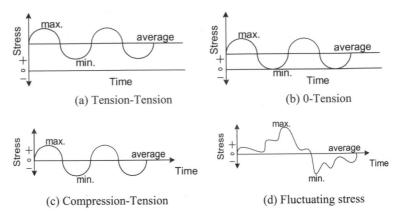

Fig. 25.5 Common fatigue load patterns

- Strength of weld metal does not play any big role on fatigue performance of the weld joints as under severe stress conditions (which generally exist in weld joint owing to the presence of notches and discontinuities). Fatigue strength and life are marginally affected by strength of weld metal.

Information required for designing for fatigue loading

- The fatigue life (number of load cycles) for which a weld joint is to be designed, e.g. 2×10^6 cycles
- Class of the weld joint based on type of the weld, penetration and other conditions
- Allowable stress range is obtained on the basis of class of weld and fatigue life required
- Value of the maximum and minimum service load expected on weld joint

Example A T joint of steel plates is subjected to 350 kN load is developed using intermitted 4 double fillet welds of length each of 40 mm at an interval of 100 mm. Allowable shear strength of the weld metal is 100 MPa. Determine the leg length of the weld.

Solution

Weld load-resisting cross-sectional area: throat thickness × length of each weld × No. of weld.

: weld throat thickness × 40 × 4 × 2.

: weld throat thickness × 320.

Load-carrying capability: load-resisting cross-sectional area × allowable shear stress.

350,000: weld throat thickness × 320 × 100.

Weld throat thickness: 350,000/320 × 100.

Weld throat thickness: 10.93 mm.

Leg length of weld: weld throat thickness × $(2)^{1/2}$.

Leg length of weld: 10.92 × 1.414 = 15.45 mm.

Example

A full penetration butt weld joint made of two steel plates each of 10 mm × 100 mm × 300 mm is subjected load fluctuation from 150 to 350 kN load. Determine if weld is safe for 10^6 or 10^5 load cycle.

Solution

Assuming class of the weld for given service condition is E.

Load-resisting cross-sectional area of weld: length × width: 10 × 100 = 1000 mm^2.

Assuming throat thickness of weld is equal to min. thickness of plate.

Max stress: 350,000/1000: 350 MPa.

Min. stress: 150,000/1000: 150 MPa.

Stress range: 200 MPa.

The allowable stress range for 10^6 and 10^5 load cycle for E class is obtained from standard table/plots 200 and 320 MPa respectively. Thus, the weld is safe for 10^5 and 10^6 load cycles.

Questions for Self-assessment

a. What is objective of designing of fillet and butt weld joint for static loading?
b. What are the important steps of weld joint design for static loading?
c. Describe methodology for developing butt and fillet weld joint design for static loading.
d. Why does approach of weld joint design for fatigue loading differ from that for static loading?
e. Describe procedure of developing butt and fillet weld joint design for fatigue loading.
f. How do we select a class of weld to be designed for fatigue loading?
g. What information must be collected for designing weld joint for fatigue loading?

Further Reading

1. Radaj R (1990) Design and analysis of fatigue resistant welded structures. Woodhead Publishing
2. Parmar RS (2002) Welding engineering & technology, 2nd edn. Khanna Publisher, New Delhi
3. Hicks J (1999) Weld joint design, 3rd edn. Abington Publishing, England
4. Maddox SJ (1991) Fatigue strength of welded structures. Woodhead Publishing

5. Welding handbook, American Welding Society (1987). 8th edn. vols. 1 & 2, USA.
6. Kou S (2003) Welding metallurgy, 2nd edn. Wiley, USA

Chapter 26
Design of Welded Joints

Fatigue of Weld Joints: Mechanism, Stages, Parameters

The fluctuations in magnitude and direction of the load more adversely affect the life and performance of a mechanical component compared to that under static loading condition. This adverse effect of load fluctuations on the life of a mechanical component is called fatigue. Reduction in the life of the mechanical components subjected to fatigue loads is mainly caused by premature fracture due to early nucleation and growth of cracks in the areas of high stress concentration. The stress concentration may occur either due to abrupt change in cross section or the presence of discontinuities in form of cracks, blowholes, weak materials, etc., in the weld joint.

26.1 Fracture Under Fatigue Loading

The fracture of the mechanical components under fatigue load conditions generally takes place in three steps: (a) nucleation of cracks or crack-like discontinuities if not present, (b) stable growth of crack and (c) catastrophic and unstable fracture. Number of fatigue load cycles required to complete each of the above three stages of the fatigue eventually determines the fatigue life of the component (Fig. 26.1). Each stage of fatigue fracture ranging from crack nucleation to catastrophic unstable fracture is controlled by different properties such as surface properties, mechanical and metallurgical properties of the component. Any of the factors related to material properties, geometry of the component, loading and service condition which can delay the completion of any of the above three stages of the fatigue will increase the fatigue life. The microphotographs of actual samples of a typical weld joint of high strength aluminium alloy are shown in Fig. 26.2. The different stages of fatigue life can be seen with clarity; however, many a times fatigue fracture surface does not exhibit the stages of fracture with clear demarcation and distinction. It is evident from the zones on the fracture surfaces corresponding to different zones that these can be varying significantly depending on the resistance to the crack growth during a particular stage of fatigue fracture (Fig. 26.2a–c). The metals with good ductility

© The Author(s), under exclusive license to Springer Nature Singapore Pte Ltd. 2022   363
D. K. Dwivedi, *Fundamentals of Metal Joining*,
https://doi.org/10.1007/978-981-16-4819-9_26

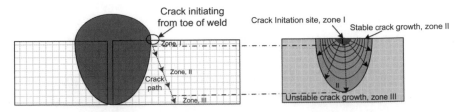

Fig. 26.1 Schematic showing stages of fatigue fracture of a weld joint

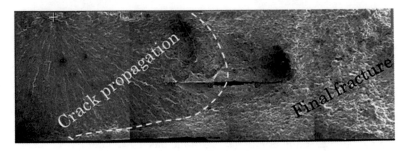

Fig. 26.2 SEM microphotograph of the fatigue fracture surface of a weld joint

show lateral contraction as well primarily due to yielding at the later stages of fatigue fracture.

26.2   Factors Affecting the Stages of Fatigue Fracture

26.2.1   Surface Crack Nucleation Stage

Surface crack nucleation stage is primarily influenced by surface properties such as roughness, hardness, yield strength and ductility of the mechanical component subjected to fatigue provided there is not stress raiser causing stress localization. Cracks on the smooth surface of the component are nucleated by micro-level deformation occurring at the surface due to slip under the influence of fluctuating loads. Repeated fluctuation of loads results in surface irregularities of micron level which act as stress raiser and site for stress concentration. Continued slip at certain crystallographic planes due to fluctuating load cycle finally produces crack-like discontinuities at the surface (Fig. 26.3). It is generally believed that first crack nucleation stage takes about 10–20% of total fatigue life cycle of the mechanical component. Since the mechanism of fatigue crack nucleation is based on micro-level slip deformation at the surface therefore factors like surface irregularities (increasing stress concentration), high ductility, low yield strength and low hardness would facilitate the micro-level surface deformations and thereby lower the number of fatigue load

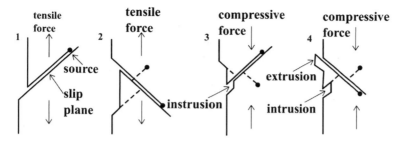

Fig. 26.3 Schematic of fatigue crack nucleation mechanism

cycles required for completing the crack nucleation stage (Fig. 26.3). Hence, for enhancing the fatigue life attempts are always made to improve the surface finish (so as to reduce stress concentration due to surface irregularities if any by grinding, lapping, polishing, etc.), increase the surface hardness and yield strength and lower the ductility using various approaches, namely shot peening, carburizing, nitriding and other hardening treatment.

Surface nucleation stage in case of welded joints becomes very crucial as almost all the weld joints generally possess poor surface finish and weld discontinuity of one or another kind which can act as a stress raiser. Further, the development of residual stresses in weld joints can also promote or discourage the surface nucleation stage depending upon the type of loading. Residual stresses similar (type) to that of external loading facilitate the crack nucleation while those of opposite kinds tend to discourage the crack nucleation. These are reasons why welding of base metal in general lowers the fatigue life up to 90% depending upon the type of the weld joints, loading conditions and surface conditions of weld.

26.2.2 Stable Crack Growth Stage

A crack nucleated in the first stage may be propagating or non-propagating type depending upon the extent of load fluctuation for a given component. A fatigue loading with limited load fluctuations or high stress ratio (ratio of minimum stress and maximum stress) especially in case of fracture tough materials may lead to the existence of non-propagating crack.

However, growth of a propagating crack is primarily determined by stress range (difference of maximum and minimum stress) and material properties such as ductility, yield strength and microstructural characteristics (size, shape and distribution of hard second phase particles in the matrix). An increase in stress range in general increases the rate of stable crack growth in second stage of fatigue fracture. Increase in yield strength and reduction in ductility increase the crack growth rate primarily due to the reduction in extent of plastic deformation (which reduces blunting of crack tip so the crack remains sharp tipped) experienced by material ahead

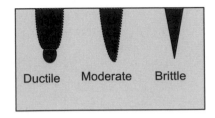

Fig. 26.4 Schematic showing three different types of crack morphologies as per hardness and ductility of metal subjected to fatigue

of crack tip under the influence of external tensile/shear load. Increase in blunting of crack tip due to plastic deformation lowers the stress concentration at the crack tip and thereby reduces the crack growth rate. A combination of high yield strength and low ductility causes limited plastic deformation at crack tip which in turn results in high stress concentration at the crack tip. High stress concentration at the crack tip produces rapid crack growth which reduces number of fatigue load cycles (fatigue life) required for completion of second stage of fatigue fracture of the component (Fig. 26.4).

All the factors associated with loading pattern and material which increase the stable crack growth rate, reduce the number of fatigue load cycle required for fatigue fracture. High stress range in general increases the stable crack growth rate. Therefore, attempts are made by design and manufacturing engineers to design the weld joints in such a way that a) stress range on the weld joint during service (if possible) is reduced and b) reduce crack growth rate by developing weld joints of fracture tough material (having requisite ductility and yield strength).

The fracture mechanics principles have also been applied in fatigue studies to understand the conditions required for different stages of fatigue. The fracture mechanics considers the materials properties, crack size and applied stress for suggesting the conditions for the growth of crack under fatigue loading. One of the common terms in fracture toughness is stress intensity factor indicating the stress intensity near the tip of crack and is extensively used to predict the crack propagation and fracture conditions in case of homogeneous, linear <u>elastic</u> material for providing a failure criterion in case of high strength, low ductility and <u>brittle</u> materials. Stress intensity factor (K) under uni-axial stress condition is given by $\sigma(\pi c)^{1/2}$ where σ is applied stress (MPa); π is the constant; c is length (in m) of surface crack and half crack length in case of sub-surface cracks inside the body).

For a given crack length, under varying load conditions stress intensity factor varies from maximum to minimum level as per external stresses. The variation in stress intensity factor is called stress intensity factor range (ΔK). A minimum stress intensity factor range needed for commencement of propagation of a crack is called threshold stress intensity factor (ΔK_{th}) as shown in Fig. 26.5. The Paris law shows the relationship between the stress intensity factor range (ΔK) and crack growth rate (dc/dN) per load cycle in second stage of fatigue fracture and is expressed as below.

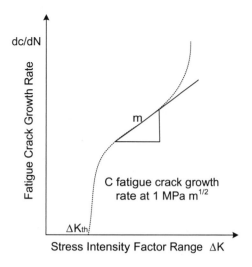

Fig. 26.5 Schematic showing relationship between fatigue crack growth rate and stress intensity factor range

$$\text{Crack growth rate } (\mathrm{d}c/\mathrm{d}N) = C(\Delta K)^m$$

where c is the crack length, N is the number of load cycles, m is slope of curve in second stage of crack growth (Fig. 26.3) C is constant and shows the crack growth rate per load cycle corresponding 1 MPa m$^{1/2}$ stress intensity factor range.

26.2.3   Sudden Fracture (Unstable Crack Growth)

Third stage of fatigue fracture corresponds to unstable rapid crack growth causing abrupt facture. This stage commences only when load-resisting cross-sectional area of the component (due to stable crack growth in second stage of fatigue fracture) is reduced to such an extent that it becomes unable to withstand maximum stress being applied during service. Hence, under such condition material failure occurs largely due to overloading of the remaining cross section. The mode of fracture at the third stage of fatigue failure may be ductile or brittle depending upon type of the material. Metals of high fracture toughness allow second stage stable crack growth (of fatigue fracture) to a greater extent which in turn delays the commencement of third stage unstable crack propagation (Fig. 26.6). Conversely for a given load, material of high fracture toughness (high strength and high ductility) can withstand to a very low load-resisting cross-sectional area prior to the commencement of third stage of fatigue fracture than that of low fracture toughness.

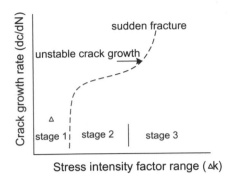

Fig. 26.6   Stages of fatigue fracture including unstable fatigue crack growth rate followed by fracture at high stress intensity factor range

Fatigue resistance and Metal properties

This section describes residual fatigue life concept and effect of various service load-related parameters on fatigue performance of the weld joints. Further, relationship between crack growth and number of load cycle has also been elaborated.

26.3   Crack Growth and Residual Fatigue Life

Once the fatigue crack nucleated (after the first stage), it grows with the increase in number of fatigue load cycles. Slope of the curve showing the relationship between crack size and number of fatigue load cycles indicates the fatigue crack growth rate does not remain constant (Fig. 26.7). The fatigue crack growth rate (slope of curve) continuously increases with increase in number of fatigue load cycles. Initially in second stage of the fatigue fracture, fatigue crack growth rate (FCGR) increases gradually in stable manner. Thereafter, in third stage of fatigue fracture, FCGR increases

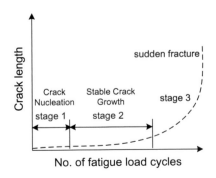

Fig. 26.7   Schematic of crack length versus number of fatigue load cycles relationship

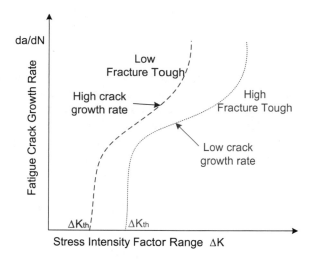

Fig. 26.8 Schematic showing the effect of fracture toughness of metal on crack growth behaviour

at very high rate with increase in number of fatigue load cycles as evident from the increasing slope of the curve in third stage.

This trend of crack size versus number of fatigue load cycle remains same even under varying service conditions of weld joints made of different materials. Moreover, the number of load cycles required for developing a particular crack size (during the second and third stage of fatigue facture) varies with factors related to service conditions, material and environment. For example, increase in stress range during fatigue loading of high strength and low ductility welds decreases the number of load cycles required to complete the second as well as third stages of fatigue fracture (Fig. 26.8). Conversely, unstable fatigue crack prorogation (increasing FCGR) occurring in third stage of fatigue fracture is attained earlier. Increase in fatigue crack size in fact decreases the load-resisting cross section (residual cross-sectional area) of weld joint which in turn increases stresses accordingly for given load fluctuations. Therefore, the above-trend of crack size vs. number of fatigue load cycles is mainly attributed to increasing true stress range for a given load fluctuation which will actually be acting on actual load-resisting cross-sectional area at any moment.

Residual fatigue life is directly determined by load-resisting cross-sectional area left due to fatigue crack growth (FCG) at any stage of fatigue life. Increase in crack length and so reduction in load-resisting cross-sectional area in general lowers the number of cycle required to complete the fatigue fracture. Thus, leftover fatigue life, i.e. residual fatigue life, of a component subjected to fluctuating load gradually decreases with increase in fatigue crack size.

26.4 Factors Affecting the Fatigue Performance of Weld Joints

There are many factors related to service load condition, material and service environment affecting one or other stage (singly or in combination) of the fatigue fracture. The fatigue behaviour of welded joints is not different from that of unwelded base metal except that weld joints need fewer number of load cycles due to many unfavourable features such as stress raisers, residual stresses, surface and sub-surface discontinuities, hardening/softening of HAZ, irregular and rough surface of the weld in as-welded conditions (if not ground and flushed) besides in-homogeneity in respect of composition, metallurgical, corrosion and mechanical properties. Therefore, in general, fatigue performance of the weld joints is usually found offer lower than the base metal. However, this trend is not common in friction stir welded joint of precipitation hardenable aluminium alloys as these develop more ductile weld nugget than heat affected zone which is generally softened due to reversion in as-welded conditions. The extent of decrease in fatigue performance (strength/life) is determined by severity of above-mentioned factors present during the service besides the weld joint configuration and whether joint is load-carrying or non-load-carrying type. Reduction in fatigue performance of a weld joint can be as low as 0.15 times of fatigue performance of corresponding base metal depending up on the joint configuration and other welding-related factors. The following section describes the influence of various service, material, environment and welding procedure-related parameters on the fatigue performance of weld joints.

26.4.1 Service Load Conditions

Service conditions influencing the fatigue performance of the weld joints mainly include fatigue load and trend of its variation. Fluctuation of the load during the service can be in different ways. The fatigue load fluctuations are characterized with the help of different parameters, namely type of stress, maximum stress, minimum stress, mean stress, stress range, stress ratio, stress amplitude, loading frequency, etc. Following section presents the influence of these parameters in systematic manner on fatigue. These parameters help to distinguish the type of stresses and extent of their variation.

26.4.1.1 Type of Stress

For nucleation and propagation of the fatigue cracks, existence of tensile or shear stress is considered to be mandatory. The presence of only compressive stress does not help in easy nucleation and propagation of the cracks. Therefore, fatigue failure

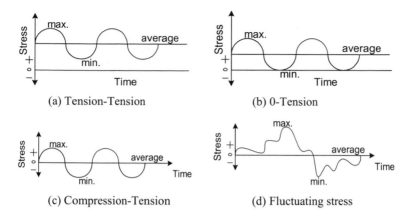

Fig. 26.9 Common fatigue load cycles

tendency is either reduced or almost eliminated when fatigue load is only of compressive type. As a customary, tensile and shear stress are taken as positive while compressive stress is taken as negative stress in calculations. These sign conventions play a major role when fatigue load fluctuation is characterized in terms of stress ratio and stress range (Fig. 26.9).

26.4.1.2   Maximum Stress

It is maximum level of stress generated by fluctuating load and significantly influences the fatigue life (strength) of the component. Any discontinuity present in weld joints remains non-propagating type until maximum tensile/shear stress (due to fatigue loading) does not become more than certain limit. Thereafter, only increase in maximum stress in general lowers the fatigue life, i.e. number of cycles required for fracture. Increase of rate of crack growth in different stages of fatigue fracture occurring at high level of maximum stress leads to the reduction in number of load cycles required to complete each of the three stages of the fatigue fracture (Fig. 26.10).

26.4.1.3   Stress Range

It is the difference between maximum and minimum stress induced by fatigue load acting on the component of a given load-resisting cross-sectional area. Difference of maximum and minimum stresses gives the stress range directly if nature of stress remains same (tensile–tensile, compressive–compressive, shear–shear during loading. However, in case when load fluctuation changes nature of load from tensile and compressive, shear and compressive or vice versa then it becomes mandatory to use sign conventions with magnitude of stress according to the type of loading for calculating the stress range. For example, stress range for tensile stress variation

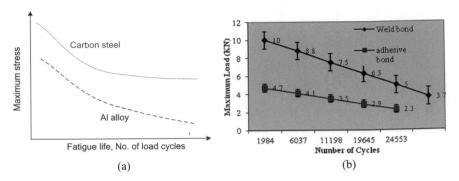

Fig. 26.10 a schematic showing the effect of maximum stress on fatigue life and **b** effect of maximum load on fatigue life of weld bond and adhesive joints.

from 130 to 230 MPa becomes 100 MPa while that in case of 250 MPa tensile stress and 100 MPa compressive stress results in 350 MPa and calculated as under: [(+250) − (−100)] MPa.

Zero stress range indicates that maximum and minimum stresses are of the same value, and there is no fluctuation in magnitude of the load means load is static in nature; therefore, material will not be experiencing any fatigue. Conversely, for premature failure of material owing to fatigue it is necessary that material is subjected to enough fluctuations in stresses during the service. The extent of fluctuation in stress (due to fatigue) is measured in terms of stress range. In general, increase in stress range lowers the fatigue life.

Most of the weld joint designs of real engineering systems for fatigue load conditions are therefore generally based on stress range or its derivative parameters such as stress amplitude (which is taken as half of the stress range) and stress ratio (ratio of minimum to maximum stress). In general, an increase in stress range decreases the fatigue life as evident from the fatigue behaviour of friction stir weld joints in different temper conditions (Fig. 26.11).

26.4.1.4   Stress Ratio

It is obtained from ratio of minimum stress to maximum stress. Lower value of stress ratio indicates greater fluctuation in fatigue load. For example, stress ratio of 0.1, 0.2 and 0.5 is commonly used for evaluating the fatigue performance of weld joints as per requirement (Fig. 26.12). Stress ratio of 0.1 indicates that maximum stress is 10 times of minimum stress. Stress ratio of zero value suggests that minimum stress is zero while stress ratio of −1 indicates that the load fluctuates equally on tensile/shear and compressive side. The decrease in stress ratio for tensile and shear fatigue loads (say from 0.9 to 0.1) adversely affects the fatigue performance.

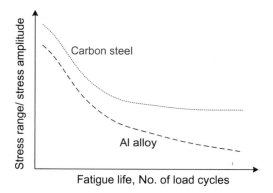

Fig. 26.11 Schematic showing effect of stress range on fatigue life of weld joints

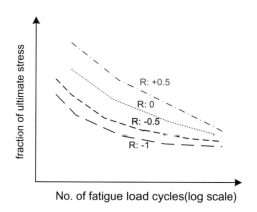

Fig. 26.12 Schematic showing effect of stress ratio (R) on fatigue life (N) for given stress conditions

26.4.1.5   Mean Stress

Mean stress is average of maximum and minimum stress. The influence of mean stress on the fatigue life mainly depends on the stress amplitude and nature of mean stress. Nature of mean stress indicates the type of stress. The effect of nature of mean stress (i.e. compressive, zero and tensile stress), on the fatigue life is more severe at low stress amplitude than the high stress amplitude. It can be observed that in general for a given stress amplitude, mean tensile stress results in lower fatigue life than the compressive and zero mean stress (Fig. 26.13). Further, increase in tensile mean stress decreases the number of load cycle required for fatigue crack nucleation and prorogation of the cracks which in turn lowers the fatigue life.

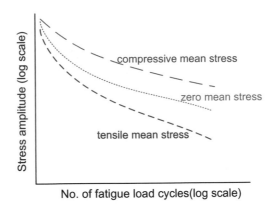

Fig. 26.13 Schematic showing effect of type of stress on S–N curve

26.4.1.6 Frequency of Fatigue Loading

Frequency of the fatigue loading is number of times a fluctuating load cycle repeats in unit time and is usually expressed in terms of Hz which indicates the number of fatigue load cycles per second. Frequency of fatigue loading has little influence on fatigue performance.

Fatigue Resistance and Materials Properties

This section describes the influence of various physical, mechanical and metallurgical characteristics on fatigue performance of the weld joint. Further, the influence of environmental conditions, namely temperature, vacuum and corrosion, on fatigue behaviour of the weld joint has also been elaborated.

26.4.2 Material Characteristics

The performance of an engineering component under fatigue load conditions is significantly influenced by various properties of weld/HAZ/base material such as physical properties, mechanical, corrosion and metallurgical properties.

26.4.2.1 Physical Properties

Many physical properties, such as melting point, thermal diffusivity and thermal expansion coefficient, of the base or filler metal can be important for the development of the sound weld joint. It is felt that probably thermal expansion coefficient of base metal is one of physical properties which can appreciably affect the fatigue performance of a sound weld joint as it directly influences the magnitude and type of residual stress developed due to weld thermal cycle experienced by the base metal during welding. Tensile residual stresses are usually left in weld metal and nearby HAZ which adversely affect the fatigue life of weld joint, and therefore, attempts are made to develop compressive residual stress in weld joints using localized heating or deformation-based approaches.

26.4.2.2   Mechanical Properties

Mechanical properties of the weld joint such as hardness, yield and ultimate tensile strength, ductility and fracture toughness significantly affect the fatigue strength of the weld joint. The extent of influence of an individual mechanical property on fatigue performance primarily depends on the way by which it affects the one or other stage of the fatigue fracture. For example, high ductility, low hardness and low yield strength reduce the number of load cycles for the crack nucleation stage while high ductility, moderate tensile strength and high fracture toughness delay second stage of fatigue fracture, i.e. lower stable crack growth rate, and both these stages constitute to about 90% of the fatigue life.

It is generally believed that under the conditions of high stress concentration as in case of welded joints (especially in fillet weld and weld with severe discontinuities and stress raisers and those used in corrosive environment), the mechanical properties such as tensile strength does not affect the fatigue performance appreciably (Fig. 26.14). Therefore, design and production engineers should not rely much on tensile strength of electrode material for developing fatigue-resistant weld joints. Moreover, in case of full penetration, ground, flushed, defect-free butt weld joints, mechanical properties, namely ductility, hardness tensile strength and fracture toughness, can play an important role in determining the fatigue performance primarily due to the absence of stress raisers. Moreover, the effect of these properties on each stage of fatigue fracture has been described in respective sections of fatigue fracture mechanism.

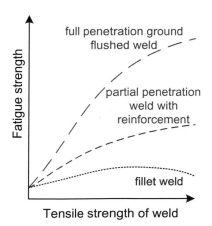

Fig. 26.14 Schematic diagram showing the fatigue strength vs. tensile strength relationship for different conditions of the weld

26.4.2.3   Metallurgical Properties

Metallurgical properties such as microstructure and segregation of elements in weld joint influence the fatigue performance. Microstructure indicates the size, shape and distribution of grains besides the type and relative amount of various phases present in the structure. Due to varying cooling conditions experienced by weld metal and heat affected zone during welding significant structural in-homogeneity is observed in the weld metal and HAZ. The mode of weld metal solidification continuously varies from planar at fusion boundary to cellular, dendritic then equiaxed at weld centreline owing variation in cooling conditions which in turn results in varying morphologies of grains in weld metal. Similarly, size of grains also varies from coarsest at fusion boundary to finest at weld centreline. A combination of a welding process, welding parameters (deciding net heat input), section size and base metal composition eventually determines mode of solidification and so the final grain and phase structure.

Needle shape phases offer more adverse effect on the fatigue life than spherical and cuboids shape micro-constituents (Fig. 26.15). In general, fine and equiaxed grains result in longer fatigue life than coarse and columnar dendritic grains as crack nucleation and stable crack growth stages of fatigue fracture are delayed. Therefore,

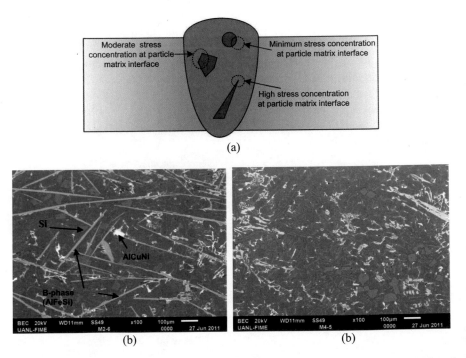

(a)

(b)   (b)

Fig. 26.15 a Schematic showing constituents of different morphologies, **b** micrographs of aluminium–silicon alloy showing needle micro-constituents of Al-Si-Fe and **c** fine and Chinese script morphologies and polyhedral shape primary silicon particles

attempts are made to have fine equiaxed grain structure in weld metal using various approaches such as controlled alloying, external excitation forces and arc pulsation.

26.4.3   Environment

Fatigue performance of a weld joint is significantly governed by the service environment conditions such as corrosion, high temperature and vacuum. In general, all these special environments affect the fatigue performance in either way (positively or negatively).

26.4.3.1   Corrosion Fatigue

The performance of a component subjected to fatigue loading in corrosive environment during the service is termed as corrosion fatigue. Corrosion means localized removal of materials either from plane smooth surface or from the tip of preexisting discontinuity. Localized corrosion from smooth surface facilitates easy nucleation of crack during first stage of fatigue fracture by forming small pits and crevices while removal of material from crack tip by corrosion accelerates the crack growth rate during second stage of fatigue fracture (Fig. 26.16). A synergic effect of stable crack growth during second stage of stable crack growth and material removal from crack tip reduces the fatigue life drastically. Moreover, how far corrosion will affect fatigue life; it depends on corrosion media for a given metal of weld, e.g. steel weld joints perform very more badly in saline environment (halide ions) than in dry atmospheric conditions.

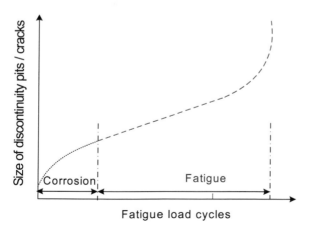

Fig. 26.16 Schematic showing contribution of corrosion and fatigue under corrosion-fatigue conditions

26.4.3.2   Effect of Temperature

Effect of slight variation in temperature on fatigue performance of the weld joint is marginal. Low temperature generally increases the hardness and tensile strength and lowers the ductility. Increase in hardness and strength delays the crack nucleation stage during first stage of fatigue fracture; however, a combination of high strength and low ductility increases the stable crack growth rate in second stage of fatigue fracture. Carbon steel and mild steel weld joints below the ductile to brittle transition temperature (DBTT) lose their toughness which in turn increases the stable fatigue crack growth rate in second stage of the fatigue fracture. On the other hand, moderate increase in temperature lowers the strength and increases the ductility. This combination of strength and ductility reduces the number of load cycles required for nucleation of the fatigue crack in first stage of fatigue fracture while this combination of strength and ductility increases crack tip blunting tendency due to easy deformation of the material ahead of the crack tip which in turn reduces stable crack growth rate during second stage of fatigue fracture in fatigue dominated temperature zone. Therefore, influence of slight increase in temperature on the fatigue life is not found to be very decisive and significant. However, too low temperatures can lower the fatigue performance appreciably due to large variation in material properties such as hardness, ductility and strength. High-temperature facilitating recovery and creep of metal will be adversely affecting the fatigue life predominantly due to dominance of creep over fatigue due to metallic failure (Fig. 26.17).

26.4.3.3   Effect of Vacuum

The fatigue performance of weld joints in vacuum is found much better than in the normal ambient conditions (Fig. 26.18). This improvement in fatigue performance is mainly attributed to the absence of any surface oxidation and any other reaction with atmospheric gases.

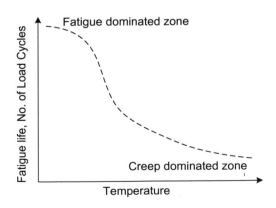

Fig. 26.17 Schematic showing effect of temperature on fatigue life

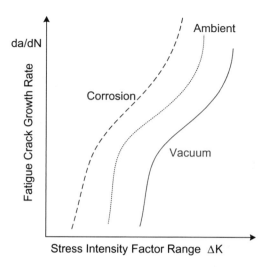

Fig. 26.18 Schematic showing effect of exposure environment on fatigue crack growth rate

Questions for Self-assessment

(a)   What is fatigue loading of weld joint?
(b)   Explain the mechanism of fatigue fracture of weld joints?
(c)   Describe factors affecting different stages of fatigue fracture.
(d)   How do material properties affect stable crack growth rate?
(e)   Explain the fatigue crack growth rate vs. stress intensity factor range relationship.
(f)   Explain the crack growth and number of fatigue load cycle relationship using suitable schematic diagram.
(g)   What is residual fatigue life?
(h)   Enlist the factors affecting the fatigue performance of weld joints?
(i)   Describe the effect of following service load-related parameters on fatigue life of weld joints.

   i.   Type of stress
   ii.   Maximum stress
   iii. Stress range
   iv.   Stress ratio
   v.   Mean stress
   vi.   Frequency of loading

(j)   How do materials properties affect the fatigue performance of a weld joint?
(k)   Explain the role of physical properties of metal to be welded on fatigue life?
(l)   Describe effect of the following mechanical properties on the fatigue performance of the weld joints

   1.   Tensile strength

2. Hardness
3. Percentage elongation
4. Fracture toughness

(m) How does microstructure of weld joints affect the fatigue strength?
(n) What is effect of morphology of micro-constituents of the weld joints?
(o) Describe the effect of the following types of environment on fatigue life of weld joints

1. Temperature
2. Vacuum
3. Corrosion.

Further Reading

1. Radaj R (1990) Design and analysis of fatigue resistant welded structures. Woodhead Publishing
2. Parmar RS (2002) Welding engineering & technology, 2nd edn. Khanna Publisher, New Delhi
3. Hicks J (1999) Weld joint design, 3rd edn. Abington Publishing, England
4. Maddox SJ (1991) Fatigue strength of welded structures. Wood-head Publishing
5. Welding handbook, American Welding Society (1987), 8th edn., vols. 1 & 2, USA

Chapter 27
Design of Welded Joints

Fatigue Strength and Welding: Welding Procedure, Improving the Fatigue Strength

27.1 Welding and Fatigue

There are many aspects related with welding which influence the fatigue performance of a sound (defect-free) weld joint such as welding procedure, weld bead geometry, weld joint configuration and residual stress in weld joint (Fig. 27.1). These parameters affect the fatigue performance in four ways (a) how stress raiser in form of weld discontinuities are induced or eliminated, (b) how do residual stresses develop due to weld thermal cycle experienced by the metal during the welding, (c) how are mechanical properties such as strength, hardness, ductility and fracture toughness of the weld joint influenced and d) how is the microstructure of the weld and HAZ affected by the welding-related parameters. In general, all those welding-related factors increasing the stress raisers, tensile residual stresses, embrittlement and coarse needle shape grain structure in weld joint adversely affect the fatigue life.

27.2 Welding Procedure

Welding procedure includes the entire range of activities from edge preparation, selection of welding process and their parameters (welding current, voltage and speed), welding consumable (welding electrode, filler, flux, and shielding gas), post-weld treatment, etc., needed for the development of a weld joint. The following sections describe effect of various steps of welding procedure on the fatigue performance of the weld joints.

© The Author(s), under exclusive license to Springer Nature Singapore Pte Ltd. 2022
D. K. Dwivedi, *Fundamentals of Metal Joining*,
https://doi.org/10.1007/978-981-16-4819-9_27

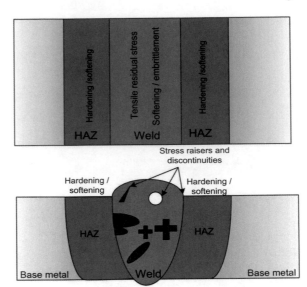

Fig. 27.1 Schematic showing few aspects related with weld affecting fatigue life

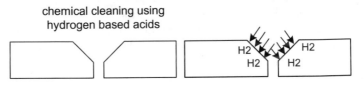

Fig. 27.2 Hydrogen-based chemical cleaning approach causes charging of hydrogen in weld

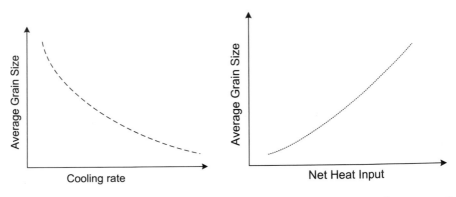

Fig. 27.3 Schematic diagram showing effect of heat input on cooling rate and grain structure of the weld

27.2.1 *Edge Preparation*

There are two main aspects of edge preparation which can influence the fatigue performance of a weld joint (a) cleanliness of faying surface and (b) cutting of faying surface of base metal to be welded by thermal cutting processes. Surface and edge of the plates to be welded must be cleaned to remove the dirt, dust, paint, oil, grease, etc., present on the surface either by mechanical or chemical methods. Use of chemical approach for cleaning the surface using hydrogen-containing acid (sulphuric acid, hydrochloric acid, etc.) sometimes introduces hydrogen in base metal which in long run can diffuse in the weld and HAZ. The presence of hydrogen in HAZ of hardenable steels facilitates crack nucleation and propagation (by HIC) besides making weldment brittle (Fig. 27.2). Further, improper cleaning sometimes leaves impurities on faying surface, which, if are melted or evaporated during the welding then these impurities can induce inclusions in the weld metal. The presence of inclusions in the weld metal acts as stress raiser for easy nucleation and growth of cracks. These in turn weakens the joint and lowers fatigue performance. Cutting of hardenable steel plates by thermal cutting methods such as gas cutting also hardens the cut edge. These hardened edges can easily develop cracks in HAZ under the influence of the residual tensile stresses caused by weld thermal cycle associated with welding.

27.2.2 *Welding Process*

Welding process affects the fatigue performance of a weld joint in two ways (a) net heat input per supplied during welding affects cooling rate and the so weld structure and (b) soundness/cleanliness of the weld. Arc welding processes use heat generated by an arc for melting of the faying surfaces of the base metal. Heat generation from welding arc (VI) of a process depends on welding current (A) and welding arc voltage (V) while net heat supplied to base metal for melting is determined by welding speed (S). Therefore, net heat (kJ/mm) supplied to the faying surfaces for melting is obtained from ratio of arc heat generated (VI) and welding speed (S). Net arc heat supplied to base metal falls over an area per the arc diameter at the surface of base metal. Net heat input per unit area of the base metal affects the amount of the heat required for melting. Higher the net heat input, lower is cooling rate (Fig. 27.3). High cooling rate results in finer grain structure and better mechanical properties hence improved fatigue performance while low cooling rate coarsens the grain structure of weld which in turn adversely affects the fatigue life. However, high cooling rate in case of hardenable steel tends to develop cracks and harden the HAZ which may deteriorate the fatigue performance of the weld joints.

Each arc welding process offers a range for net heat input which in turn affects the cooling rate and so the grain structure and fatigue performance accordingly. For example, shielded metal arc welding possess provides lower net heat input per unit area than gas tungsten arc welding for developing sound weld joints.

Impurities in the form of inclusion in weld metal are introduced due to the inter-actions between the molten weld metal and atmospheric gases. However, the extent of contamination of the weld metal by atmospheric gases depends on the shielding method associated with the particular welding process to protect the "molten weld". Each method has its own approach/mechanism of protecting the weld metal. GTA welding offers minimum adverse effect of atmospheric gases and weld thermal cycle so it results in the cleanest weld in terms of lowest oxygen and nitrogen content in the weld metal as compared to other welding process. On contrary, SAW welding results in high oxygen concentration in weld while self-shielded arc welding process produces weld joints with large amount of oxygen and nitrogen as impurities in the weld metal. These gases in turn result in more concentration of inclusions and porosity in the weld. These discontinuities degrade the fatigue performance. Therefore, the selection of welding process affects the fatigue performance appreciably.

27.2.3 Welding Consumables

Depending upon the welding process being used for the fabrication of a fusion weld joint, variety of welding consumables such as welding electrode, filler wire, shielding gas, flux, etc., are applied. The extent up to which the factors related with welding consumables influence the fatigue performance is determined by the fact that how following characteristics related with welding are affected by welding consumables:

(a) net heat input
(b) cleanliness of the weld metal
(c) residual stress development
(d) microstructure and chemical composition
(e) mechanical properties of the weld joints.

Effect of each of the above aspects related with welding has already been described under separate headings in previous section. In the following section, influence of welding consumable on each of the aspects will be elaborated. The summarized effect of heat input and edge preparation on weld characteristics is presented in Fig. 27.4. Increasing in heat input increases the residual stress while microstructure and mechanical properties are badly compromised. On the other hand, inappropriate edge preparation and poor control over the weld encourage the stress raisers and weld discontinuities.

Electrode

Electrode diameter and its material affect the arc heat generation (due to variation in area over which heat is applied and amount of heat generated (as per welding current and arc voltage) which in turn governs weld thermal cycle and related parameters such as cooling rate, solidification rate, peak temperature and width of HAZ. Large

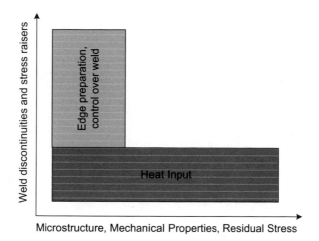

Fig. 27.4 Schematic showing the way heat input and edge preparation can affect the various characteristics of weld joints

diameter electrodes use high welding current which in turn results in high net heat input. Composition of the electrode material affects the solidification mechanism of the weld metal, residual stress in weldment and mechanical properties of the weld metal. Electrode material similar to that of base metal results in epitaxial solidification and otherwise heterogeneous solidification through nucleation and growth mechanism is observed. The difference in thermal expansion coefficient and yield strength of electrode metal with respect to base metal determines the magnitude of residual stress in weld and HAZ region. Larger is the difference in thermal expansion coefficient of two (base metal and weld metal) higher will be the residual stresses. Further, low yield strength weld metal results in lower residual stresses than high yield strength metal. The development of tensile residual stresses in general lowers fatigue life of weld joints (Fig. 27.5). Further, fatigue performance of the weld joints is determined by the way the solidification mode, microstructure and residual stress affect the mechanical properties of weldment. The equiaxed solidification mode, fine grain structure, compressive residual stresses improve the fatigue performance of the weld joints.

Coating material and flux

The presence of low ionization potential elements like Na, K, Ca (in large amount) lowers the heat generation as easy emission of free electrons from these elements in coating material in the arc gap improve the electrical conductivity by increasing the charge particle density which in turn reduces the electrical resistance of arc column. Therefore, heat generation for a given current setting is reduced. Additionally, the basicity index of the flux or coating material on the electrode affects the cleanliness of the weld. In general, flux or coating material having basicity index greater than 1.2 results in cleaner a weld than that of low basicity index. Thickness of the

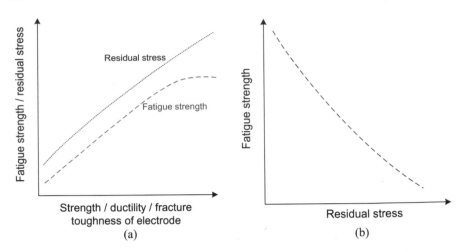

Fig. 27.5 Schematic showing effect of **a** mechanical properties of electrode on fatigue and residual stress and **b** residual stress on fatigue strength

coating material on the core wire in SMA welding affects the contamination of the molten weld pool by influencing the inactive gas (amount) shielding capability from atmospheric gases. Thicker is flux/coating on the core wire better is protection due to release of large amount of inactive protective gases from thermal decomposition of coating material and so cleaner is the weld. Increase in thickness of flux layer in SAW lowers the cooling rate of weld metal during the solidification and increases the protection from atmospheric contamination. Effect of both these factors on fatigue performance of the weld is expected to be different; e.g. low cooling should adversely affect the mechanical properties and fatigue performance while cleaner weld should offer better fatigue performance owing to the absence of stress raisers in the form of inclusions (Fig. 27.6).

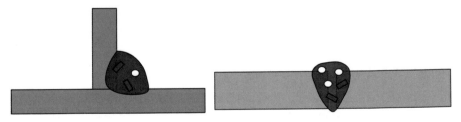

Fig. 27.6 Schematic showing stress raisers in a) filler and b) butt weld joint deteriorating fatigue strength

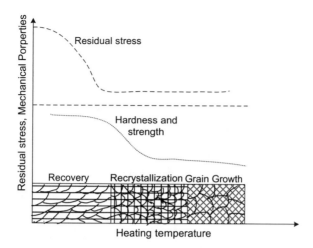

Fig. 27.7 Schematic showing effect of heating temperature due to varying heat input on structure and mechanical properties

Shielding gas

The effect of shielding gas (helium, argon, carbon dioxide, and mixture of these gases with oxygen, helium and hydrogen) on fatigue performance of the weld joint is determined by two factors:

(a) **Effect of shielding gas on the arc heat generation**: The shielding gas affects the heat generation in the arc gap due to the difference in ionization potential of different shielding gases. The variation in heat input in turn affects the cooling rate and so resulting microstructure and mechanical properties of the weld. High ionization potential shielding gas in general burns the arc hotter which in turn leads to lower net heat input and higher cooling rate. The high cooling rate results in finer structure and improved mechanical properties and so enhanced fatigue performance of the weld joint (Fig. 27.7). Similarly, addition of oxygen, hydrogen and helium in argon also increases the arc heat generation and penetration capability of the arc.

(b) **Effect of shielding gas on the cleanliness of the weld**: Shielding capability of each of the above-mentioned gases to protect the molten weld pool from atmospheric gases is found to be different. Helium and argon provide more effective shielding than carbon dioxide and other gases. Hence, He and Ar result in better fatigue performance of the weld joints. Carbon dioxide tends to decompose in arc environment to produce Co and O_2. The presence of oxygen arc zone contaminates the weld metal.

27.2.4   Post-Weld Heat Treatment

Weld joints are given variety of heat treatments (normalizing, tempering, stress relieving, Q &T, T6 treatment) for different purposes ranging from just relieving the residual stress to manipulating the microstructure in order to obtain the desired combination of the mechanical properties. In general, post-weld heat treatment relieves the residual stresses and improves the mechanical properties, which in turn result in improved fatigue performance of the weld joints (Fig. 27.8). However, improper selection of type of PWHT and their parameters, like heating rate, maximum temperature, soaking time and then cooling rate, can deteriorate the microstructure and mechanical properties by inducing unfavourable softening or hardening of HAZ, embrittlement, tensile residual stresses and cracks in the HAZ. Thus, unfavourable PWHT can adversely affect the fatigue performance of the weld joint.

Improving Fatigue Performance

This section presents various approaches commonly used for enhancing the fatigue performance of weld joints, namely reducing stress raiser, improving mechanical properties and inducing compressive residual stresses. Methods of improving fatigue behaviour of weld joints based on the above approaches have been elaborated.

27.3   Improving the Fatigue Performance of the Weld Joints

The fatigue performance of welded joints can be improved using single or multipronged approach including enhancing the load-carrying capability of the weld joint by improving the mechanical properties of the weld, reducing the stress raisers, developing favourable compressive residual stresses. The basic principles of these

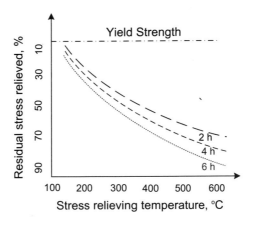

Fig. 27.8   Effect of stress-relieving post-weld heat treatment temperature and time on residual stress relieved

approaches have been presented in the following sections. The fatigue life is improvement using any of the above approaches is based on delaying the completion of one or more of three stages of fatigue fracture under a given set of fatigue loading conditions.

27.3.1 Increasing Load-Carrying Capacity of the Weld Joint

Increase in load-carrying capacity in terms of strength, ductility and toughness improves the tolerance to fatigue loads because of increase in number of load cycles required to complete each stage of the fatigue fracture (Fig. 27.9). Improvement in mechanical properties of the weld joint with right combination of strength, ductility and toughness delays the nucleation and crack growth stages of the fatigue fracture. Load-carrying capability of the weld joints can be enhanced by selecting proper electrode or filler metal and proper welding procedure so to obtain the desired microstructure and mechanical properties of the weld joints. Efforts are made to achieve the fine equiaxed grain structure in weld with minimum adverse effect of weld thermal cycle on the heat affected zone. These factors are influenced by electrode material composition, net heat input during welding and the presence of nucleating agents in weld metal to promote heterogeneous nucleation so as to achieve fine equiaxed grain structure in the weld metal. Inoculation involving addition of the element like Ti, V, Al and Zr is commonly used in steel and aluminium weld to realize the fine equiaxed grain structure. Additionally, application of external excitation techniques such magnetic arc oscillation, arc pulsation and gravitational force method can also be used for grain refinement of weld metal. Selection of proper welding parameters (welding current, speed) and shielding gas also aid in refinement of the grain structure of the weld by reducing the net heat input for developing weld joints. In general, fine equiaxed grain structure is known to enhance the load-carrying capacity of weld joints and fatigue performance of the weld joints. Post-weld heat treatment,

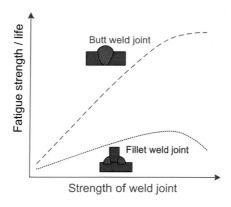

Fig. 27.9 Schematic showing effect of tensile strength of weld joint on fatigue strength/life

namely normalizing, improves the fatigue performance of weld joints by (a) refinement of the grain structure and (b) relieving the residual stress. Surface and case hardening treatments like carburizing and nitriding are also known to improve the fatigue performance of the weld joints in two ways (a) increase the surface hardness up to certain depth and (b) inducing compressive residual stresses.

27.3.2 Reducing Stress Raisers

First stage of fatigue fracture, i.e. crack nucleation, is largely influenced by the presence of the stress raisers on the surface of component subjected to fatigue loading. These stress raisers in the weld joints are mostly found in the form of ripples present on the surface of weld in as-welded condition, sharp change in cross section at the toe of the weld, cracks in weld metal and heat affected zone, inclusions in weld, too high bead angle, excessive reinforcement of the weld bead, crater and under-fill (Fig. 27.10).

In order to reduce adverse effect of stress raisers on fatigue performance of weld joints, it is necessary that stress raisers in form of poor weld bead geometry and weld discontinuities are reduced as much as possible by the selection of the proper welding parameters, consumable, manipulation of welding arc and placement of molten weld metal (Fig. 27.11). The presence of inclusions and defects in the weld metal can be reduced by remelting of small amount of weld metal near toe of the weld using tungsten inert gas arc heat (Fig. 27.12). This process of partial remelting weld bead to remove discontinuities and inclusions especially near the toe of the weld is called TIG dressing. TIG dressing is reported to increase the fatigue life by 20–30% especially under low stress fatigue conditions. The TIG dressing also disturbs the system of residual stress favourably by remelting a small portion of weld and HAZ.

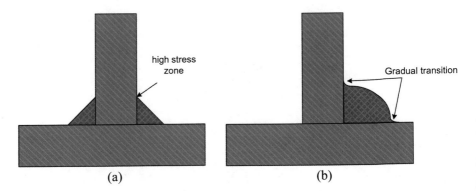

Fig. 27.10 Reducing stress concentration at toe of the weld **a** toe with sudden change in cross section causing high stress concentration and **b** providing some fillet at the toe of the weld by grinding

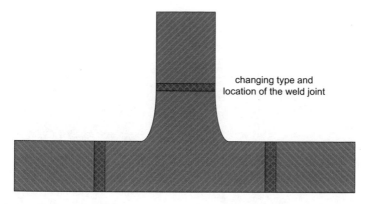

Fig. 27.11 Schematic diagram showing change on joint configuration from fillet to butt joints

Controlled removal of material from the toe of the weld by machining or grinding in order to give suitable fillet so as to avoid abrupt change in cross section of the weld is another method of enhancing the fatigue life of weld joints by reducing the severity of stress raisers.

Further, attempts should be made to reduce the weld bead angle as low as possible so that transition in cross-sectional area from the base metal to the weld bead is gradual to produce a weld joint without stress raisers (Fig. 27.10). Weld joints with machined, ground and flushed weld bead offer minimum stress concentration effect and hence maximum fatigue life. Additionally, efforts can be made to relocate the stress raisers away from the high stress areas by redesigning the components. For example, fillet weld can be replaced with butt weld by relocating the weld through redesign of component (Fig. 27.11).

27.3.3 Developing Compressive Residual Stress

The development of compressive residual stress for improving the fatigue performance of the weld joints is based on simple concept of lowering the effective applied tensile stresses. This residual compressive stress to some extent neutralizes/cancels the magnitude of externally applied tensile stress. Therefore, this method is found effective only when fatigue load is tensile in nature. The magnitude of maximum stress is lower than yield strength. Moreover, this method marginally affects the fatigue performance of the weld joints under low cycle fatigue conditions when fluctuating loads and corresponding stresses are more than yield strength of weld joint. Improvement in fatigue performance of the weld joint by this method can vary from 20–30%. There are many methods, namely shot peening, overloading, spot heating, and post-weld heat treatment, which can be used to induce compressive

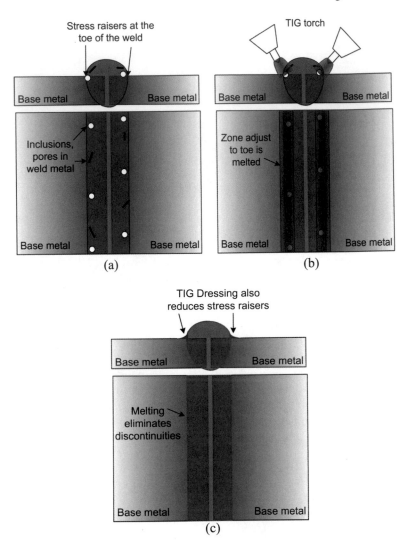

Fig. 27.12 Schematic of TIG dressing process **a** weld joint with discontinuities, **b** passing TIG arc along the weld toe and **c** remover weld discontinuities and increased radius at the weld toe

residual stress. All these methods are based on principles of differential dimensional/volumetric change at the surface layer with respect to that core of the weld by application of either localizing heating or stresses beyond yield point.

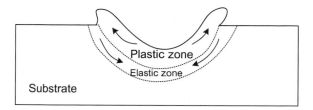

Fig. 27.13 Schematic of shot peening mechanism

27.3.3.1 Shot Peening

In case of shot peening, high-speed steel balls are directed towards the surface of the weld joint on which compressive residual stress is to be developed. Impact of shots produces indentation through localized plastic deformation at the surface layers of weld and HAZ while metal layers just below the plastically deformed surface layers are subjected to elastic deformation. Material further deeper from the surface remains unaffected by shots and plastic deformation occurring at the surface (Fig. 27.13). Elastically deformed layers tend to regain their dimensions while plastically elongated surface layers resist any comeback. Since both plastically and elastically elongated layers are metallurgically bonded together therefore elastically elongated under-surface metal layer tends to put plastically elongated surface layer under compression while elastically elongated under-surface layers come under tension as balancing stresses. Thus, residual compressive stresses are induced at shot-peened surface. The presence of tensile residual stress below the surface is not considered to be much damaging for fatigue life as mostly fatigue failures commence at the surface.

27.3.3.2 Overloading

This method helps to reduce the residual stresses by (a) developing the opposite kind of elastic stresses and (b) relieving the locked in strain using plastic deformation by overloading the component under consideration. Stages of this method regarding external loading and residual stress in as-welded and stress relieved conditions are shown in Fig. 27.14 (Fig. 27.15).

27.3.3.3 Shallow Hardening

Shallow hardening improves the fatigue performance in two ways (a) increase in the hardness of surface and near surface layers which in turn delays crack nucleation stage of fatigue fracture and (b) development of residual compressive stress at the surface reduces adverse effect of the external tensile stresses on all stages of fatigue fracture hence improves the fatigue performance. However, under external compressive loading conditions, residual compressive stresses will deteriorate the fatigue performance of welds.

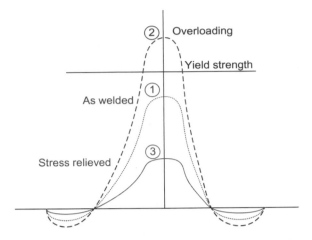

Fig. 27.14 Schematic of overloading approach using external loading methods

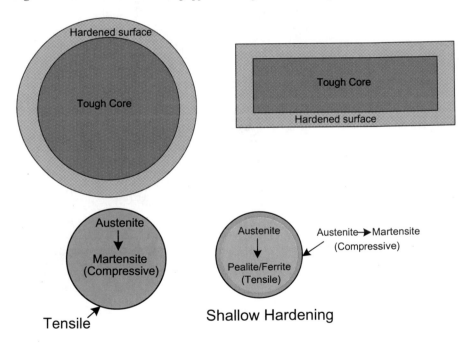

Fig. 27.15 Schematic of shallow hardening approach. It is showing type of residual stresses developed in case of through section and shallow hardening of steel

Questions for Self-assessment

(a) What is effect of various parameters related with welding on fatigue strength of weld joints?

(b) How does edge preparation influence the fatigue behaviour of the weld joints?

(c) Why does the fatigue performance of the weld joint developed using different processes vary?

(d) Describe role of following welding consumables on fatigue strength of the weld joints?

   i. Electrode
   ii. Coating materials
   iii. Shielding gas

(e) Does post-weld heat treatment affect the fatigue behaviour of weld joints? If yes, how?

(f) What are different approaches used for improving the fatigue performance of the weld joints?

(g) What is effect of stress raisers on fatigue life? Explain how fatigue life can be increased by reducing the stress raisers in weld joints?

(h) How does the development of residual compressive stress help in reducing fatigue failure?

(i) Explain the principle of following methods of inducing compressive residual stresses for improving the fatigue life of the weld joints?

   i. Shot peening
   ii. Overloading
   iii. Shallow hardening.

Further Reading

1. Hicks J (1999) Weld joint design, 3rd edn. Abington Publishing, England
2. Maddox SJ (1991) Fatigue strength of welded structures. Woodhead Publishing
3. Parmar RS (2002) Welding engineering & technology, 2nd edn. Khanna Publisher, New Delhi
4. Welding handbook, American Welding Society (1987), 8th edn., vols. 1 & 2, USA

Part IX
Inspection and Testing of Weld Joints

Chapter 28
Inspection and Testing of Weld Joint

Destructive Testing: Tensile, Bend, Hardness, Toughness, Fracture Toughness. Non-destructive Testing: Dye Penetrant, Magnetic Particle, Ultrasonic Testing

28.1 Introduction

To produce quality weld joints, it is necessary to keep an eye on what is being done in three different stages of the welding

- Before welding such as cleaning, edge preparation and baking of electrode to ensure sound and defect-free weld joints
- During welding various aspects such as manipulation of heat source, selection of input parameters (pressure of oxygen and fuel gas, welding current, arc voltage, welding speed, shielding gases and electrode selection) affecting the heat input and so melting, solidification and cooling rates besides protection of the weld pool from atmospheric contamination
- After welding steps, if any, such as removal of the slag, peening, post-welding treatment.

Selection of proper method and parameters of each of above steps and their meticulous execution in different stages of production of a weld joint determine the quality of the weld joint. Inspection is mainly carried out to assess ground realties in respect of progress of the work or how meticulously things are being implemented. Testing helps to: (a) assess the suitability of the weld joint for a particular application in light requirements of the service, (b) take decision on whether to go ahead (with further processing or accept/reject the same) at any stage of welding and (c) quantify the performance parameters related to soundness and performance of weld joints.

Testing methods of the weld joint are broadly classified as destructive testing and non-destructive testing. Destructive testing methods damage the test piece to more or less extent. The extent of damage on (destructive) tested specimens sometime can be up to complete fracture (like in tensile or fatigue testing), thus making it unuseable for the intended purpose, while in case of non-destructive tested specimen, the extent of damage on tested specimen is either none or negligible which does not adversely affect their usability for the intended purpose in anyways for most of the general-purpose applications.

D. K. Dwivedi, *Fundamentals of Metal Joining*,
https://doi.org/10.1007/978-981-16-4819-9_28

Weld joints are generally subjected to destructive tests such as hardness, toughness, bend and tensile test for developing the welding procedure specification and assessing the suitability of weld joint for a particular application.

Visual inspection reflects the quality of external features of a weld joint such as weld bead profile indicating weld width and reinforcement, bead angle and external defects such as craters, cracks and distortion only.

28.2 Destructive Test

28.2.1 Tensile Test

Tensile properties of the weld joints, namely yield and ultimate strength and ductility (%age elongation and %age reduction in area), can be obtained either in ambient condition or in special environment (low temperature, high temperature, corrosion, etc.) depending upon the requirement of the application using tensile test which is usually conducted at constant crosshead speed (ranging from 0.0001 to 10,000 mm/min). Tensile properties of the weld joint are obtained in two ways (a) taking specimen from transverse direction of weld joint consisting of base metal–heat affected zone–weld metal–heat affected zone–base metal and (b) all weld metal specimen as shown in Fig. 28.1a, b.

Tensile test results must be supported by respective engineering stress and strain diagram indicating modulus of elasticity, elongation at fracture, yield and ultimate strength (Fig. 28.2). Test results should include information on the following point about test conditions:

- Type of sample (transverse weld and all weld specimen)
- Strain rate (mm/min)
- Temperature or any other environment in which test was conducted if any

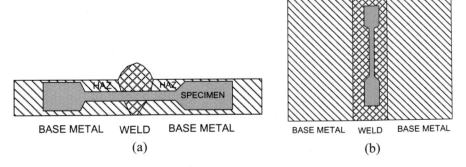

BASE METAL   WELD   BASE METAL   BASE METAL   WELD   BASE METAL

(a)   (b)

Fig. 28.1 Schematic of tensile specimens from **a** transverse section of weld joints and **b** all weld specimen

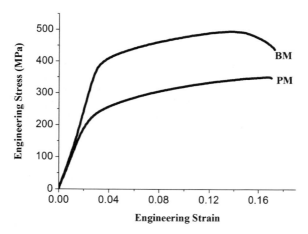

Fig. 28.2 Typical stress–strain diagram of aluminium alloy in as received (BM) and friction stir processed (PM) condition

- Topography, morphology, texture of the fracture surface indicating the mode of fracture and respective stress state
- Location of the fracture.

28.2.2 Bend Test

Bend test is one of the most important and commonly used destructive tests to determine the ductility and soundness (for the presence of porosity, inclusion, penetration and other macro-size internal weld discontinuities, if any) of the weld joint produced under one set of welding conditions. Bending of the weld joint can be done from face or root side depending upon the purpose, i.e. whether face or root side of the weld is to be assessed. The root side bending shows the lack of penetration and lack of fusion at the root, if any. Further, bending can be performed using simple compressive/bending load and die of standard size for free and guided bending, respectively (Figs. 28.3 and 28.4). Moreover, free bending can lead to face or root bending, while guided bending is performed by placing the weld joint over the die as needs (face or root is subjected to elongation and severity of deformation). Thus, guided bending is better and performed in well-controlled condition as shown in Fig. 28.4.

For bend test, the load is increased until cracks start to appear on face or root of the weld for face and root bend test, respectively, and angle of bend at this stage is used as a measured of ductility of weld joints. The higher the bend angle (needed for crack initiation), greater is the ductility of the weld joint. Fracture surface of the joint from the face/root side due to bending reveals the presence of internal weld discontinuities, if any.

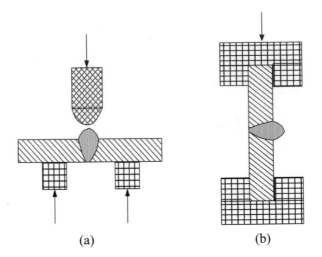

(a)   (b)

Fig. 28.3 Schematics of free bend tests

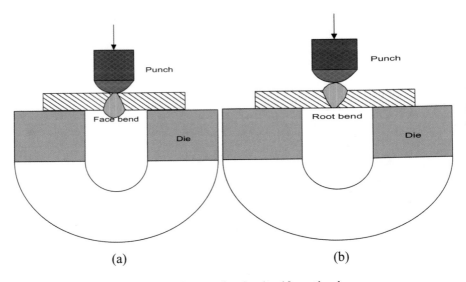

(a)   (b)

Fig. 28.4 Schematics of guided bend tests: **a** face bend and **b** root bend

28.2.3 *Hardness Test*

Hardness is defined as resistance to indentation and is commonly used as a measure of resistance to abrasion or scratching. For the formation of a scratch or causing abrasion, a relative movement is required between two or more interacting bodies. Out of two, one body must penetrate/indent into other body. Indentation is the penetration of a pointed object (harder) into other object (softer) under the influence of external

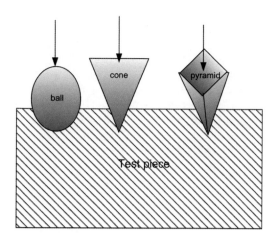

Fig. 28.5 Schematic diagram showing indentation using different indenters corresponding to different hardness test methods

load. Resistance to the penetration of the pointed object (indenter) into the softer one depends on the hardness of the sample. The load is applied on the surface of the test component through the indenter, and then, effect of indenter is checked/measured at the surface of the body.

All methods of hardness testing are based on the principle of applying the standard load through the indenter (a pointed object) and measuring the penetration in terms of diameter/diagonal/depth of indentation as per method of the hardness test (Fig. 28.5). High penetration of an indenter at a given standard load suggests low hardness. Various methods of hardness testing can be compared on the basis of the following three criteria: (1) type of indenter, (2) magnitude of load and (3) measurement of indentation.

Parameter	Brinell	Rockwell		Knoop	Vickers
Load	500–2000 kg	Minor: 10 kg Major: 60–200 kg as dictated by scale to be used (A-C)		10–3000 g	
Indenter	Ball	Ball or cone		Cone	Pyramid
Measurement	Diameter	Depth		Diagonal	Diagonal

Penetration due to applied normal load can be influenced by unevenness on the surface and presence of hard surface films such as oxides, lubricants, dust and dirt, if any. Therefore, surface of the test piece should be cleaned and polished before hardness test. In case of Brinell hardness test, full load is applied directly for causing indentation to measure the hardness, while in case of Rockwell hardness test, minor load (10 kgf) is applied first before applying major load (60, 90 and 140 kgf as per scale to be used, i.e. HRA, HRB, HRC). Minor load is applied to ensure the firm metallic contact between the indenter and sample surface by breaking surface films and impurities, if any, present on the surface. Minor load is not expected to cause indentation. Indentation is caused by major load only. Therefore, cleaning and polishing of the surface films become mandatory for accuracy in hardness test results in case of Brinell test method as major load is applied directly.

Steel ball of different diameters (D) is used as an indenter in Brinell hardness test. Diameter of indentation (d) is measured to calculate the projected area and determine the hardness. Brinell hardness test results are expressed in terms of pressure generated due to load (P). It is calculated by the ratio of load applied and projected contact area. Load in the range of 500–3000 kg can be applied depending upon the type of material to be tested. Higher load is applied for hardness testing of hard materials as compared to soft materials.

$$\text{BHN} : \frac{2P}{\pi D \left[D - \left(D^2 - d^2 \right) \right]^{1/2}}$$

In case of Rockwell hardness test, first minor load of 10 kgf is applied, then major load of 60–140 kgf is applied on the surface of the workpiece through the indenter and the same is decided by scale (A, B, C and D) to be used as per type of material to be tested. Minor load is not changed. Out of above 4 scales, B and C scales are commonly used. Different indenters and major loads are required for each scale. Steel ball and diamond cone are two types of indenters used in Rockwell testing. B scale uses hardened steel ball and major load of 90 kg, whereas C scale uses diamond cone and major load of 140 kg; accordingly, hardness is written in terms of HRB and HRC, respectively.

Vickers hardness test uses square pyramid shape indenter of diamond and load ranging from 1 to 120 kg. Average length (L) of two diagonals of square indentation is used as a measure of hardness. The longer the average diagonal length, lower is the hardness. Vickers hardness number (VHN) or diamond pyramid hardness (DPH) is the ratio of load (P) and apparent area of indentation given by the relation: DPH: $1.854P/L^2$.

Hardness test provides very quick indications (less than 10 s) of tensile strength and mechanical properties. However, hardness testing suffers from (a) need of conducting a number of tests to arrive at a value in conclusive manner and (b) proper sample surface preparation is needed for consistent results.

Mechanical Testing of Weld Joints

The following section describes three important destructive testing methods of welded joints, namely toughness test, fatigue test and fracture toughness test. Additionally, concept of the fracture toughness and conditions required for fracture toughness test for different stress conditions has also been presented. Further, few non-destructive testing methods have also been presented.

28.2.4 *Toughness Testing*

In actual practice, mechanical components during the service are invariably subjected to various kinds of loads, namely static and dynamic loads which are classified on the basis of the rate of change in magnitude of load and direction. Dynamic loads are characterized by high rate of change in load magnitude and direction. Reverse happens in case of static loads. During the hardness and tensile tests, load is increased very slowly that corresponds to the behaviour of material under more or less static loading condition. A very wide range rate of loading can be used for the tensile test. Rate of loading governs the strain rate and so the rate of hardening which can affect mechanical behaviour of material. For example, material showing ductile behaviour at low rate of loading can exhibit brittle behaviour under high rate of loading condition.

The toughness test simulates service conditions often encountered by mechanical components used in transportation, agricultural and construction equipment. A material of high impact resistance is said to be a tough material. Toughness is the ability of a material to resist both fracture and deformation. Conversely, toughness indicates combination of strength and ductility. To be tough, a material must exhibit both fairly good strength and ductility to resist cracking and deformation under impact loading. Notches are made intentionally in impact test specimens to increase the stress concentration so as to increase tendency to fracture as most of the mechanical components have stress raisers. To withstand an impact force, a notched material must be tough. Despite indicating the behaviour of materials under impact conditions, toughness which is generally measured in terms of Nm and joules is not for design of mechanical component. Nowhere in design of mechanical components and weld joint design, the toughness value is used for calculation purpose however, it helps is selection of proper material selection at the design stage.

To study the behaviour of material under dynamic load conditions (at high rate of loading), toughness test is frequently conducted. There are two methods used for toughness testing, namely Izod and Charpy tests, based on the common principle of applying the load at high rate and measuring the amount of energy absorbed (Nm or Joule) in breaking the sample due to impact (Fig. 28.6). However, there are some differences also in these two methods in terms of sample size and shape, method of holding of the sample and maximum energy content of pendulum that hits the sample during the test.

S. No.	Toughness test	Sample	Holding
1	Izod	Held vertically on anvil as cantilever	Cantilever type and notch face the pendulum

(continued)

(continued)

S. No.	Toughness test	Sample	Holding
2	Charpy	Held horizontally on anvil as simply supported beam	Simply supported type and notch are opposite side of pendulum impact (not facing pendulum)

Standard sample for both testing methods have a notch. The sample for each test method is mounted on the machine in specific ways i.e. notch faces to pendulum in case Izod test while pendulum hits the sample from back of the notch in Charpy test (Fig. 28.7).

Since most of the mechanical components are invariably designed with notches and stress raisers, it becomes important to know about the behaviour of material with notch under impact loading. Hence, toughness test is usually conducted on a sample with notch. Moreover, un-notched samples can also be used for the toughness test and the results of the test are expressed accordingly.

Results of impact tests are expressed in terms of either amount of energy absorbed (Nm) or amount of energy absorbed per unit cross-sectional area (Nm/cm^2) by standard sample. It may be noted that values of toughness are not directly used for design purpose but these only indicate the ability of the material to withstand against shock/impact load, i.e. load applied at very high rate. These tests are useful for comparing the resistance to impact loading of different materials or the same material in different processing conditions such as heat treatment, procedure and mechanical working. Resistance to the impact loading of a material appreciably depends on the surrounding temperature (Fig. 28.8). Therefore, temperature at which toughness test is conducted must be reported with results. Toughness test report should include

- Energy absorbed (in Nm or Joules) or energy absorbed per unit area
- Method of test, i.e. Izod or Charpy
- Notched or un-notched samples and its cross section if standard sample is not used
- Temperature of the specimen during test.

28.2.5 Fatigue Behaviour of Weld Joint

The fatigue performance of the metallic components in general is determined in two ways (a) endurance limit, i.e. indicating the maximum stress, stress amplitude or stress range for infinite life (typically more than 2 millions of load cycles), and (b) number of load cycle a joint can withstand for a set of loading conditions as desired. Two types of samples are generally prepared for fatigue studies as per ASTM 466 (Fig. 28.9a, b). Reduced radius sample generally ensures fracture from weld joint or any specific location of interest (Fig. 28.10a, b). The fatigue performance is appreciably influenced by various variables related to fatigue test, namely stress ratio, type of stress (tension–tension, reverse bending, tension–compression, zero-tension),

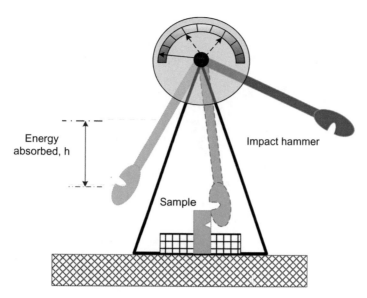

Fig. 28.6 Schematic showing principle of toughness test

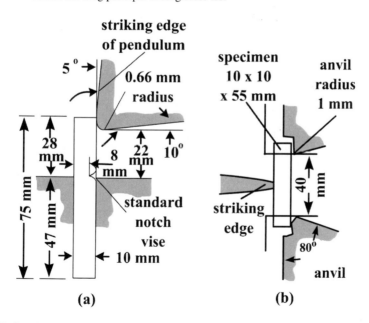

Fig. 28.7 Standard specimens for **a** Izod and **b** Charpy impact tests

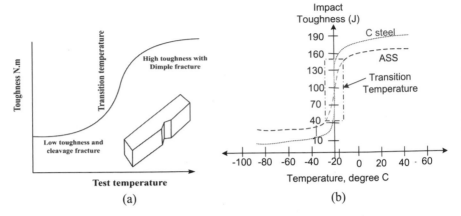

Fig. 28.8 DBTT: **a** schematic diagram showing ductile to brittle transition behaviour and **b** toughness versus temperature relation for common structural steel

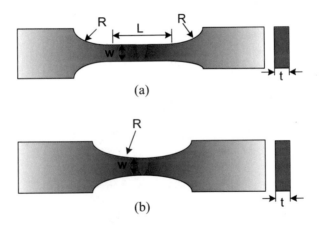

Fig. 28.9 Standard specimen for fatigue testing

maximum stress, stress range, loading frequency and surrounding environmental conditions such as temperature, corrosion, vacuum and tribological aspects, if any. Each and every parameter to be used for the fatigue test must be carefully selected and recorded with results while reporting. The fatigue test results should include the following.

- Type of loading: axial pulsating/reverse bending/tension–compression
- Maximum stress
- Stress ratio (ratio of minimum stress to maximum stress)
- Temperature: ambient/vacuum/corrosion
- Frequency of pulsating load: load cycles per min
- Type of sample.

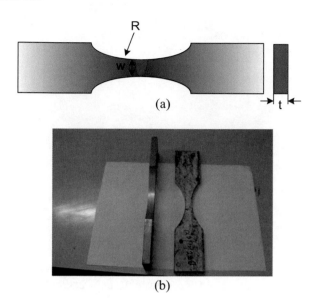

Fig. 28.10 Fatigue test sample: **a** schematic diagram of standard fatigue test sample with continuous radius between ends and **b** photograph of typical specimen

To conduct a fatigue test, first step is to perform the tensile test on the weld joint for establishing the yielding strength of metal. The maximum stress for fatigue test (in the begining) is taken as 0.9 times of yield strength of the component. For plotting the stress amplitude–number of cycle (S–N) curve, fatigue test is first conducted with maximum applied tensile load corresponding to 0.9 times of yield strength of weld joint under study to determine the number of load cycle required for fracture and then in the same way test is repeated at 0.85, 0.8, 0.75, 0.7 …. times of yield strength of weld joint until endurance limits or desired fatigue life is achieved (Fig. 28.11). Typical dimensions of a standard specimen as per ASTM 466 are as under.

- Continuous radius (R): 100 mm
- Width (W): 10.3 mm
- Thickness (T): 11 mm (as received)
- Gripping length: 50 mm.

28.2.6 Fracture Toughness

The resistance to crack growth is known as fracture toughness and is measured using various parameters such as stress intensity around the crack tip (K), opening of crack mouth also called crack tip opening displacement (CTOD) and energy requirement for growth of crack (J or G). The mechanical properties, namely yield

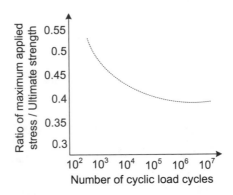

Fig. 28.11 Typical data on fatigue test showing peak stress/ultimate stress versus number of cycle relationship for structure steel

strength and ductility and thickness of the weld joint, under study primarily dictate the suitable parameter to be used for determining the fracture toughness. The fracture toughness parameter, namely stress intensity factor (K), is commonly used for weld joint of heavy sections of high strength and low ductility material developing plain strain conditions, while crack tip opening tip displacement and energy-based methods (G and J integrals) are used for comparatively thinner sections made of low strength and high ductility material and those developing plain stress condition under external loading. Measurement of fracture toughness using any of above parameters is performed using two types of samples: (a) compact tension specimen (CT) and (b) three-point bending specimen (TPB). Schematics of two types of specimen are shown in Fig. 28.12. In general, during these tests, applied external load is increased until strain/crack opening displacement/energy versus load relationship becomes nonlinear. This critical value of load (P) is used for calculations of fracture toughness using relevant formulas.

$W = 2B$, $a = B$, $W - a = B$ and radius of hole $r = 0.25B$ where B is plate thickness.

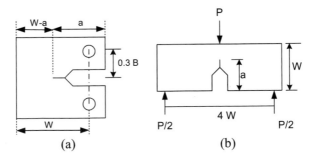

Fig. 28.12 Schematic of fracture toughness specimens using **a** compact tension and **b** three-point bending approaches

Although different standards have been published for determining K, CTOD and J integral, and these tests are very similar, and generally values of these three parameters used as a measure of the fracture toughness can be established from one type of test.

In general, stress intensity factor (K) decreases with increase in specimen thickness. This trend continues up to a limit of thickness; thereafter, K becomes independent of the plate thickness. The corresponding value of K is called critical stress intensity factor (K_c) and occurs in *plane strain condition*. K_{IC} is used for the estimation of the critical stress need to apply to a specimen with a given crack length for catastrophic fracture to take place.

$$\sigma_C \leq K_{IC}/(Y(\pi\, a)^{1/2})$$

where K_{IC} is the stress intensity factor, measured in MPa*m$^{1/2}$, σ_C is the critical stress applied to the specimen, a is the crack length for edge crack or half crack length for internal crack and Y is a geometry factor.

28.3 Non-destructive Testing (NDT)

To determine the presence of surface and sub-surface imperfections, non-destructive testing of weld joints can be carried out using variety of techniques as per needs. Apart from the visual inspection, many non-destructive testing methods including dye penetrant test (DPT), magnetic particle test (MPT), eddy current test (ECT), ultrasonic test (UT), radiographic test (RT), etc., are used in manufacturing industry for assessing the soundness of weld joints. In the following section, principle and capability of some non-destructive testing methods have been described.

28.3.1 Dye Penetrant Test

This is one of the simplest non-destructive testing methods primarily used for detecting the presence of surface defects only. In this method, a thin low viscosity and low surface tension liquid containing suitable dye is applied on the surface to be tested (Fig. 28.13). The thin liquid penetrates (by capillary action) into fine cavities, pores and cracks, if any, present on the surface. Excess liquid present at surface is wiped out. Then, suitable developer like talc or chalk powder is sprinkled over the surface. Developer sucks out thin liquid with dye wherever it is present inside the surface discontinuities on the surface of weld joints. Dye with liquid changes colour of developer and indicates location, size and distribution of surface defects.

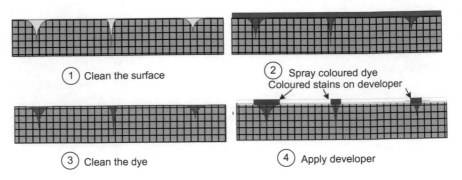

① Clean the surface   ② Spray coloured dye
                              Coloured stains on developer

③ Clean the dye   ④ Apply developer

Fig. 28.13 Schematic showing four steps of dye penetrant test

28.3.2 Magnetic Particle Testing

This method is mainly used for assessing the surface and near-sub-surface defects in the welded joints of magnetic material. The magnetic particle testing is based on the simple concept of difference in the flow of magnetic line of forces through magnetic and non-magnetic materials. Magnetic flux flows easily through metal from south to north pole. The component to be evaluated is magnetized using electrical energy or suitable permanent magnetic. The electro-magnetization is performed using suitable yoke which is applied across the location/area to be tested. The presence of any discontinuity in the form of crack, porosity, near-surface defects in the path of flow of these lines results in leakage of magnetic flux by forming two additional poles. The magnetic powder particles (either in dry or suspended in thin liquid form) are sprinkled over the surface of components magnetized for testing. The magnetic particles tend to migrate towards the location, wherefrom leakage of magnetic flux is taking place and then gets piled up there (Fig. 28.14). The particles align along discontinuities on the surface, near or shallow sub-surface discontinuities. The location and pattern of piled-up magnetic powder particles suggest the location, size, type of discontinuity present on the surface or near-surface region. Hazy pile of powder

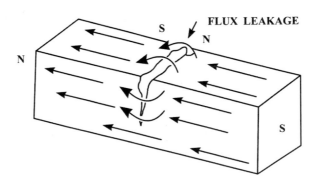

Fig. 28.14 Principle of magnetic particle test

particle indicates existence of the sub-surface defect. Formation of very thin line of powder particles suggests the presence of crack with details of size and location of crack. However, this method of testing is found fit for ferromagnetic metal only.

28.3.3   Ultrasonic Testing

This is one of the most popular, quick, cost-effective and capable methods of NDT as it not only indicates the presence of discontinuities but also provides information about location of discontinuities using ultrasonic vibrations. Ultrasonic vibrations have capability to penetrate into the metals and are reflected as soon as they come across a change of medium, e.g. metal to air, air to metal, metal to discontinuities, etc. This reflection characteristic of ultrasonic vibrations from the interfaces of change in medium is mainly exploited for detecting the presence or absence of discontinuities. Application of ultrasonic vibrations in a sound metal system at the sources produces two interfaces: a) at top surface due to change of medium from air to metal and b) at the bottom surface due to change of medium from metal to air. More than two interfaces occur only if there are discontinuities of metal system being tested. The ultrasonic vibrations are used in two ways for NDT: a) transmission and b) reflection of vibrations to evaluate the soundness of the weld joints in consideration. Both the methods are very effective for parallel surface components, e.g. plates and sheets.

28.3.3.1   Transmission Approach

The transmission approach of ultrasonic testing uses two separate devices, namely transmitter and receiver of vibration. Transmitting probe generates and sends the ultrasonic vibrations, and receiver gets these vibrations at other end. Therefore, transmission approach needs access to both the sides of the components to be tested. Inputs from transmitting and receiving probes are given to oscilloscope (Fig. 28.15). Metal system without discontinuities shows the two peaks in oscilloscope, i.e. one from the top surface and another from the bottom surface. In the presence of discontinuity in the metal being tested, ultrasonic vibrations are reflected so they do not reach up to the receiving end and so no signal is received. Under this condition, only one peak is observed in the oscilloscope and the absence of another peak from bottom surface suggests the presence of discontinuity in the metal being tested. One by one entire surface area of the component to be tested is scanned using transmitting and receiving probes. However, transmission approach is not very useful due to two reasons: (a) requirement of access to both sides of component to be tested and (b) difficulty in placement of receiving probe in line of transmitting probe sending ultrasonic vibrations, especially in case of components having thick sections.

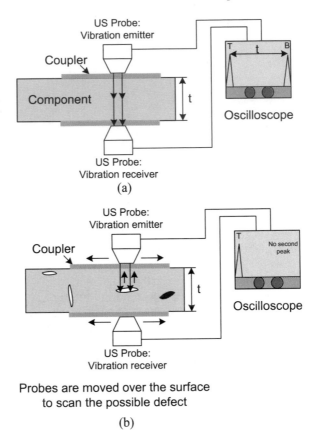

Fig. 28.15 Transmission type of ultrasonic testing of **a** plate without discontinuities and **b** plate with large crack like discontinuity perpendicular to ultrasonic vibrations

28.3.3.2   Reflection Approach

The reflection approach uses single probe which acts as a transmitter as well as receiver of ultrasonic vibrations (Fig. 28.16a–d). In metal system without discontinuities, application of ultrasonic vibrations results in the two peaks in oscilloscope, i.e. one from the top surface and another from the bottom surface like transmission approach (Fig. 28.16a). In the presence of discontinuity in the metal being tested, ultrasonic vibrations are reflected. Vibrations reflected from the discontinuity show additional peaks between the surface and bottom peaks in the oscilloscope. Relative location of the intermediate peaks (between the top and bottom surface peaks) suggests the distance of discontinuity from the surface (Fig. 28.16b-d). The reflection approach overcomes both limitations of transmission approach as it uses single probe so it does not require (a) access of both sides of the component to be tested and (b) alignment of transmission and receiving probes.

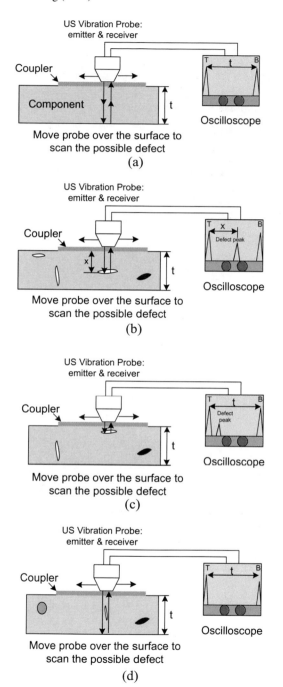

Fig. 28.16 Reflection type of ultrasonic testing: **a** sound plate free from discontinuities, **b** plate with discontinuities in middle of thickness, **c** plate with discontinuity adjustment to the surface and **d** plate with discontinuities almost parallel to ultrasonic vibrations

28.3.3.3   Pitch-Catch Method

In this method, ultrasonic vibrations are transmitted using 45 and 60 degrees to the surface of the material to be tested (Fig. 28.17). Reflected vibrations from the other reflecting surface or discontinuity are used to identify the presence and location of discontinuity in weld joints and other parallel sided surfaces.

Coupler

For effective transmission of ultrasonic vibrations from the transmitting/source probe to the metal surface, generally a fluid mostly in the form of water or low viscosity liquid called coupler is used. The coupler ensures proper contact and transmission of vibration from source probe to metal surface with minimum losses. Water is considered as the best coupling media because it is readily available has low viscosity

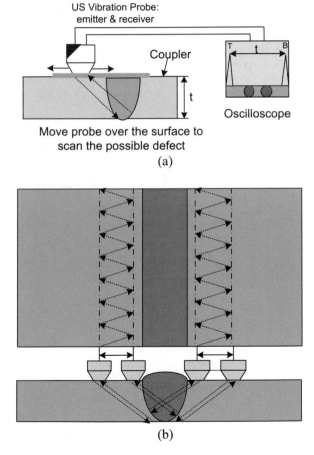

Fig. 28.17 Schematic of pitch-catch method of ultrasonic testing: **a** use of oblique probe for weld joint testing and **b** pattern of movement of the probe for weld

and is relatively safe to use with most construction materials. In the pitch-catch method, a water-based gel has proven to be the most practical coupling agent.

Questions for Self-assessment

(a) What are stages of inspection in welding and what should be looked into at each stage?

(b) How destructive testing is different from non-destructive testing methods?

(c) What are methods of hardness testing? Describe methodology of Brinell, Rockwell, Vickers and Knoop hardness testing.

(d) What information can be obtained from tensile test?

(e) Describe method of tensile testing of weld joints. Draw schematic diagram of engineering stress and stress curve and show yield point, ultimate point, fracture point and modulus of elasticity.

(f) What information must be provided with results of tensile test?

(g) Define toughness and how can it be used in engineering design?

(h) Name various methods of toughness testing along with basic principles of the same.

(i) Distinguish the Charpy and Izod toughness testing methods?

(j) What information related to test must be provided with results of toughness test?

(k) What is fatigue and how can fatigue strength of ferrous and non-ferrous metals be obtained?

(l) What is significance of the following terms in fatigue test: stress amplitude, stress ratio, loading pattern, loading frequency?

(m) Describe step-by-step procedure of fatigue testing.

(n) How do we express fatigue test results? What information must be provided with results of fatigue test?

(o) What is fracture toughness and how can it be obtained for a material.

(p) What are parameters commonly used for measuring fracture toughness of hard and brittle material, high-strength low ductility material, low strength and ductile metals?

Further Reading

American Welding Society (1983) Inspection and testing of weld joints Welding handbook, 7th edn. American Welding Society, USA

Kou S (2003) Welding metallurgy, 2nd edn. Willey, USA

Lancaster JF (2009) Metallurgy of welding, 6th edn. Abington Publishing, England

American Society for Metals (1993) Metals handbook-welding, brazing and soldering, 10th edn., vol. 6. American Society for Metals, USA

Parmar RS (2002) Welding engineering & technology, 2nd edn. Khanna Publisher, New Delhi

Little R (2001) Welding and welding technology, 1st edn. McGraw Hill

Cary H (1988) Welding technology, 2nd edn. Prentice Hall

Nadkarni SV (2010) Modern arc welding technology. Ador Welding Limited, New Delhi

Parmar RS, Welding process and engineering. Khanna Publisher, New Delhi
American Society for Metals (1993) Metals Handbook-mechanical testing and evaluation, vol. 8. American Society for Metals, USA
American Welding Society (2017) Welding handbook, 8th edn., vols. 1 & 2. American Welding Society, USA
Dwivedi DK (2018) Surface engineering. Springer, New Delhi
Dwivedi DK (2013) Production and properties of cast Al-Si alloys. New Age International, New Delhi

Part X
Weldability of Metals

Chapter 29
Weldability of Metals: Characteristics of Metals and Weldability

Weldability of Steel

29.1 Understanding Weldability

Weldability is considered as ease of accomplishing a satisfactory weld joint and can be determined from the quality, effort and cost required for developing a weld joint. Quality of the weld joint, however, can be determined by many factors, but the joint must fulfil the service requirements. The characteristics of the metal determining the quality of weld joint includes tendency of cracking, hardening and softening of HAZ, oxidation, evaporation, structural modification and affinity to gases. While efforts required for producing a sound weld joint are determined by properties of metal system in consideration, namely melting point, thermal expansion coefficient, thermal and electrical conductivity, defects inherent in base metal and surface conditions. All the factors adversely affecting the weld quality and increasing the efforts (and skill required) for producing a satisfactory weld joint will in turn be decreasing the weldability of the metal.

In view of above, it can be said that weldability of metal is not an intrinsic property as it gets influenced by (a) all steps related with welding procedure, (b) purpose of the weld joints and (c) fabrication conditions. Welding of a metal using one process may shows poor weldability (like Al welding with SMA welding process) and good weldability when the same metal is welded with some other welding process (Al welding with TIG/MIG). Similarly, a steel weld joint may perform well under normal atmospheric conditions (0–45 °C), but the same may exhibit very poor toughness and ductility at sub-zero temperature conditions. Steps of the welding procedure, namely preparation of surface and edge, preheating, welding process, welding parameters, post-weld treatment such as relieving the residual stresses, can influence the weldability of metal appreciably. Therefore, weldability of a metal is considered as a relative term.

© The Author(s), under exclusive license to Springer Nature Singapore Pte Ltd. 2022
D. K. Dwivedi, *Fundamentals of Metal Joining*,
https://doi.org/10.1007/978-981-16-4819-9_29

29.2   Weldability of Metals by Fusion Welding Processes

The weldability of a metal (for a given welding procedure, purpose, fabrication condition) by the fusion welding processes is influenced by its many properties such as chemical properties (compositions, affinity of atmospheric gases with molten metal, solubility of gases in liquid and state of metal), physical properties (melting temperature, solidification temperature range, thermal conductivity, thermal expansion coefficient), mechanical properties (yield strength, toughness and ductility) and dimensional properties (thickness, cross section) including geometry affecting choice of joint configuration.

Chemical composition of the metal is one such characteristics which affects almost all above-mentioned characteristics except dimensional properties. Therefore, change in chemical composition of base metal affects the ease of welding. The following section presents different metal properties and ways by which these affect the ease of welding.

29.2.1   Composition

Metals having unfavourable elements like Pb, S, P and other low melting temperature constituents increase the solidification cracking and liquation cracking tendency while high concentration of hardening elements (like C, W, V) increases the cracking and embrittlement tendency. Composition of steel directly affects the embrittlement and cracking tendency. Few empirical equations showing relation between cracking tendency and chemical composition are given below.

(1)   **Hot Cracking Sensitivity**

$$HCS = \frac{(S + P + Si/25 + Ni/100) \times 1000}{3Mn + Cr + Mo + V}$$

If HCS < 4 then steel is not sensitive to hot cracking.

(2)   **Unit Crack Susceptibility (for SAW)**

$$UCS = 230C + 90S + 75P + 45Nb - 12.3Si - 4.5Mn - 1$$

Low risk if UCS $\leq$ 10, and high risk if UCS > 30.

(3)   **Cold cracking using Hydrogen Control Approach** (for steels in Zones—I and III of Graville diagram Fig. 29.1)

$$CE = C + \frac{Mn}{6} + \frac{Cr + Mo + V}{5} + \frac{Ni + Cu}{15}.$$

Cracking Parameter P_W : $Pcm + (H_D/60) + K/(40 \times 10^3)$,

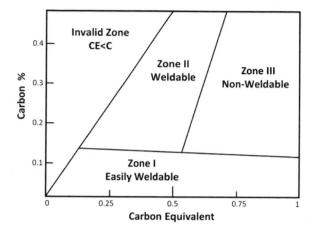

Fig. 29.1 Graville diagram showing weldability of steel corresponding different carbon % and carbon equivalent

$$Pcm = C + \frac{Si}{30} + \frac{Mn + Cu + Cr}{20} + \frac{Ni}{60} + \frac{V}{15} + 5B$$

Weld restraint, $K = K_o \times h$, with h = combined thickness in mm and $K_o \approx 69$.
Preheat temperature to avoid cold cracking T(°C) = 1440P$_W$ − 392.

(4)   **Prediction of Reheat Cracking**

$$P_{sr} = Cr + Cu + 2Mo + 10V + 7Nb + 5Ti - 2$$

$$\Delta G = Cr + 3.3Mo + 8.1V + 10C - 2$$

If ΔG, $P_{sr} > 0$, material susceptible to cracking.

(5)   **Temper Embrittlement Prediction**

$$J = (Si + Mn)(P + Sn) \times 10^4$$

If steel in not sensitive if $J \leq 180$,
    In case of weld metal $P_E = C + Mn + Mo + Cr/3 + Si/4 + 3.5(10P + 5Sb + 4Sn + As)$.
To avoid embrittlement $P_E \leq 3$.

29.2.2   Affinity of Weld Metal with Atmospheric Gases

Metals like Al, Mg, Ti show affinity with gases (oxygen and nitrogen) and impurities by forming compound through chemical reactions (Fig. 29.2a). Most of the metals

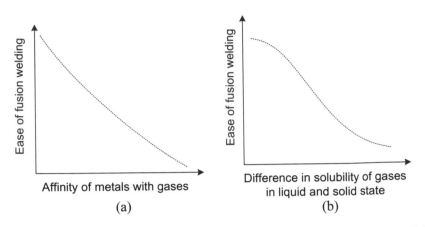

Fig. 29.2 Schematic showing influence of atmospheric gases on ease of fusion welding **a** affinity of metals with gases and **b** difference in solubility of gases in liquid and solid state

at high temperature during fusion welding (if not protected) tends to react with atmospheric gases and impurities to form their oxides, nitrides, sulphides, hydrides which can be observed in the form inclusions in the weld metal.

Solubility of Gases

Gases dissolved in metals affects fusion welding in two ways (a) forming porosity and (b) increasing embrittlement and cracking. A large difference in solubility of gases liquid and solid state of a metal increases the possibility of porosity formation due to high gas entrapment tendency (Fig. 29.2b), while gases like oxygen, hydrogen dissolved metal in solid state can promote blistering (Cu) and cracking (steels), respectively.

29.2.3   Melting Temperature

Ease of fusion welding is also affected by melting temperature of the metal as determines the net heat input and power density of heat source needed to facilitate welding. High melting point temperature metal (like Ti, W, Co) is found to be more difficult to weld by fusion welding than low melting temperature metals (Al, Mg) provided other characteristics remain identical (Fig. 29.3). Fusion welding of medium melting temperature metals is found be comparatively earlier as compared to those of both extremes (very high or very low). Low melting temperature metals make control of weld pool difficult and sensitive for melt through especially in case of thin sections while high melting temperature metals impose difficulty in melting of faying surfaces. Ease of joining by solid state welding processes (relying/involving on large scale deformation) also decreases with increase in melting temperature of parent metals primarily due to increased yield strength and reduction in ductility.

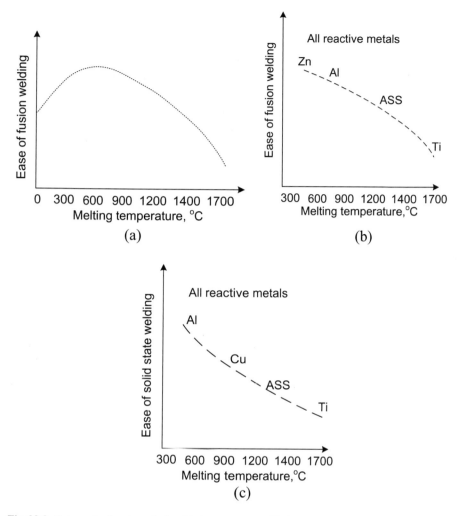

Fig. 29.3 Schematic showing relationship between ease of fusion welding and melting temperature of different types of metals **a** entire range of metals, **b** common metals and **c** reactive metals

29.2.4 Solidification Temperature Range

Solidification temperature range is indicated by difference of solidus and liquidus temperature. The solidification temperature range (> 50 °C) increases the solidification cracking (SC) and liquation cracking (LC) tendency (Fig. 29.4). Therefore, metals (pure, eutectic composition) solidifying wither at a single temperature or with a narrow solidification temperature range exhibit limited weld centreline cracking (SC) and one adjacent to the fusion boundary (LC).

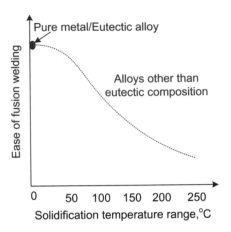

Fig. 29.4 Schematic showing relationship between ease of fusion welding and solidification temperature range

29.2.5 Thermal Conductivity

Thermal conductivity of metals determines how far and how fast heat will be dissipated award from the weld pool to the base metal during the fusion welding. These are factors affecting the net heat input requirement for the fusion welding and width of heat affected zone so ease of fusion welding. High thermal conductivity in general increases the requirement of heat input for welding and width of heat affected zone (Fig. 29.5). Therefore, microstructure, mechanical properties and residual stress behaviour of fusion weld joints are affected by thermal conductivity. In general, therefore, an increase in thermal conductivity decreases the ease of fusion welding of metals.

29.2.6 Thermal Expansion Coefficient

Linear thermal expansion coefficient of a metal directly affects the residual stress and distortion tendency of the weld joint. An increase in thermal coefficient increases the differential localized expansion and contraction experienced by the parent metal during fusion welding which in turn increases the residual stresses under identical welding conditions (Fig. 29.6). Therefore, metals like aluminium, austenitic stainless steel having high thermal expansion coefficient show greater residual stress and distortion tendency as compared to metals like mild steel and ferritic steel having low thermal expansion coefficient.

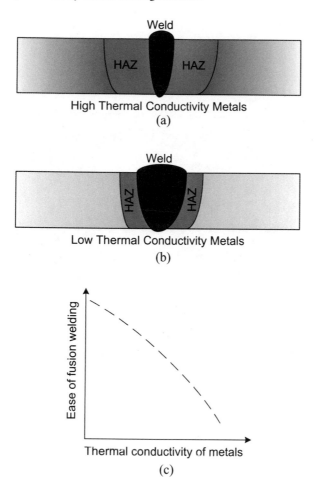

Fig. 29.5 Schematic showing **a** effect of high thermal conductivity on weld zone and HAZ, **b** effect of low thermal conductivity on weld zone and HAZ and **c** generic relationship between ease of fusion welding and thermal conductivity

29.2.7 Yield Strength

The fusion welding of a metal is not directly affected by its yield strength but ease of fusion welding is certain influenced. The maximum residual stress induced in a fusion weld joint depends on the yield strength of weld metal (Fig. 29.7). The autogenous welding and welding by matching filler/electrode results in maximum residual stress approximately similar to the base metal yield strength which in turn causes many issues related with cracking of weld metal and heat affected zone. Therefore, low yield strength filler/electrode (Ni alloys, austenitic stainless steel) is

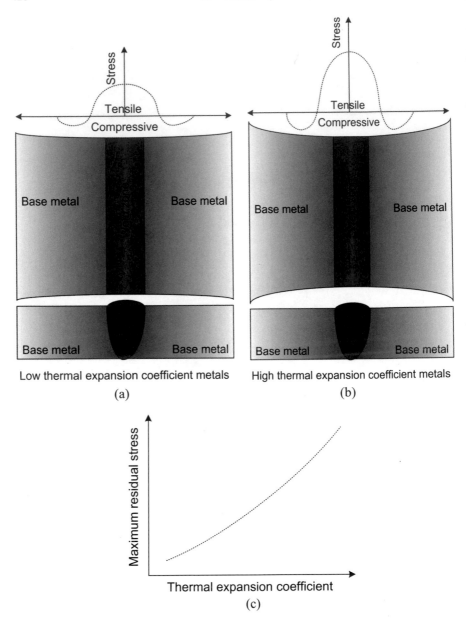

Fig. 29.6 Schematic showing effect of thermal expansion coefficient of metal on residual stress and distortion tendency **a** low thermal expansion coefficient metal, **b** high thermal expansion coefficient metal and **c** residual stress and thermal expansion behaviour of metal

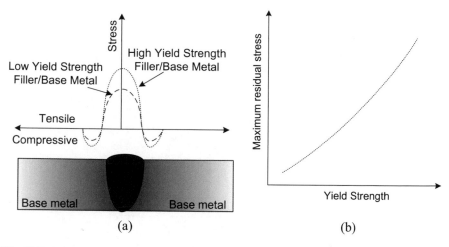

Fig. 29.7 Schematic showing effect of yield strength of base/filler/electrode metal on **a** distribution of residual stress and **b** maximum residual stress

sometime intentionally selected for fusion welding of high hardenability base metals and cast irons.

29.2.8 Toughness and Ductility

Toughness and ductility of base metal affect the cracking tendency of weld joints under the influence of residual stresses. Base metals having low toughness (say < 5 J) and poor ductility (< 4–5%) show high tendency of cracking and embrittlement of the weld joint (Fig. 29.8). Therefore, fusion welding of such metals needs extra care (with regard to the selection of process, filler, pre- and post-heating) which in turn adversely affects their weldability.

29.2.9 Thickness

Thickness of metal has significant effect on the weldability. Neither very thick (> 25 mm) nor thin (< 1 mm) are considered easy to weld due to unique issues related to them (Fig. 29.9). Weldability of very thick plates is considered poor due to the need of edge preparation (like double V groove), requirement large of weld passes to complete the weld, high residual stress, possibility of plain strain condition, needs of firm and well-designed fixture and need of post weld heat treatment to restore the properties of parent metal, while thin sheet impose difficulties related to control of molten weld pool, melt through, distortion tendency, difficulty in holding of the jobs.

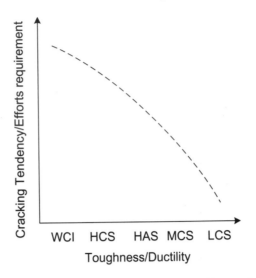

Fig. 29.8 Schematic showing general effort requirement/cracking tendency as function of toughness and ductility for different ferrous metals

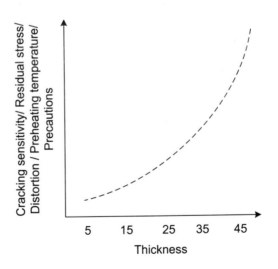

Fig. 29.9 Schematic showing general trend of ease of fusion welding (as dictated by various factors) as function of thickness of metals

29.2.9.1   Geometry of the Base Metal

The geometry of the base metal affects the ease of holding, access/application of heat for welding, and control over the molten weld pool. The welding of flat plates/sheets is considered to be easier than cylindrical, conical and other complex

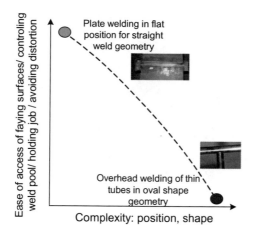

Fig. 29.10 Schematic showing general trend of ease of fusion welding (as dictated by various factors) as function of complexity shape/position

geometries primarily due to difficulty in control over the weld pool and need of especial attachments and fixtures for welding (Fig. 29.10).

29.3   Weldability of Metals by Solid State Joining Processes

The weldability of a metal (for a given welding procedure, purpose, fabrication condition) by the solid state joining processes is influenced by comparatively fewer properties such as chemical properties (compositions, affinity of atmospheric gases in solid state), physical properties (thermal conductivity, thermal expansion coefficient), mechanical properties (yield strength, work hardening and ductility) and dimensional properties (thickness, cross section) including geometry affecting choice of joint configuration.

29.3.1   Composition

The effect of composition on weldability of a metal by solid state joining processes depends how far it affects (a) reactivity of parent metal with atmospheric gases on heating to high temperature, (b) thermal conductivity and thermal expansion behaviour, (c) yield strength, ductility and work hardening behaviour. Effect of these factors will be elaborated in following sections.

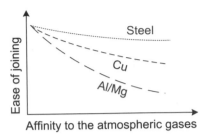

Fig. 29.11 Schematic showing relation between ease of joining of metal by solid state joining processes and affinity to atmospheric gases

29.3.2 Affinity with Atmospheric Gases

A pre-requisite for solid state joining is a clean surface of base metals to be joined for ensuring a direct metal to metal contact needed for metallurgical continuity through localized plastic deformation, diffusion and metallurgical transformation during solid state joining. Metals like Al, Mg, stainless steels form very thin transparent oxide layer at the surface which acts as barrier in atomic diffusion across the interface and developing metallurgical bond during solid state joining. Further, alloying elements like aluminium and chromium in iron increases oxide formation tendency both at room and at elevated temperatures which in turn affects the cleanliness of surfaces needed for perfect metallic intimacy for developing sound joint free from inclusions (Fig. 29.11). Surfaces of parent metals to be joined by solid state joining processes therefore must be cleaned using appropriate mechanical, chemical and ultrasonic cleaning methods so as to develop a sound weld joints.

29.3.3 Thermal Conductivity and Thermal Expansion Coefficient

The influence of thermal conductivity and thermal expansion coefficient on joining base metals by solid state joining is largely similar to that of fusion welding processes; however, the effect is somewhat less due to comparatively limited role of heat in solid joining. Still there are many solid state joining processes like friction stir welding wherein a lot of heat is generated inherently by friction and deformation. Solid state joining needs either macro- or microscale interfacial plastic deformation. The limited interfacial yielding during solid state joining can occur due to many reasons like very high yield strength of metal, very low ductility, high resistance to thermal softening, very limited rise temperature of the metal due to high thermal conductivity despite of enough heat generation (Fig. 29.12). Joining of high thermal conductivity metal like copper by solid state joining process such friction stir welding and even joining of Al and Cu by resistance welding is found to be difficult because heat is

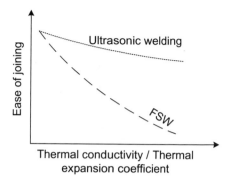

Fig. 29.12 Schematic showing relation between ease of joining of metal by solid state joining processes (USW, FSW) and thermal conductivity and thermal expansion behaviour

transferred rapidly from the weld zone to the underlying base metal leading to limited temperature rise for faying surface causing poor flowability of metal needed for solid state joining.

29.3.4 Yield Strength and Ductility

These two mechanical properties significantly determine the ease of plastic deformation. Low yield strength, low resistance to thermal softening and high ductility of metal facilitate interfacial yielding needed for solid state joining which in turn makes them good weldable by solid state joining processes (Fig. 29.13). The weldability of metals by the solid state joining processes (like explosive welding, ultrasonic welding, friction stir welding based on significant interfacial plastic deformation to facilitate mechanical interlocking and metallurgical bonding) is predominantly affected by the yield strength and ductility. In general, high ductility and low yield strength metal offer good weldability.

29.3.5 Work Hardening

Yield strength and ductility of metals considered to be important for solid state joining are affected by plastic deformation. Plastic deformation of metal either at room temperature or at marginally high temperature in general increases the yield strength and reduces the ductility due to work hardening behaviour of metals. However, such variation in yield strength and ductility depends on stacking fault energy (SFE) associated with metal affecting the work hardening tendency. Metals like stainless steel, steel, cobalt alloys, many super alloys having comparatively low SFE show high work hardening tendency during solid state joining (SSJ). These metals during

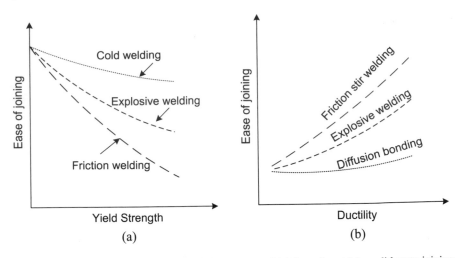

Fig. 29.13 Schematic showing relation between ease of joining of metal by solid state joining processes and **a** yield strength and **b** ductility

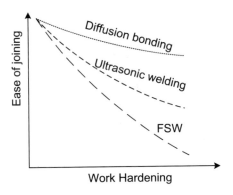

Fig. 29.14 Schematic showing relation between ease of joining of metal by solid state joining processes (USW and FSW) and thermal conductivity and thermal expansion behaviour

SSJ cause significant increase in yield strength and reduction in the ductility which can make their joining difficult (Fig. 29.14).

29.4 Weldability of Steels

The weldability of ferrous metals is considered to be complex due to allotropic behaviour of Fe–C system, transformation hardening behaviour, many metallurgical transformation leading to embrittlement, formation of a range phases exhibiting mechanical properties from very low to very high hardness, strength, toughness and ductility. The localized metallurgical transformation during welding and joining makes it further complex.

29.4.1 Weldability of Steel and Composition

Weldability of steels can be judged by two parameters (a) cleanliness of weld metal and (b) properties of HAZ. Cleanliness of weld metal is related with presence of inclusion in the form of slag or gases, whereas HAZ properties are primarily controlled by hardenability of the steel. Proper shielding of arc zone and degassing of molten weld metal can be used to control first factor. Proper shielding can be done by inactive gases released by combustion of electrode coatings in SMA or inert gases (Ar, He, Co_2) in case of TIG, MIG welding. Hardenability of steel is primarily governed by the composition. All the factors increasing the hardenability adversely affect the weldability because steel becomes more hard, brittle and sensitive to fracture/cracking; therefore, it needs extra care. So, more precautions should be taken to produce a sound weld joint (Fig. 29.15).

Addition of all alloying elements (C, Mn, Ni, W, Cr, etc.) except cobalt increases the hardenability which in turn decreases the weldability. To find the combined effect of alloying elements on hardenability/weldability, carbon equivalent (CE) is determined. The most of the carbon equivalent (CE) equations used to evaluate weldability depends type of steel, i.e. alloy steel or carbon steel.

- Common CE equation for low alloy steel is as under:

$$CE = C + Mn/6 + (Cr + Mo + V)/5 + (Ni + Cu)/15$$

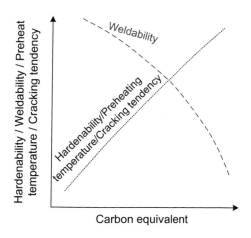

Fig. 29.15 Schematic showing relation between ease of welding of steel (in terms of different factors) as a function of carbon equivalent

(elements are expressed in weight percent amounts).
- For low carbon steels and micro-alloy steels, CE is obtained using following equation: $CE = C + Si/25 + (Mn + Cr)/16 + (Cr + Ni + Mo)/20 + V/15$
- From the Welding Journal, for low carbon, micro-alloyed steels, Ito-Besseyo carbon equivalent:

$$Ceq = C + Si/30 + (Mn + Cu + Cr)/20 + Ni/60 + Mo/15 + V/10 + 5 * B$$

Since the effect of different alloying elements on hardenability of steel is different, therefore, their influence on weldability will also be different. In general, high the CE steel needs high preheat temperature to produce crack-free weld joint. The following point can be kept in mind as broad guidelines for welding steel.

- $CE < 0.45$No preheat required,
- $0.45 < CE < 0.72$00–500 °C of preheat may be used
- $CE > 0.7$Cannot be welded.

Thickness of plate to be welded affects the cooling rate which in turn influences the hardening and cracking tendency. To take into account the thickness of plate above criteria is modified to get compensated carbon equivalent (CCE) relation.

$$CCE = CE + 0.00425t$$

where t is the thickness of plate in mm.

- $CCE < 0.4$No preheat required,
- $0.4 < CCE < 0.7$200–500 °C of preheat may be used
- $CCE > 0.7$Cannot be welded.

From the weldability point of view, steels can be placed in five categories based on chemical composition, mechanical properties, heat treatment conditions and high temperature properties: (a) carbon steel, (b) high strength low alloy steel, (c) quench and tempered steel, (d) heat treatable steel and (e) Cr–Mo steel. These steels need to be welded in different forms such as sheets, plates, pipes and forgings. In case of steel welding, it is important to consider thickness of base metal as it affects the heat input, cooling rate and restraint conditions during welding.

29.4.2 Different Types of Steel and Welding

Carbon steel generally welded in as rolled condition (besides annealed and normalized one). The weldable carbon steel is mostly composed of carbon about 0.25%, Mn up to 1.65%, Si up to 0.6% with residual amount of S and P below 0.05%. High strength low alloy steel (HSLA) is designed to have yield strength in range of 290–550 MPa using alloying concentration lesser than 1% in total. These can be welded in conditions same as that of carbon steel. Quench and tempered (Q and T) steels can be a carbon steel or HSLA steel category that are generally heat treated to impart yield strength in range of 350–1030 MPa. Heat treatable steels generally contain carbon more than carbon steel or HSLA steels to increase their response to the heat treatment. However, presence of high carbon in these steels increases the hardenability which in turn decreases the weldability owing to increased embrittlement and cracking tendency of heat affected zone. Further, PWHT of heat treatable steel weld joints is done to enhance their toughness and induce ductility because of the presence of high carbon in these heat-treatable steels. Cr–Mo steels are primarily designed to have high resistance to corrosion, thermal softening and creep at elevated temperature (up to 700 °C). Therefore, these are commonly used in petrochemical industries and thermal power plants. Weld joints of Cr–Mo steels are generally given PWHT to regain ductility, toughness and corrosion-resistance and reduce the residual stresses.

29.5 Common Problems in Steel Welding

29.5.1 Cracking of HAZ Due to Hardening

The cooling rate experienced by the weld metal and HAZ during welding generally exceeds the critical cooling (CCR) which in turn increases the chances of martensitic transformation. It is well known from the physical metallurgy of the steels that this transformation increases the hardness and brittleness and generates tensile residual stresses. This combination of high hardness and tensile residual stresses makes the steel prone to the cracking.

29.5.2 Cold Cracking

Another important effect of solid state transformation is the cold cracking. It is also termed as delayed/hydrogen-induced cracking because these two factors (delay and hydrogen) are basically responsible for cold cracking. Applied/residual tensile stress vs. time relationship for failure by cold cracking is shown in Fig. 29.16. It can be observed that increase in stress decreases the time required for initiation and complete fracture by cold cracking.

Origin of this problem lies in the variation of solubility of hydrogen in the steel with the temperature. Reduction in temperature decreases solubility of hydrogen in solid state due to change in crystal structure from F. C. C. to B. C. C. High temperature transformation (like austenite to pearlite or bainite) allows escape of some of excess hydrogen (beyond the solubility) by diffusion. But in case of low temperature transformation (austenite into martensite), when rate of diffusion reduces significantly, hydrogen cannot escape and is trapped in steel as solid solution. Dissolved

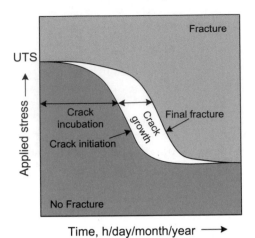

Fig. 29.16 Schematic showing stress versus time relationship for fracture by cold crack

hydrogen has more damaging effect in the presence of martensite and the same has been explained below.

Hydrogen dissolved in atomic state at low temperature tends to diffuse out gradually towards the vacancies and other cavities. At these locations, atomic hydrogen converts into diatomic H_2 gas and with time, continued diffusion of hydrogen towards these discontinuities as this gas starts to build up pressure in the cavities (Fig. 29.17). If the pressure exceeds the fracture stress of metal, cavities expands by cracking. Cracking of metal increases the volume which in turn reduces the pressure. Due to continuous diffusion of hydrogen towards the cavities after some time again as pressure exceeds the fracture stress, and crack propagates further. This process of building up on pressure and propagation of cracks is repeated until compete fracture

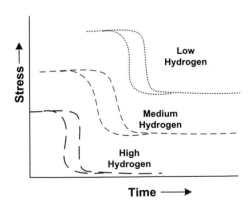

Fig. 29.17 Schematic showing effect of hydrogen concentration on cold cracking at different stress levels

takes place without external load. Since this type cracking and fracture takes place after some time of welding, hence, it is called delayed cracking. Delay for complete fracture depends on the following factors:

- Hardenability of steel
- Amount of hydrogen dissolved in atomic state
- Magnitude of residual tensile stress.

Hardenability of steel affects the critical cooling rate. Steel of high hardenability promotes the martensitic transformation; therefore, it has high hardness and brittleness. High hardness increases the cracking tendency, whereas soft and ductile metals reduce it. Crack tips are blunted in case of ductile metals, so they reduce the cracking sensitivity and increases the stress level for fracture. As a result, crack propagation rate is reduced in case of ductile and low strength metal. Therefore, steels of low hardenability/strength will therefore minimize the cold/delayed cracking. This is evident from high cold crack tendency of high yield strength steel than low strength steels (Fig. 29.18).

Larger the amount of dissolved hydrogen, faster will be the delayed/hydrogen-induced cracking.

Remedy

- Use of low hydrogen electrodes.
- Preheating of plates to be welded.
- Use of austenitic electrodes.

Use of low hydrogen electrodes will reduce the hydrogen content in weld metal. Preheating of the plate will reduce the cooing rate, which will allow longer time for gases to escape during the liquid-to-solid state and solid–solid transformation. It

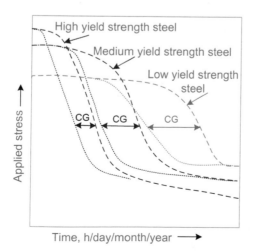

Fig. 29.18 Schematic showing effect of strength of steel on cold cracking behaviour

may also reduce the cooling rate below the critical cooling rate so that martensitic transformation can be avoided and austenite can be transformed into softer phases and phase mixtures like pearlite, bainite, etc. These soft phases further reduce the cracking tendency. The use of austenitic electrode also avoids the martensite formation and provides mainly austenite matrix in weld zone. Austenite is a soft and tough phase having high solubility (%) for hydrogen. All these characteristics of austenite reduce the cold/delayed cracking.

Questions for Self-assessment

a.   Define weldability of metals? Describe the metal properties affecting weldability.
b.   Explain the factors determining the weldability of metals.
c.   How does composition of steel affect its weldability?
d.   What is carbon equivalent and how is it related with weldability of steel?
e.   Describe following common problems associated with welding of steel in respect of factors contributing towards their development, mechanism and remedial method

  i.   Porosity
  ii. Hardening and embrittlement
  iii. Cold cracking tendency.

Further Reading

American Welding Society (1987) Welding handbook, 8th edn, vol 1 & 2, USA
American Society for Metals (1993) Metals handbook-welding, brazing and soldering, 10th edn, vol 6, USA
Avner SH (2009) Introduction to physical metallurgy, 2nd edn. McGraw Hill, New Delhi
Dwivedi (2013) Production and properties of cast Al-Si alloys. New Age International, New Delhi (2013)
Dwivedi DK (2018) Surface engineering. Springer, New Delhi
Kou S (2003) Welding metallurgy, 2nd edn. Wiley, USA
Lancaster JF (1999) Metallurgy of welding, 6th edn. Abington Publishing, England
Nadkarni SV (2010) Modern arc welding technology. Ador Welding Limited, New Delhi

Chapter 30
Weldability of Metals: Weldability of Aluminium Alloys: Porosity, HAZ Softening and Solidification Cracking

30.1 Need of Aluminium Welding

Welding of the aluminium is considered to be slightly difficult than the steel due to high thermal and electrical conductivity, high thermal expansion coefficient, refractory aluminium oxide (Al_2O_3) formation tendency and low stiffness (Fig. 30.1). However, increase in applications of aluminium alloys in all sectors of industry is a driving force for technologists to develop viable and efficient technologies for joining of aluminium without much adverse effect on their mechanical, chemical and metallurgical performances desired for longer life. The performance of weld joints of an aluminium alloys to a great extent is determined by its composition, alloy temper condition and method of manufacturing besides welding-related parameters. All the three aspects are usually included in aluminium alloy specification. Aluminium alloy may be produced either only by cast or by casting and subsequent forming process (which are called wrought alloys). Welding of wrought aluminium alloys is more common, and therefore, in this chapter, discussions are related to wrought aluminium alloys. Depending upon the composition, aluminium alloy are classified from 1XXX through 9XXX series. Some of aluminium alloys (1XXX, 3XXX, 4XXX and 5XXX) are non-heat-treatable, and others (2XXX, 6XXX and 7XXX series) are heat treatable.

30.1.1 Strengthening of Non-heat-Treatable Aluminium Alloys and Welding

The strength of the non-heat-treatable aluminium alloys is mostly dictated by solid solution strengthening and dispersion hardening effects of alloying elements such as silicon, iron, manganese and magnesium. Magnesium is the most effective alloying element in solid solution strengthening, and therefore, 5XXX series aluminium

D. K. Dwivedi, *Fundamentals of Metal Joining*,
https://doi.org/10.1007/978-981-16-4819-9_30

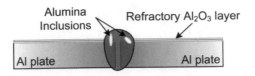

Fig. 30.1 Schematic showing formation of alumina and resulting in oxide inclusions

alloys have relatively high strength even in annealed condition. Most of the non-heat-treatable aluminium alloys are work hardenable. Heating of these alloys during welding (due to weld thermal cycle) lowers prior work hardening effect and improves the ductility which in turn can lead to loss of strength of HAZ (Fig. 30.2). Moreover, high-strength solid solution alloys of 5XXX series such as Al–Mg and Al–Mg–Mn are found suitable for welded construction structures as they offer largely uniform mechanical properties in the various zones of a welded joint.

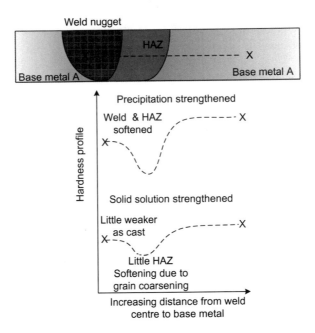

Fig. 30.2 Schematic of transverse section of the weld joint showing different zones, distribution of hardness of weld and HAZ of solid solution and precipitation-strengthened aluminium alloys

30.1.2 *Strengthening of Heat-Treatable Aluminium Alloys and Welding*

The most of heat-treatable aluminium alloys (2XXX, 6XXX and 7XXX series) are strengthened by solid solution formation, work hardening and precipitation strengthening depending upon the alloy condition and manufacturing history. Strength of these alloys in annealed condition is either similar or slightly better as compared to non-heat-treatable alloys mainly due to presence of alloying elements such as copper, magnesium, zinc and silicon. Generally, heat-treatable aluminium alloys are precipitation hardened. The precipitation hardening involves solutionizing followed by quenching and aging either at room temperature (natural aging) or elevated temperature (artificial aging).

Three most common precipitation-hardenable aluminium alloys namely Al–Cu (2XXX series), Al–Mg–Si (6XXX series) and Al–Zn–Mg (7XXX series) are primarily hardened by forming phases, namely Al_2Cu, Mg_2Si and Zn_2Mg, respectively, besides many complex intermetallic compounds developed during aging process. Therefore, presence and loss of these precipitates significantly affect the mechanical performance (hardness, tensile strength and % elongation) of weld joints of these alloys. However, the existence of theses hardening precipitates is influenced by weld thermal cycle experienced by base metal and weld metal during welding. In general, all factors decreasing the heat input (either due to low welding current, increase in welding speed or use of low heat input welding processes such as electron beam, pulse TIG) would reduce the width of heat affected zone associate adverse effects such as the possibility of partial melting of low melting point phases (eutectic) present at grain boundary, overaging, grain growth, reversion or dissolution of precipitates or a combination of few or all.

$$Al - Cu - Mg(e.g., 2024): SS \rightarrow GP \rightarrow S^{'}(Al_2CuMg) \rightarrow S(Al_2CuMg)$$

$$Al - Mg - Si(e.g., 6061): SS \rightarrow GP \rightarrow \beta^{'}(Mg_2Si) \rightarrow \beta(Mg_2Si)$$

$$Al - Zn - Mg(e.g., 7005): SS \rightarrow GP \rightarrow \eta^{'}(Zn_2Mg) \rightarrow \eta(Zn_2Mg)$$

In the solution heat-treated condition, heat-treatable alloys exhibit lower cracking tendency than in the aged condition mainly due to more uniform microstructure and lesser restraint imposed by base metal. Welding of heat-treatable aluminium alloy in aged condition leads to reversion (loss/dissolution of precipitates) and overaging (coarsening of precipitates by consuming fine precipitates) effect which in turn softens the HAZ to some extent (Fig. 30.2). However, under influence welding thermal cycle, alloying elements are dissolved during heating and form heterogeneous solid solution, and subsequently rapid cooling results in supersaturation of these elements in aluminium matrix. Thus, solutionizing and quenching influence the heat affected zone. Thereafter, aging of some of the alloys like Al–Zn–Mg occurs slowly even at room temperature which in turn help to attain strength almost similar to that of base metal, while other heat-treatable alloy like Al–Cu and Al–Mg–Si alloys do not show appreciable age hardening at room temperature. Hence, Al–Zn–Mg alloys are preferred when post-weld heat treatment is neither possible nor feasible.

30.2 Weldability of Aluminium Alloys

Weldability of aluminium alloys like any other metal system must be assessed in light of purpose (application of weld joint considering service conditions), welding procedure being used and welding conditions in which welding need to be performed. Weldability of aluminium may be very poor when joined by shielded metal arc welding or gas welding, but the same may be very good when joint is made using tungsten inert gas or gas metal arc welding process. Similarly, other aspects of welding procedure such as edge preparation, welding parameters, pre and post-weld heat treatment can significantly dictate the weldability of aluminium owing to their ability to affect the soundness of weld joints and mechanical performance. Thus, all the factors governing the soundness of the aluminium weld, the mechanical and metallurgical features determine the weldability of aluminium alloy system. In general, aluminium is considered to be of comparatively lower weldability than steels due to various reasons (a) high affinity of aluminium towards atmospheric gases, (b) high thermal expansion coefficient, (c) high thermal and electrical conductivity, (d) poor rigidity and (e) high solidification temperature range. These characteristics of aluminium alloys in general make them sensitive from defect formation point of view during welding. The extent of undesirable effect of above characteristics on performance of the weld joints is generally reduced using two approaches (a) effective protection of the weld pool contamination from atmospheric gases using proper shielding method and (b) reducing influence of weld thermal cycling using higher energy density welding processes. Former approach mainly deals with using various environments (vacuum, Ar, He, or their mixtures with hydrogen and oxygen) to shield the weld pool from ambient gases, while later one has led to the development of newer welding processes such as laser, pulse variants of TIG and MIG, and friction stir welding.

30.3 Typical Welding Problems in Aluminium Alloys

30.3.1 Porosity

Porosity in aluminium weld joints can be of two types (a) hydrogen-induced porosity and (b) interdendritic shrinkage porosity and both are caused by entirely different factors. Former one is caused by the presence of hydrogen in the weld owing to unfavourable welding conditions such as improper cleaning, moisture in electrode,

Fig. 30.3 Schematic showing different types of porosities in weld metal

shielding gases and oxide layer, presence of hydro-carbons in form of oil, paint, grease, etc. (Fig. 30.3).

In presence of hydrogen porosity in the weld metal mainly occurs due to high difference in solubility of hydrogen in liquid and solid state of aluminium alloy (Fig. 30.4). During solidification of the weld metal, the excess hydrogen is rejected at the advancing solid–liquid interface in the weld which in turn leads to the development of hydrogen-induced porosity especially under high solidification rate conditions as high cooling rate experienced by the weld pool increases tendency of entrapment of hydrogen (Fig. 30.5).

Excessive hydrogen porosity can severely reduce strength, ductility and fatigue resistance of aluminium welds due to two reasons (a) reduction in effective load resisting cross-sectional area of the weld joints and (b) loss of metallic continuity owing to the presence of gas pockets which in turn increases the stress concentration at the weld pores. It also reduces the life of aluminium welds. Therefore, to control hydrogen-induced porosity in aluminium following approaches can be used (a) proper cleaning of surfaces, baking of the electrodes to drive off moisture and

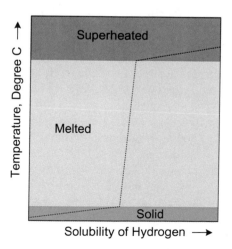

Fig. 30.4 Schematic showing general trend of variation in solubility of hydrogen in aluminium alloy

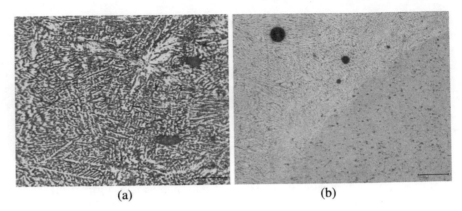

Fig. 30.5 Micrographs showing **a** dendritic and **b** gas porosity in aluminium welds (100X)

remove the impurities from weld surface (b) addition of freon to the shielding gas and (c) churning the weld pool during weld solidification using suitable electromagnetic fields.

Interdendritic porosity in weld mainly occurs due to poor fluidity of molten weld metal and rapid solidification. Preheating of plates and increasing heat input (using high current and low welding speed) help in reducing the interdendritic porosity.

30.3.2 Inclusion

In general, presence of any foreign constituent (one which is not desired) in the weld can be considered as inclusion, and these may be in the form of gases, thin films and solid particles. High affinity of aluminium with atmospheric gases increases the tendency of formation of oxides and nitrides (having density similar to that of aluminium) especially when (a) protection of weld pool is not enough, (b) proper cleaning of filler and base metal has not been done, (c) shielding gases are not pure enough and therefore making oxygen and hydrogen available to molten weld pool during welding and (d) gases are present in dissolved state in aluminium itself and tungsten inclusion while using GTA welding. Mostly, inclusion of oxides and nitrides of aluminium is found in weld joints in case of unfavourable welding conditions. Presence of these inclusions disrupts the metallic continuity in the weld therefore these provide site for stress concentration and become a source of weakness leading to the deterioration in mechanical and corrosion performance of the weld joints (Fig. 30.6). Ductility, notch toughness and fatigue resistance of the weld joints are very adversely affected by the presence of the inclusion. To reduce the formation of inclusion in weld, it is important to give proper attention to (a) avoid sources of atmospheric gases, (b) developing proper welding procedure specification (selection of proper electrode, welding parameters, shielding gases and manipulation of during

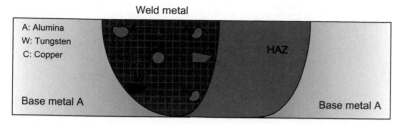

Fig. 30.6 Schematic showing different type of inclusions and other impurities in weld joints

welding) and (c) manipulation of GTAW torch properly so as to avoid the formation of tungsten inclusion.

30.3.3 Solidification Cracking

This is an interdendritic type of cracking mostly observed along the weld centreline in very last stage of solidification primarily due to two factors (a) development of tensile residual stresses and (b) presence of low melting point phases in interdendritic regions of solidifying weld is called solidification cracking (Fig. 30.7).

The solidification cracking mainly occurs when residual tensile stress developed in weld (owing to contraction of base metal and weld metal) goes beyond the strength of solidifying weld metal. Moreover, the contribution of solidification shrinkage of weld metal in development of the tensile residual stress is generally marginal. All the factors, namely thermal expansion coefficient of weld and base metal, melting point, weld bead profile, type of weld, degree of constraint, thickness of workpiece, etc., affecting the contraction of the weld will govern the residual stresses and so

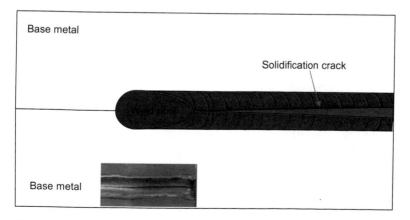

Fig. 30.7 Schematic of solidification cracking in aluminium weld

solidification cracking tendency. The presence of tensile or shear stress is mandatory for cracking means no residual tensile stress no cracking. Residual stresses in weld joint cannot be eliminated but can be minimized by developing proper welding procedure.

Increase in degree of restraint during welding in general increases solidification cracking tendency due to increased residual tensile stresses. Similarly, concave fillet weld bead profile results higher solidification cracking tendency than those of convex weld bead profile. In same line, other related materials characteristics of base metal such as increase in thickness of plate, high thermal expansion of coefficient and wider solidification temperature in general increases the residual stresses and so solidification cracking tendency.

Apart from the residual tensile stresses, strength and ductility of weld metal in terminal stage of solidification also predominantly determine the solidification cracking tendency. In general, all the factors such as composition of the weld metal, microstructure, segregation tendency, wider solidification temperature range and higher fluidity of low melting point phases (owing to reduction in surface tension and viscosity) of molten weld metal increase the solidification cracking.

30.3.3.1 Composition of Aluminium Alloy

Presence of all alloying element (silicon, copper, magnesium, zinc) in such a quantity that increases the solidification temperature range tends to increase the solidification cracking tendency. In general, addition of these elements in aluminium first widens the solidification temperature range, and then after reaching maximum, it decreases gradually. It can be observed that addition of these elements at certain level results in maximum range of solidification temperature and that corresponds to highest solidification cracking tendency. It can be noticed from Fig. 30.8 that solidification cracking is lower with both very low and high concentration of alloying element owing to varying amount of low melting point eutectic and other phases. A very limited amount of low melting point phases obviously increases resistance to solidification cracking due to high strength of solidified weld metal in terminal stage of solidification while in case of aluminium alloy (such as eutectic or near to the eutectic composition) or those with high concentration of alloying elements having large amount of low melting point phases to facilitate healing of cracks by the backfill of incipient cracks which in turn decreases the solidification cracking tendency (Fig. 30.8).

Therefore, selection of filler metal for welding of aluminium alloys is done in such a way that for given dilution level, concentration of alloying element in weld metal corresponds to minimum solidification temperature so as to reduce the solidification cracking possibility. In general, application of Al-5%Mg and Al-(5–12%) Si fillers are commonly used to avoid solidification cracking during welding of aluminium alloys.

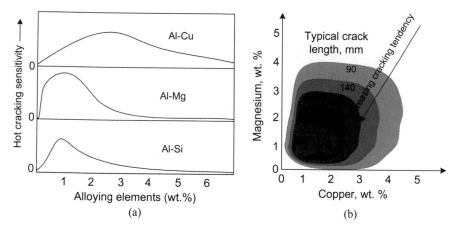

Fig. 30.8 Influence of alloying elements on solidification cracking tendency in aluminium alloy containing **a** singly Si, Mg, Cu, and **b** combined effect of Cu and Mg

The influence of microstructure of weld metal on solidification cracking depends on the way it affects the segregation tendency owing to variation in size and orientation of grains. In general, fine grain structure results is large grain boundary area and hence more uniform distribution of low melting point phases and reduced segregation of alloying element (Fig. 30.9). Further, fine equiaxed grain structure provides better heeling of incipient crack through back fill by liquid metal available at last to solidify due to improved fluidity of melt through the micro-channel present between already solidified metal. Conversely for a given solidification, cracking sensitive alloy composition coarse columnar grain structure having abutting orientation encourages the cracking tendency as compared to fine equiaxed and axially grain (Fig. 30.10). Moreover, the morphology of low melting point phases as governed by their surface tension and viscosity in liquid state near last stage of solidification also affects the solidification cracking sensitivity. In general, low melting point phases having low surface tension and low viscosity (so high fluidity) solidify in form of thin films

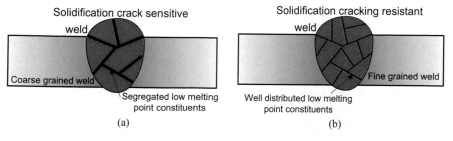

Fig. 30.9 Schematic showing **a** coarse grain promoting alloy element segregation and so solidification cracking tendency and **b** a fine grain resulting in more uniform distribution of alloy element so reducing solidification cracking tendency

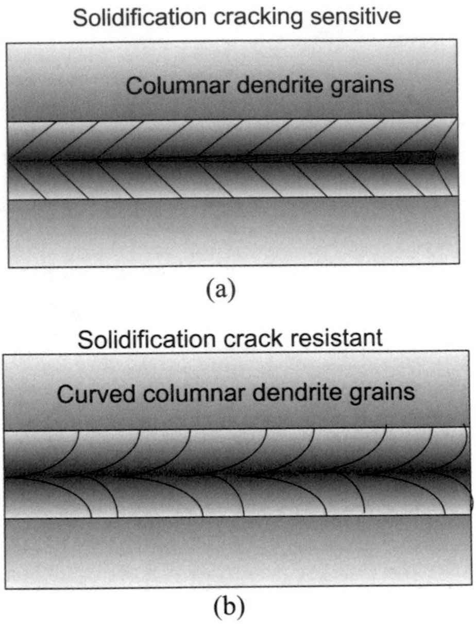

Solidification cracking sensitive

(a)

Solidification crack resistant

(b)

Fig. 30.10 Schematic diagram showing influence of grain structure of weld **a** columnar grains and **b** curved grains

and layer in interdendritic regions which are considered to be more crack sensitive than those of globular morphology formed in case of high surface tension and high viscousity low melting point molten metal.

30.3.3.2 Control of Solidification Cracking

- Changing composition of the weld metal so as to reduce the solidification temperature range and increase the amount of low melting point eutectic phases and phase mixtures to facilitate heating of incipient cracks
- Refinement of the grain structure: The microstructure of weld metal can be controlled in many ways such as addition of grain refiner, use of external electromagnetic or mechanical forces and selection of proper welding parameters such as heat input (VI) and welding speed or use pulse current for welding. Addition of grain refiner (Ti, B, Zr, etc.) in aluminium weld metal so as to facilitate the development of fine and equiaxed grain structure and reduce columnar grain structure. Similarly, low heat input leads to development of fine equiaxed grains, and low welding speed produces curved grain associated with pear drop shaped weld pool. Mechanical vibrations and electromagnetic stirring of weld pool also help to refine the grain structure avoid the abutting columnar grains.
- Reducing tensile residual stresses developing in weld joints using any of the approaches such as controlling weld bead geometry, selection of weld joint design,

welding procedure and low strength filler can help in reducing the solidification cracking.

Questions for Self-assessment

a. What are properties of aluminium alloy that make it somewhat difficult to weld?
b. What is softening of HAZ in heat-treatable aluminium alloy weld joints?
c. Explain the mechanism of solidification cracking using suitable sketch.
d. What are factors affecting the solidification cracking tendency.
e. Explain the effect of metallurgical aspects on solidification cracking of aluminium weld joints.
f. How does alloy composition affect the solidification cracking behaviour of aluminium weld joints?

Further Reading

Avner SH (2009), Introduction to physical metallurgy, 2nd edn. McGraw Hill, New Delhi
Dwivedi DK (2013) Production and properties of cast Al-Si alloys. New Age International, New Delhi
Dwivedi DK (2018) Surface engineering. Springer, New Delhi
Kou S (2003) Welding metallurgy, 2nd edn. Willey, USA
Lancaster JF (1999) Metallurgy of welding, 6th edn. Abington Publishing, England
Mandal NR (2005) Aluminium welding, 2nd edn. Narosa Publications
Metals Handbook (1993) Welding, brazing and soldering, 10th edn, vol 6. American Society for Metals, USA
Nadkarni SV (2010) Modern arc welding technology. Ador Welding Limited, New Delhi
Welding Handbook (1987) 8th edn, vols 1 and 2. American Welding Society, USA

Printed in the United States
by Baker & Taylor Publisher Services